ALGAL GENETIC RESOURCES

Cosmeceuticals, Nutraceuticals, and Pharmaceuticals from Algae

ALGAL GENETIC RESOURCES

*Cosmeceuticals, Nutraceuticals,
and Pharmaceuticals from Algae*

Edited by
Jeyabalan Sangeetha, PhD
Devarajan Thangadurai, PhD

AAP APPLE
ACADEMIC
PRESS

First edition published 2023

Apple Academic Press Inc.
1265 Goldenrod Circle, NE,
Palm Bay, FL 32905 USA

4164 Lakeshore Road, Burlington,
ON, L7L 1A4 Canada

CRC Press
6000 Broken Sound Parkway NW,
Suite 300, Boca Raton, FL 33487-2742 USA

2 Park Square, Milton Park,
Abingdon, Oxon, OX14 4RN UK

Library and Archives Canada Cataloguing in Publication

Title: Algal genetic resources : cosmeceuticals, nutraceuticals, and pharmaceuticals from algae / edited by Jeyabalan Sangeetha, PhD, Devarajan Thangadurai, PhD.
Names: Sangeetha, Jeyabalan, editor. | Thangadurai, D. (Devarajan), 1976- editor.
Series: Current advances in biodiversity, conservation, and environmental sciences (Series)
Description: First edition. | Series statement: Current advances in biodiversity, conservation, and environmental sciences | Includes bibliographical references and index.
Identifiers: Canadiana (print) 20210381558 | Canadiana (ebook) 20210381582 | ISBN 9781774637487 (hardcover) | ISBN 9781774637494 (softcover) | ISBN 9781003277095 (ebook)
Subjects: LCSH: Microalgae. | LCSH: Cyanobacteria.
Classification: LCC QK568.M52 A44 2022 | DDC 579.8—dc23

Library of Congress Cataloging-in-Publication Data

Names: Sangeetha, Jeyabalan, editor. | Thangadurai, D. (Devarajan), 1976- editor.
Title: Algal genetic resources : cosmeceuticals, nutraceuticals, and pharmaceuticals from algae / edited by Jeyabalan Sangeetha, Devarajan Thangadurai.
Other titles: Current advances in biodiversity, conservation, and environmental sciences (Series)
Description: First edition. | Palm Bay, FL : Apple Academic Press ; Boca Raton, FL : CRC Press, 2022. | Series: Current advances in biodiversity, conservation, and environmental sciences | Includes bibliographical references and index. | Summary: "This book focuses on the current and potential applications of microalgae and cyanobacteria in pharmaceuticals, nutraceuticals, and cosmeceuticals. The book deals with the very recent and advanced techniques and technologies in algal cultivation and extraction for its application. The chapters discuss the biological importance, properties, and uses of algal metabolites and microalgae-based compounds in drug development, in food nutrition enhancement, and in the development of cosmetics with medicinal properties. The chapter authors cover a range of diverse topics on algal biological resources, such as Algae as a nutraceutical and functional food ingredient The extraction of food bioactive compounds from microalgae Spirulina-derived nutraceuticals and their applications in the food industry Anticancer compounds from freshwater microalgae Cosmetic applications of microalgal and cyanobacterial pigments More This unique book, Algal Genetic Resources: Cosmeceuticals, Nutraceuticals, and Pharmaceuticals from Algae, will enlighten readers on the vast usefulness of microalgae and cyanobacteria as an important resource for the cosmeceutical, pharmaceutical and nutraceutical industries for their broad biotechnological potential industrial applications. The volume will be a valuable reference for scientists and researchers in these areas as well as for advanced students and faculty in ecology, phycology, botany, agriculture, biotechnology, microbiology, environmental biotechnology, plant science, and life sciences"-- Provided by publisher.
Identifiers: LCCN 2021057328 (print) | LCCN 2021057329 (ebook) | ISBN 9781774637487 (hardback) | ISBN 9781774637494 (paperback) | ISBN 9781003277095 (ebook)
Subjects: MESH: Cyanobacteria | Microalgae | Dietary Supplements | Phytochemicals--therapeutic use | Biotechnology--methods
Classification: LCC QR99.63 (print) | LCC QR99.63 (ebook) | NLM QW 131 | DDC 579.3/9--dc23/eng/20220106
LC record available at https://lccn.loc.gov/2021057328
LC ebook record available at https://lccn.loc.gov/2021057329

ISBN: 978-1-77463-748-7 (hbk)
ISBN: 978-1-77463-749-4 (pbk)
ISBN: 978-1-00327-709-5 (ebk)

ABOUT THE CURRENT ADVANCES IN BIODIVERSITY, CONSERVATION AND ENVIRONMENTAL SCIENCES BOOK SERIES

SERIES EDITOR

Jeyabalan Sangeetha, PhD

Assistant Professor, Central University of Kerala, Kasaragod, Kerala, India

Devarajan Thangadurai, PhD

Associate Professor, Karnatak University, Dharwad, Karnataka, India

BOOKS IN THE SERIES

Biodiversity and Conservation: Characterization and Utilization of Plants, Microbes, and Natural Resources for Sustainable Development and Ecosystem Management

Jeyabalan Sangeetha, PhD, Devarajan Thangadurai, PhD, Hong Ching Goh, PhD, and Saher Islam, PhD

Beneficial Microbes for Sustainable Agriculture and Environmental Management

Jeyabalan Sangeetha, PhD, Devarajan Thangadurai, PhD, and Saher Islam, PhD

Algal Genetic Resources: Cosmeceuticals, Nutraceuticals, and Pharmaceuticals from Algae

Jeyabalan Sangeetha, PhD, and Devarajan Thangadurai, PhD

ABOUT THE EDITORS

Jeyabalan Sangeetha, PhD

Assistant Professor, Central University of Kerala, Kasaragod, South India

Jeyabalan Sangeetha, PhD, is an Assistant Professor at Central University of Kerala, Kasaragod, South India. She has edited/coedited several books in her research areas, which include environmental toxicology, environmental microbiology, environmental biotechnology, and environmental nanotechnology. She earned her BSc in Microbiology and PhD in Environmental Science from Bharathidasan University, Tiruchirappalli, Tamil Nadu, India. She holds an MSc in Environmental Science from Bharathiar University, Coimbatore, Tamil Nadu, India. She is the recipient of a Tamil Nadu Government Scholarship and a Rajiv Gandhi National Fellowship of the University Grants Commission, Government of India for her doctoral studies. She served as the Dr. D.S. Kothari Postdoctoral Fellow and UGC Postdoctoral Fellow at Karnatak University, Dharwad, South India, during 2012–2016.

Devarajan Thangadurai, PhD

Associate Professor, Karnatak University, Dharwad, South India;
Editor-in-Chief, Biotechnology, Bioinformatics and
Bioengineering and Acta Biologica Indica

Devarajan Thangadurai, PhD, is an Associate Professor at Karnatak University in South India and Editor-in-Chief of the international journals *Biotechnology, Bioinformatics and Bioengineering* and *Acta Biologica Indica*. He has authored/edited over 30 books with national and international publishers and has visited 24 countries in Asia, Europe, Africa, and the Middle East for academic visits, scientific meetings, and international collaborations. He received his PhD in Botany from Sri Krishnadevaraya University in South India as a CSIR Senior Research Fellow with funding from the Ministry of Science and Technology, Government of India. He served as a Postdoctoral Fellow at the University of Madeira, Portugal; University of Delhi, India; and ICAR National Research Centre for Banana, India. He is the recipient of a Best Young Scientist Award with a Gold Medal from Acharya Nagarjuna University, India, and the VGST-SMYSR Young Scientist Award of the Government of Karnataka, Republic of India.

CONTENTS

CONTRIBUTORS

Iffat Zareen Ahmad
Natural Products Laboratory, Department of Bioengineering, Integral University, Dasauli, Lucknow, Uttar Pradesh–226026, India

Vishal Aparadh
Department of Botany, S.P.K. College, Sawantwadi–416510, Maharashtra, India

Sangeetha Arunachalam
Department of Food Technology, Kongu Engineering College, Perundurai, Erode–638060, Tamil Nadu, India

Malik Badshah
Department of Microbiology, Faculty of Biological Sciences, Quaid-i-Azam University, Islamabad–45320, Pakistan

Andressa M. Baseggio
Department of Food and Nutrition, University of Campinas (UNICAMP), Campinas, SP 13083-862, Brazil

Usman Ali Chaudhry
Infection Control and Disease Prevention Center, Ministry of Health, Tabuk, Kingdom of Saudi Arabia

Jorge Alberto Vieira Costa
Laboratory of Biochemical Engineering, College of Chemistry and Food Engineering, Federal University of Rio Grande, Rio Grande, Brazil

Nivas Desai
Department of Botany, S.P.K. College, Sawantwadi–416510, Maharashtra, India

Uttam Dethe
Department of Botany, S.P.K. College, Sawantwadi–416510, Maharashtra, India

Andrêssa S. Fernandes
Department of Food Science and Technology, Federal University of Santa Maria (UFSM), Roraima Avenue, 1000, 97105-900, Santa Maria, RS, Brazil

Dattatray Gaikwad
Sub-Center, Babasaheb Ambedkar Marathwada University, Aurangabad, Osmanabad–413501, Maharashtra, India

A. Catarina Guedes
CIIMAR-Interdisciplinary Center of Marine and Environmental Research, University of Porto, Novo Edifício do Terminal de Cruzeiros do Porto de Leixões. Av. General Norton de Matos, s/n, 4450-208 Matosinhos, Portugal

Fariha Hasan
Department of Microbiology, Faculty of Biological Sciences, Quaid-i-Azam University, Islamabad–45320, Pakistan

Eduardo Jacob-Lopes
Department of Food Science and Technology, Federal University of Santa Maria (UFSM),
Roraima Avenue, 1000, 97105-900, Santa Maria, RS, Brazil

Mehwish Jaffer
Department of Plant Sciences, Lahore College for Woman University, Lahore–54000, Pakistan

Pannaga Pavan Jutur
Omics of Algae Group, Industrial Biotechnology, International Center for Genetic Engineering and
Biotechnology, Aruna Asaf Ali Marg, New Delhi–110067, India

Senthilkumar Kandasamy
Department of Chemical Engineering, Kongu Engineering College, Perundurai, Erode–638060,
Tamil Nadu, India

Rupali Kaur
Center of Biotechnology, Nehru Science Center, University of Allahabad, Prayagraj,
Uttar Pradesh–211002, India

Samiullah Khan
Department of Microbiology, Faculty of Biological Sciences, Quaid-i-Azam University,
Islamabad–45320, Pakistan

Ameet Kumar
Department of Microbiology, Faculty of Biological Sciences, Quaid-i-Azam University,
Islamabad–45320, Pakistan

Manonmani Kumaraguruparaswami
Department of Food Technology, Kongu Engineering College, Perundurai, Erode–638060,
Tamil Nadu, India

Cristiane Reinaldo Lisboa
Laboratory of Biochemical Engineering, College of Chemistry and Food Engineering, Federal
University of Rio Grande, Rio Grande, Brazil

Graciliana Lopes
CIIMAR-Interdisciplinary Center of Marine and Environmental Research, University of Porto,
Novo Edifício do Terminal de Cruzeiros do Porto de Leixões. Av. General Norton de Matos,
s/n, 4450-208 Matosinhos, Portugal; FCUP-Faculty of Sciences, University of Porto,
Rua do Campo Alegre s/n, 4169-007 Porto, Portugal

Mário R. Maróstica Jr.
Department of Food and Nutrition, University of Campinas (UNICAMP), Campinas, SP 13083-862,
Brazil

Clara Martins
Coimbra Collection of Algae (ACOI), Department of Life Sciences, University of Coimbra,
Coimbra–3000-456, Portugal

Etiele Greque De Morais
Center for Marine Sciences, Faculty of Sciences and Technology, University of Algarve, Algarve, Portugal

Michele Greque De Morais
Laboratory of Microbiology and Biochemistry, College of Chemistry and Food Engineering,
Federal University of Rio Grande, Rio Grande, Brazil

Dhivya Nallamuthu
Department of Food Technology, Kalasalingam Academy of Research and Education,
Virudhunagar–626126, Tamil Nadu, India

Tatiele C. Do Nascimento
Department of Food Science and Technology, Federal University of Santa Maria (UFSM), Roraima Avenue, 1000, 97105-900, Santa Maria, RS, Brazil

Pricila P. Nass
Department of Food Science and Technology, Federal University of Santa Maria (UFSM), Roraima Avenue, 1000, 97105-900, Santa Maria, RS, Brazil

Aneela Nawaz
Department of Microbiology, Faculty of Biological Sciences, Quaid-i-Azam University, Islamabad–45320, Pakistan

Asha Arumugam Nesamma
Omics of Algae Group, Industrial Biotechnology, International Center for Genetic Engineering and Biotechnology, Aruna Asaf Ali Marg, New Delhi–110067, India

Fernando Pagels
CIIMAR-Interdisciplinary Center of Marine and Environmental Research, University of Porto, Novo Edifício do Terminal de Cruzeiros do Porto de Leixões. Av. General Norton de Matos, s/n, 4450-208 Matosinhos, Portugal; FCUP-Faculty of Sciences, University of Porto, Rua do Campo Alegre s/n, 4169-007 Porto, Portugal

Umesh Pawar
Department of Botany, S.P.K. College, Sawantwadi–416510, Maharashtra, India

Md. Akhlaqur Rahman
Department of Biotechnology, S.S. Khanna Girls' Degree College, Prayagraj, Uttar Pradesh–211003, India

Mohammed Rehmanji
Omics of Algae Group, Industrial Biotechnology, International Center for Genetic Engineering and Biotechnology, Aruna Asaf Ali Marg, New Delhi–110067, India

Lília Santos
Coimbra Collection of Algae (ACOI), Department of Life Sciences, University of Coimbra, Coimbra–3000-456, Portugal

Lucielen Oliveira Santos
Laboratory of Biotechnology, College of Chemistry and Food Engineering, Federal University of Rio Grande, Rio Grande, Brazil

Haram Sarfraz
Natural Products Laboratory, Department of Bioengineering, Integral University, Dasauli, Lucknow, Uttar Pradesh–226026, India

Shabnum Shaheen
Department of Plant Sciences, Lahore College for Woman University, Lahore–54000, Pakistan

Samuel Silvestre
CICS-UBI-Health Sciences Research Center, University of Beira Interior, Av. Infante D. Henrique, Covilhã–6200-506, Portugal; CNC-Center for Neuroscience and Cell Biology, University of Coimbra, Coimbra–3004-517, Portugal

Sangeetha Gandhi Sivasubramaniyan
Department of Food Technology, JCT College of Engineering and Technology, Coimbatore–641105, Tamil Nadu, India

Deepak Subramani
Department of Food Technology, Kongu Engineering College, Perundurai, Erode–638060,
Tamil Nadu, India

Shanthy Sundaram
Center of Biotechnology, Nehru Science Center, University of Allahabad, Prayagraj,
Uttar Pradesh–211002, India

Sukannya Suresh
Omics of Algae Group, Industrial Biotechnology, International Center for Genetic Engineering and
Biotechnology, Aruna Asaf Ali Marg, New Delhi–110067, India

Vitor Vasconcelos
CIIMAR-Interdisciplinary Center of Marine and Environmental Research, University of Porto,
Novo Edifício do Terminal de Cruzeiros do Porto de Leixões. Av. General Norton de Matos,
s/n, 4450-208 Matosinhos, Portugal; FCUP-Faculty of Sciences, University of Porto,
Rua do Campo Alegre s/n, 4169-007 Porto, Portugal

Bruna Da Silva Vaz
Laboratory of Microbiology and Biochemistry, College of Chemistry and Food Engineering,
Federal University of Rio Grande, Rio Grande, Brazil

Leila Q. Zepka
Department of Food Science and Technology, Federal University of Santa Maria (UFSM),
Roraima Avenue, 1000, 97105-900, Santa Maria, RS, Brazil

ABBREVIATIONS

$^1O_2$	quenching singlet oxygen
aa	amino acids
AA	arachidonic acid
ACCase	acetyl-CoA carboxylase
ACE	angiotensin I-converting enzyme
ADP	adenosine diphosphate
AEPS	aqueous extracellular polysaccharides
AFPs	antifreeze proteins
AGS	human gastric cancer cell line
AI	angiotensin I
AIDS	acquired immunodeficiency syndrome
AII	angiotensin II
ALA	α-linolenic acid
ALT	alanine aminotransferase
AMA1	apical major antigen
AMPK	adenosine monophosphate-activated protein kinase
APC	allophycocyanin
APTS	aqueous two-phase system
Asp	aspartate
ATP	adenosine triphosphate
BKT	beta-carotene ketolase
B-PE	B-phycoerythrin
BPH-1	non-tumoral human prostate epithelial cell
BV	biliverdin
$CaCl_2$	calcium chloride
Caco-2	human colon cancer cells
CAGR	compound annual growth rate
CaSp	calcium free spirulan
CAT	catalase
CD4	cluster of differentiation 4
CDBs	conjugated double bonds
CEM	human lymphoblastoid leukemia
CFIA	Canadian Food Inspection Agency
CFU	colony forming units

CHL	chlorophyllin
COOH	carboxyl groups
COX-2	cyclooxygenase-2
CPAP	*Chlorella pyrenoidosa* antitumor polypeptide
C-PC	C-phycocyanin
C-PE	C-phycoerythrin
CPPS Ia	*Chlorella pyrenoidosa* polysaccharides Ia
CPPS IIa	*Chlorella pyrenoidosa* polysaccharides IIa
CRISPR/Cas9	clustered regulatory interspaced short palindromic repeats/CRISPR associated protein 9
Cr-SPs	*Chlamydomonas reinhardtii* ethanol-extracted sulfated polysaccharides
CSFV	classical swine flu virus
CTB	cholera toxin B
DGDG	digalactosyl diacylglycerol
DHA	docosahexaenoic acid
DMSP	dimethylsulphoniopropionate
DNA	deoxyribonucleic acid
DPA	docosapentaenoic acid
DPPH	2,2-diphenyl-1-picrylhydrazyl
DSSC	dye-sensitized solar cells
DW	distilled water
e^-	electron
EAAs	essential amino acids
EACC	Ehrlich ascites carcinoma cell
EAE	enzyme-assisted extraction
EDTA	ethylene diamine tetraacetic acid
EFA	essential fatty acids
EFSA	European Food Safety Authority
EMS	ethyl methane sulphonate
EPA	eicosapentaenoic acid
EPS-CP	exopolysaccharides from *C. pyrenoidosa*
EPS-CS	exopolysaccharides from *Chlorococcum* sp.
EPSs	exopolysaccharides
EPS-SS	exopolysaccharides from *Scenedesmus* sp.
EU	European Union
FA	fatty acids
FACS	fluorescence-activated cell sorting
FAMEs	fatty acid methyl esters

FAO	Food and Agricultural Organization
FDA	Food and Drug Administration
FFA	free fatty acid
FGH	flounder growth hormone
FMDV	food and mouth disease virus
fMet-tRNA^{fmet}	N-formyl-methionyl-tRNA
FSANZ	Food standards Australia New Zealand
g day^{-1}	gram per day
g	gram
G361	human malignant melanoma
GAD65	glutamic acid decarboxylase-65
GBSS	granule-bound starch synthase
GGPP	geranylgeranyl pyrophosphate
GIT	gastrointestinal tract
GLA	γ-linolenic acid
Glu	glutamate
GLUT-4	glucose transporter type 4
GM	genetic modification
GMP	good manufacturing practices
GPAT	glycerol-3-phosphate acyltransferases
GPx	glutathione peroxidase
GR	glutathione reductase
GRAS	generally regarded as safe
GSH	glutathione
GSH-Px	glutathione peroxidase
GST	glutathione transferase
GTP	guanosine triphosphate
h	hour
H$_2$	hydrogen
H$_2$O	water
HACCP	hazards analysis and critical control points
HBcAg	hepatitis B virus capsid antigen
HBsAg	hepatitis B surface Antigen
HeLa	human cervical cancer cell line
HepG2	human hepatoma cancer cell line
HIV	human immunodeficiency virus
HPH	high-pressure homogenization
HPV-16	human papillomavirus
Hs578T	human breast cancer cell line

HSV	herpes simplex virus
IBA	indole-3-butyric acid
IC_{50}	half-maximal inhibitory concentration
IFN-γ	interferon-gamma
Ig-E	immunoglobulin E
IL	interleukin
IL-1	interleukin-1
IL-10	interleukin 10
IL-12	interleukin 12
IL-13	interleukin 13
IL-1β	interleukin 1 *beta*
IL-4	interleukin 4
IL-6	interleukin 6
IL-8	interleukin 8
IL-β	interleukin-β
iNOS	inducible nitric oxide synthase
IRS	insulin receptor substrate
kDa	kilodalton
L^{-1}	liter
LC-PUFAs	long-chain polyunsaturated fatty acids
LDL	low-density lipoprotein
LDL-C	low-density lipoprotein-cholesterol
LNA	linoleic acid
LNCaP	human prostate cancer cell-derived from a metastatic site in the lymph node
LTB4	leukotriene B4
MAAs	mycosporine-like amino acids
MAE	microwave-assisted extraction
MAPK	mitogen-activated protein kinase
MCP-1	monocyte chemoattractant protein-1
MDA	malondialdehyde
mg/kg	milligram per kilogram
mg/ml	microgram/microliter
MGDG	monogalactosyl diacylglycerol
mM	millimolar
MPO	myeloperoxide
mRNA	messenger ribonucleic acid
MSP1	major surface protein
MUFA	monounsaturated fatty acids

N_2	nitrogen
NADH	nicotinamide adenine dinucleotide
nASX	natural astaxanthin
NAXA	natural algae astaxanthin association
NDIN	new dietary ingredient notification
NF-κB	nuclear transcription factor *kappa-beta*
NH_2	amino groups
NH_4^+	ammonium
nm	nanometer
NO_2^-	nitrogen dioxide
NO_3^-	nitrate
OtElo5	*Ostreococcus tauri*-Δ5 elongase
PAHs	polycyclic aromatic hydrocarbons
PB	phosphate buffer
PBP	phycobiliprotein
PBR	photobioreactor
PBS	phycobilisome
PC	phycocyanin
PE	phycoerythrin
PEB	phycoerythrobilin
PEG	polyethylene glycol
PepT1	transporter peptide
PGE 2	prostaglandin E2
pH	potential of hydrogen
Pi	phosphate
PI3K-AKT	phosphoinositide 3-kinases-AKT
pKa	negative base logarithm of the acid dissociation constant (Ka)
PMN	polymorphonuclear leukocytes
PMSF	phenylmethylsulfonyl fluoride
PS	photosystem
PSY	phytoene synthases
PUB	phycourobilin
PUFA	polyunsaturated fatty acids
PVB	phycobiliviolin
RAS	renin-angiotensin system
RDA	recommended dietary allowance
ROS	reactive oxygen species
rRNA	ribosomal ribonucleic acid

RT	radiation therapy
SFAs	saturated fatty acids
SFE/SCF	super-critical fluid extraction
SGI	simulated gastric juice
SIJ	simulated intestinal juice
SOD	superoxide dismutase
SQE	squalene epoxidase
TAG	triacylglycerol
TG	triglyceride
TGF-β	transforming growth factor-beta
TNF	tumor necrosis factor
TNF-α	tumor necrosis factor-alpha
TPA	tetradecanoylphorbol-13-acetate
tRNA	transporter ribonucleic acid
TSP	total soluble protein
UAE	ultrasound-assisted extraction
USA	United States of America
USD	United States dollar
US EPA	United States Environmental Protection Agency
US-FDA	United States Food and Drug Administration
UV	ultraviolet
VP28/WSSV	viral envelop protein of white spot syndrome virus
VSV	vesicular stomatitis virus
WAT	white adipose tissue
WHO	World Health Organization
ZEP	zeaxanthin epoxidase

SYMBOLS

α	alpha
β	beta
μg	microgram
μg/g	microgram per gram
μM	micrometer
μmol	micro moles
γ	gamma
Δ	delta
ω	omega

PREFACE

Since several billion years, algae have been ubiquitous on the Earth, with an extremely diverse group of photosynthetic organisms that have evolved successfully from its origin. The algae possess a range of forms, from the picoplankton to macrophytic organisms, which inhabit the open ocean and meadows in the ocean waters, respectively. Different sizes and simple structures of algae support the distribution not only in the aquatic ecosystem but also in other habitats, including extreme environments. These are acting as a vital source for primary productivity and nutrient cycles in the aquatic ecosystem and are an important contributor for a country's economy as it is the base for many valuable products in the commercial market. The simple form and diversified characters of algae lead to exploring it for commercial utilization.

In recent past decades, much research took place on algal genetics and genetic breeding to enhance our knowledge on algae and its genetic resources through advanced molecular tools and techniques such as sequencing and genome analysis. Though, research on algal genomics is in the infancy phase, and algal global gene expression is limited to supporting large-scale algal genome projects. The molecular tool is useful to identify the specific species for future genome studies, and the obtained information is used to understand the functional, structural, and evolutionary aspects of the algae.

Providentially, for several algal species, sequencing of algae projects was implemented and is in progress. *Cyanidioschyzon merolae* (red algae), *Thalasiosira pseudonana* (diatom), and green algae such as *Chlamydomonas reinhardtii, Volovox carteri,* and *Ostreococcus tauri* are the well-studied and advanced genome projects. The algae diversity is enormous due to its size, shape, and the chemical compounds produced during metabolic processes.

Unique and commercially valuable compounds are being produced by different types of algae. Enzymes necessary for the synthesis of omega-3 polyunsaturated fatty acids (PUFA) are lacking in human beings, plants, and animals and hence depend on external sources. Mostly, human beings are dependent on fish and fish oil for this purposes, but most of the microalgae also could be a good source for omega-3 polyunsaturated fatty acids. In most parts of the world, algae are being used as a good nutritious and functional food with many health benefits.

"Functional food," or "nutraceuticals," is a term that expresses the food that contains bioactive compounds that play a major role in human health in addition to providing basic nutrition. *Chlorella* and *Spirulina* are well-known, safe, and highly nutritious foods. Algae are not only used in the pharma and food industries but also act as a main ingredient in the cosmeceutical industry. The term "cosmeceuticals" describes products having both cosmetic and therapeutic characteristics, which supports the beneficial effect on skincare and attractiveness. Compounds obtained from algae-like carrageenan, polysaccharides, proteins, lipids, vitamins, especially A and B1, iron, copper, and other minerals are used as thickening agents, moisturizers, and antioxidants, and act as anti-aging compounds.

Because the algae and their bioactive compounds are possessing greater commercial applications in different industries like food, fodder, nutraceuticals, pharmaceuticals, biofuels, biofertilizers, and cosmetics, they are substantially contributing to a country's economic growth.

This book focuses on the potential applications of microalgae and cyanobacteria in different fields. Mainly, many researchers are working on the application of algae in the fields of pharmaceuticals and nutraceuticals. The book discusses in depth the cosmetical applications of microalgae and cyanobacterial pigments. Other chapters discuss the broad extraction and applications of algal and cyanobacterial bioactive compounds in the fields of nutraceuticals and pharmaceuticals. One of the chapters discusses the potential of freshwater microalgae as anticancer agents. The biological properties and importance of algae and microalgal products are discussed thoroughly.

After reading the book, readers may realize that microalgae and cyanobacteria are treasures for the cosmeceutical, pharmaceutical, and nutraceutical industries as their biotechnological potential are more and necessary for broad industrial applications.

—**Jeyabalan Sangeetha, PhD**
Devarajan Thangadurai, PhD

CHAPTER 1

COSMETICAL APPLICATIONS OF MICROALGAL AND CYANOBACTERIAL PIGMENTS

FERNANDO PAGELS,[1,2] GRACILIANA LOPES,[1,2]
VITOR VASCONCELOS,[1,2] and A. CATARINA GUEDES[1]

[1]CIIMAR-Interdisciplinary Center of Marine and Environmental
Research, University of Porto, Novo Edifício do Terminal de Cruzeiros
do Porto de Leixões. Av. General Norton de Matos,
s/n, 4450-208 Matosinhos, Portugal

[2]FCUP-Faculty of Sciences, University of Porto,
Rua do Campo Alegre s/n, 4169-007 Porto, Portugal

ABSTRACT

Pigments from microalgae and cyanobacteria are known for several bioactivities, and their natural feature and vibrant colors have increased market interest in recent years. Health and cosmetical applications are the most representative sectors in the microalgal industry, mainly because of the large associated-profit and a relatively low need of biomass. Carotenoids, chlorophylls, and phycobiliproteins (PBP) can be used in cosmetical formulations and can be easily obtained from microalgae and cyanobacteria, although there is still competition from synthetic equivalents and other natural sources (vascular plants). These pigments have already been seen as skin tonics, with a capacity of skin protection and prevention and slow-down of skin-associated disorders, due to their recognized antioxidant, anti-inflammatory, anti-aging, antitumoral, and photoprotector activity. This chapter gives a general overview of the potential of microalgae and cyanobacteria for the cosmetic industry, taking into account the challenges of these products from an economic perspective.

1.1   INTRODUCTION

1.1.1   NATURAL COSMETOLOGY

The use of natural products for cosmetical purposes started even before the concept of cosmetics, and for a long time, cosmetics were made from mineral materials and herbal pastes and oils (Scott, 2016). Even with the lack of records, the use of body painting in religious rituals dates back to prehistory, as shown by paintings and archaeological artifacts (Stewart, 2017). With a greater record, ancient Egypt becomes a reference for the use of cosmetics. In fact, the use of cosmetics is an important factor for the analysis of social, economic, and political condition of society through history. In ancient Egypt, for instance, the facial painting was related to nobility and clergy as a way of reasserting their social class or represents a close relation to the divine. On the other hand, it was common for the use of creams and pastes by all society as a way of cleaning the body and avoid infections (Stewart, 2017).

When it comes to the use of algae, both microalgae and seaweeds have been described and used through history as food, feed, and in traditional medicine. Since the Roman Empire, seaweeds such as *Fucus vesiculosus* have been associated with healing properties, being used in creams for both medicinal and cosmetic purposes. Reports from the use of seaweed for these purposes are also found in ancient China, Hawaii, Polynesia, Japan, Ireland, and the UK, among other countries (Anis et al., 2017). In the case of microalgae and cyanobacteria, the consumption of *Arthrospira platensis*, commonly known as Spirulina, dates from the Azteca Empire, although a wide knowledge of its bioactive properties and greater use of other microalgae only took place in the 20th century (Sánchez et al., 2003).

In an opposite way, after the industrial revolution, synthetic chemicals appeared, and the use of detergents and soaps increased. In cosmetics, surfactants appear as revolutionary ingredients, the older homemade cosmetic production declined, and a greater commercialization of industrially-produced cosmetic products began (Ashawat et al., 2009). Such products, which are still used today (in a better developed way) are subject of some controversy about their side effects, such as skin allergies and environmental hazards (Ashawat et al., 2009).

In the 21st century, the trend of green-labeled products, sustainability, and environmental consciousness led to a change on cosmetical industry demands, and to the rescue of the already known sources of bioactive products, plants, and algae (Chiu and Chuang, 2017). Even though a natural origin *per se* does not mean only health improvement and exemption of side effects, it is true that the compounds and extracts used in natural cosmetics have indeed the potential for health benefits. From the formulation, both considering active ingredients and excipients, to the label and packaging material, the cosmetic industry has been adapting the production to a greener way (Sahota, 2013). The recent concept of natural includes harvested, collected, and mined materials that might have been processed without chemical reaction, which might include biotechnological techniques of extraction and purification (Sahota, 2013). Since then, cosmetical industry search for greener ways of production using this strategy for slogans and marketing. In fact, the competitive industry is now searching for new markets and new products with other functions than beauty; with the use of cosmetics with a health purpose, a new sector of industry began, the called cosmeceuticals (Chiu and Chuang, 2017).

With the concept of cosmeceutical, the use of microalgae as a source of ingredients becomes even more potentialized. These organisms are source of phenolic compounds, pigments, polysaccharides, fatty acids (FAs), and peptides with widely described potential for health products (Guedes et al., 2011b). Microalgae and cyanobacteria have been attracting the attention of natural cosmetical industry, usually in the form of extracts, due to their bioactive potential, vibrant, and different colors and cosmetical enhancing-characteristics as moisturizing, stabilizing agents, along with their anti-inflammatory, antioxidant, anti-aging, and whitening capacities (Joshi et al., 2018; Morone et al., 2019).

1.1.2   PIGMENTS AS METABOLITES OF INTEREST

The so-called pigments refer to compounds able to absorb light in the visible spectrum, resulting in a reflection of color that can be perceived by the human eye. Microalgae and cyanobacteria are photosynthetic organisms that produce a wide array of pigments that act as accessories to photosynthesis, and thus harvest light for their metabolism (Masojídek et al., 2013).

Nowadays, besides from *Chlorella vulgaris*, usually found as raw biomass, in tablets or in extracts, one cyanobacterium and two other microalgae are commercialized due to their high production of pigments: *Arthrospira platensis* (Spirulina), known for the production of the blue pigment phycocyanin (PC); *Haematococcus pluvialis*, the largest natural producer of astaxanthin; and *Dunaliella salina*, standing out for the production of β-carotene.

The main constrain in the use of microalgae pigments in industry is due to the high cost of extraction and purification. In addition, the huge variability in pigments content and profile can be a big constrain to the cosmetic industry, as cosmetic formulations require a consistent and stable active ingredient. To overcome these constraints, the standardization of cultivation and extraction techniques, as well as a better understanding of the biological aspects of microalgae pigments are needed (Pagels et al., 2019).

Several classes of pigments are found in microalgae and in cyanobacteria, their profile often constituting a taxonomic marker (Jeffrey et al., 2012). In general, pigments are grouped into three main groups: chlorophylls, carotenoids, and phycobiliproteins (PBPs).

Chlorophylls are a group of compounds with a chemical structure formed by a cyclic tetrapyrrole ring, where a symmetric arrangement of four pyrrole rings forms a chlorin ring where a magnesium molecule is linked. In addition, a phytol group confers the molecule to a hydrophobic behavior. Differences between chlorophyll kinds (*a*, *b*, *c*, *d*, and *f*) are due to small modifications on extremities radicals (Pareek et al., 2017). These molecules represent important natural green pigments, and many microalgae species, especially green microalgae, contain a large amount of chlorophyll in their cells. However, market competition with vascular plants, namely grass, makes the use of chlorophyll from microalgae an expensive source for industrial uses (Odjadjare et al., 2017; Pagels et al., 2019). Chlorophylls α and β have been associated with antimutagenic and antioxidant capacities (Zepka et al., 2019), although to a less extend than other microalgae pigments. Apart from chlorophylls, due to the action of several enzymes inside microalgae cell, it occurs the formation of chlorophylls derivatives, such as pheophytin α and β, pheophorbide α and β, and chlorophyllin (CHL), among others. Such compounds have been also described as antioxidant, chemopreventive, and antimutagenic agents, as well as anti-inflammatory, antimicrobial, and anti-obesogenic (Freitas et al., 2019a; Zepka et al., 2019).

Carotenoids are terpene pigments, derived from a 40-carbon polyene chain, and essential for photosynthetic activity, increasing the light absorption range. There are more than 700 carotenoids described to date, being divided in two classes: the first (carotenes) containing only carbon and hydrogen (e.g., β-carotene) and the second (xanthophylls) containing also oxygen in their structure (e.g., astaxanthin) (Guedes et al., 2011a). Carotenoids can be found as yellow, orange, or red pigments, due to different molecular structures comprising different degrees of unsaturation, as well as the presence of different functional groups. They have a biological role in light-harvesting, but also as scavengers of reactive oxygen species (ROS), being one of the

main non-enzymatic antioxidant components in the photosynthetic apparatus (Zakar et al., 2016). These pigments are known for several important
biological activities, including antioxidant, anti-inflammatory, anticancer,
antimicrobial, and anti-aging capacity, as well as being precursors of vitamins, emphasizing their promising role in cosmetic formulations, not only
as excipients able to confer color or protect the final product from oxidation
or microbial contamination, but also as active ingredients able to treat skin
disorders and slow down the aging process (Pagels et al., 2019).

PBPs are a group of colored proteins present in cyanobacteria and in a few
kinds of microalgae, and responsible for light absorption and photosystem
(PS) regulation. Their molecules are composed of protein subunits and a lipidic
chromophore, called phycobilin (Schluchter et al., 2010). Four kinds of PBPs
can be found in cyanobacteria and red microalgae and their profile varies
according to the species, its classification being correlated to the maximum
absorption wavelength and structure of the molecule: phycoerythrin (PE)
(red/pink), PC (blue), phycoerythrocyanin (red/purple) and allophycocyanin
(APC) (turquoise) (Pagels et al., 2019). In terms of bioactive potential, PBPs
have been associated with antioxidant, anticancer, anti-inflammatory, antimicrobial, anti-diabetes, and antiviral activities what, together with carotenoids,
can constitute an added value for cosmetic industry applications (Pagels et
al., 2019).

Pigments have been a taxonomic marker used for algae classification
since early microalgae descriptions (e.g., Jeffrey, 1974). Cyanobacteria (blue-
green algae) are known to have chlorophyll *a* in their composition, with some
species being able to produce chlorophyll *b*, *c*, *d*, and *f* (Jeffrey et al., 2012;
Zepka et al., 2019). These organisms are able to produce a variety of carotenoids, the most representative being zeaxanthin, β-carotene, and in some
cases echinenone (Jeffrey et al., 2012). Furthermore, cyanobacteria and only
a few groups of microalgae are able to produce PBPs, which are the pigments
produced in larger amounts (Pagels et al., 2019).

Among microalgae, the commercial *Haematococcus pluvialis*, *Chlorella
vulgaris*, and *Dunaliella salina* are included into the Chlorophyta (green
algae) phylum. Chlorophyta contains chlorophyll α and β, and a vast variety
of carotenoids (including lutein, violaxanthin, neoxanthin, antheraxanthin,
β-carotene, zeaxanthin, and astaxanthin). Moreover, diatoms (e.g., *Phaeodactylum tricornutum*) are able to produce fucoxanthin, a carotenoid known
by antioxidant, anti-inflammatory, and anticancer activities, among others
(Peng et al., 2011). The multivalence of microalgae and their pigments
makes their use in natural cosmetics attractive, and also reinforced by the
increase in the consumption of natural products.

This chapter covers fundamental aspects of natural cosmetology, bioactive potential of microalgae and cyanobacteria, the use of the main groups of pigments in cosmetics and finally an economical perspective for the use of the referred microorganisms as source of natural and bioactive colorants in cosmetic industry.

1.2 SKINCARE

Skin is the largest organ in the human body and represents the main barrier of the body against microbes and environmental factors, such as light. In addition, it is fundamental in the process of thermal regulation and touch perception. In its composition, it can be distinct in three tissue layers: epidermis, dermis, and hypodermis (Graham et al., 2019). The first, epidermis, is a layer essentially made up of cells joined together that do not have any intercellular substance between them. As the region with the greatest contact with environmental stresses, it suffers the greatest damage and is most often renewed. In its basal region, it has stem cells that differentiate into keratinocytes, the main cell type in this layer. In addition to the keratinocytes, epidermis shelter melanocytes, the cells responsible for melanin synthesis and light exposure protection (D'Errico et al., 2007). Next, the dermis is the support layer of the skin, with an extracellular matrix rich in structural proteins. It is essential to support the vascularization of the skin, sensory nerves, sebaceous and sweat glands, and hair follicles. This layer, when affected, leads to more common skin disorders, such as inflammation and aging. Finally, the hypodermis, composed essentially of FAs, is responsible for thermoregulation, insulation, nutrition, and protection of the inner tissues from injuries (Graham et al., 2019).

Like any tissue and cell structure, the skin is susceptible to several diseases, disorders, and aging processes, and it is up to the pharma- and cosmeceutical to prevent or treat some of these changes. Regarding skin disorders, it is possible to point five types of situations: lesions, inflammation, abnormal pigmentation, cancer, and aging.

Lesions are injuries or damages that leave a mark in the skin, and they may or may not be associated with other disorders, such as inflammation. Lesions can be classified according to the damage caused: primary lesions, such as pustule (acne), bulla (contact dermatitis) or vesicles (herpes), among others; or secondary lesion, that is the result of irritated or manipulated primary lesions. For example, the formation of a fissure or an excoriation and later a scar or a crust (Stoicescu, 2020). The use of cosmetics in lesions deal with the attenuation of symptoms, prevention, and an anti-inflammatory action.

Inflammation in the skin can be a reflection of disease or oxidative processes. The most common examples are dermatitis, eczema, herpes, and psoriasis. The cosmetic effect can be indirect, related to antioxidant processes or direct, involving inhibitors of inflammatory enzymes. Cosmetic application can lead to remedial action or symptom reduction, as is the case with relaxing creams for rosacea (Kishore et al., 2019).

Abnormal pigmentation, such as vitiligo, lentigo, and melasma, can derive from genetic characteristics or be the consequence of metabolic changes and external stresses, and they may or may not bring problems to health (Spritz and Hearing, 2013). Pigmentation disorders are associated with changes in melanocytes, tyrosinase activity, and melanin concentration; and cosmetics can, in this case, act as a regulator of melanin production, controlling its production, in the case of anti-aging cosmetics or sunscreens or stimulating its production in the case of tanning cosmetics (Chisvert et al., 2018).

Hypertrophy is an abnormal growth of the skin, usually harmless. The most common hypertrophies are keratoma (callus), mole, skin tag and verruca. Hypertrophy can develop into a malignant form, becoming cancer. Skin cancer comes in three distinct types: basal cell carcinoma, squamous cell carcinoma, and the most dangerous, malignant melanoma. The appearance of skin cancer is usually associated with sun exposure, but can happen due to exposure to toxic substances, or due to weakness of the immune system (D´Errico et al., 2007). Herein, cosmetical products can act as preventive, protecting the skin against sun damage, or as post-exposure treatment, reducing the damage caused by the sun or even toxic substances, helping in cell recovery (Lyons et al., 2020).

Finally, one of the biggest problems related to the skin, and those cosmetics try to overcome is the aging process. The skin undergoes two types of aging: intrinsic and extrinsic aging. Intrinsic aging is chronologic and related to genetic and hormonal influence (Tobin, 2017). The main changes are the dryness of the skin, decreased elasticity, fine wrinkles and skin changes that lead to expression lines (Khavkin and Ellis, 2011). On the other hand, extrinsic aging is caused by external factors such as UV radiation, diet, chemicals, trauma, smoking, and chronical diseases. The exposure to such factors leads to deep wrinkles, laxity, roughness, hyperpigmentation, telangiectasia, impaired wound healing, and hypertrophy (Tobin, 2017). UV radiation, in particular, is the main external factor of premature aging (photo-aging), being directly responsible for DNA damage, formation of ROS, inflammation, and immunosuppression (Yaar and Gilchrest, 2007). In terms of cosmetics, sunscreens are the main protection against UV radiation, besides, antioxidants, anti-inflammatories, and tyrosinase inhibitors can decrease symptoms and delay the effects of aging (Yaar and Gilchrest, 2007).

1.3   COSMETIC COMPOSITION

In a cosmetic, the general composition includes an active ingredient and excipients (vehicle, thickening agents and additives) (Balboa et al., 2015). Microalgae and cyanobacteria are able to produce compounds that can be used with different functions within a cosmetic formulation, as schematized in Figure 1.1.

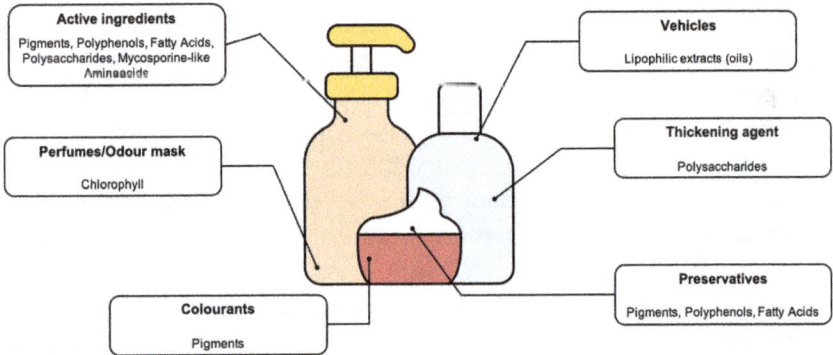

Active ingredients
Pigments, Polyphenols, Fatty Acids, Polysaccharides, Mycosporine-like Aminoacids

Vehicles
Lipophilic extracts (oils)

Perfumes/Odour mask
Chlorophyll

Thickening agent
Polysaccharides

Colourants
Pigments

Preservatives
Pigments, Polyphenols, Fatty Acids

FIGURE 1.1   Potential cosmetical applications of microalgal and cyanobacterial secondary metabolites.

The active ingredients are the main components of the cosmetic product, being responsible for its function, e.g., colorant in makeups, UV protector(s) in sunscreens and anti-aging cream and moisturizers in body lotions (Balboa et al., 2015). Compounds such as pigments, polyphenols, FAs, mycosporine-like amino acids (MAAs) and polysaccharides constitute the groups with greater potential to act as active ingredients in cosmetic formulations, due to the wide range of biological activities with interest in this industry, particularly their antioxidant, anti-inflammatory, photoprotector, and anti-aging capacity, among others (Couteau and Coiffard, 2018).

Besides, all ingredients that do not have a specific function as an active ingredient are called excipients, including the vehicle of the active ingredient, thickening agents, and additives (preservatives, colorants, and perfumes).

In the case of the vehicle, it is necessary to consider the delivery form of the cosmetical, being the component responsible for dissolving or dispersing the active agents and other cosmetic ingredients-water and oils (natural, mineral, or synthetic) (Balboa et al., 2015). Lipophilic extracts from microalgae and cyanobacteria can represent a potential ingredient as a vehicle, as the extracts give the oily characteristic and have also the presence of active ingredients (e.g., carotenoids and polyunsaturated FAs) (Couteau and Coiffard, 2018).

Regarding the thickening ingredients, the main function is to increase the viscosity of the formula, also avoiding dehydration. For this concern, polymers, and methylcellulose are the main used materials. In microalgae and cyanobacteria, the use of polysaccharides is possible, although these organisms usually produce less polysaccharides when compared to other natural sources (e.g., seaweeds) (Mourelle et al., 2017).

In terms of additives, preservatives, colorants, and perfumes are fundamental to allow the preservation of cosmetic formulations, both concerning their chemical and biological stability, to increase their shelf period, and to increase their economic value and the consumer interest and acceptance, through the manipulation of the organoleptic characteristics by means of color and odor change. To achieve this, many synthetic compounds are usually added, although natural sources are now being preferred because of its multivalent feature of acting as a preservative and as an active ingredient (Balboa et al., 2015). Preservative components are added to increase the stability and durability of the cosmetic and can be classified as antioxidants and antimicrobials. Again, microalgae and cyanobacteria are able to produce several compounds able to fit in this requirement, including pigments, polyphenols, flavonoids, and MAAs (Morone et al., 2019). Furthermore, in terms of colorants, microalgae and cyanobacteria pigments have a great advantage by the range of colors that can be found-blue, pink (PBPs), orange, yellow, red (carotenoids) and green (chlorophylls). Moreover, as perfume components, microalgae, and cyanobacteria are able to produce compounds that can mask or reduce unpleasant odors, such as chlorophyll, that can be used as deodorant (Mourelle et al., 2017) or to reduce the smell in wounds during treatments (Hosikian et al., 2010).

1.4 CAROTENOIDS IN COSMETICS

Within cells, carotenoids are responsible for absorbing light for the photosynthetic reaction, constituting one of the main barriers against oxidative stress in the PS, due to their high efficiency in quenching singlet oxygen ($^1O_2$). Thus, carotenoids are responsible for the non-photosynthetic quenching, dissipating part of the absorbed radiation in the form of heat, and consequently decreasing potential damage to the organism (Anunciato and Rocha-Filho, 2012). When extracted, the molecule can be used with a similar function in human skin, where the continuous exposition of skin to sunlight (UV radiation and high intensity light) leads to oxidative stress and a possible damage of cell membranes and DNA (Shegokar and Mitri, 2012). Carotenoids can be used in sunscreens, anti-aging, and antioxidant formulations. Moreover, the anti-inflammatory and antimicrobial capacity also associated to these pigments, can increase their interest for cosmeceutical purposes (Shegokar and Mitri, 2012).

Two strategies for carotenoids action in skin can be followed. In the first, the product is applied topically, and the delivery occurs from the external skin layers to the internal ones. The second, by oral intake, the delivery occurs by diffusion from the adipose tissue, blood, and lymph, or through glandular secretion on the skin surface. The strategy must take into account the stability of carotenoids; once these pigments are very unstable; therefore, antioxidant additives are recommended for the carotenoid-containing cosmetical formulations (Meléndez-Martínez et al., 2019). On the other hand, the use of extracts overcomes part of this issue, since they contain, apart from pigments, other compounds that increase not only the stability but also the useful life of the aforementioned pigments (e.g., polyphenols).

The uses of carotenoids from microalgae and cyanobacteria for skin health have been reported by several studies. β-Carotene from *Dunaliella salina* and astaxanthin from *Haematococcus pluvialis* are the most widely used, while *Nannochloropsis* spp. appear as potential source of canthaxanthin, and diatoms as a source of fucoxanthin. The main findings of the skin health effect of carotenoids and patents focusing their cosmetic application are summarized in Table 1.1.

The potential of β-carotene has been discussed in both *in vitro* and *in vivo* studies, including clinical trials. As carotenoids, in general, β-carotene is often associated to photoprotection mechanisms. Stahl et al. (2000) observed a decrease in UV-induced erythema formation on human skin by using β-carotene from *Dunaliella salina* as supplementation, together with α-tocopherol in an oral-intake clinical trial. The trial was performed inducing erythema in healthy patients by illuminating the skin with a solar simulator. After 8 weeks, the erythema suppression was higher in treatments with carotenoids mix. The antioxidant effect of the two compounds provided protection against the UV radiation. The supplement was provided orally, and an increase in β-carotene in the serum was observed.

TABLE 1.1 Use of Carotenoids from Microalgae and Cyanobacteria as Active Ingredients in Cosmetics

Pigments	Source	Application	References
Astaxanthin	*Haematococcus pluvialis*	Anti-aging	Tominaga et al. (2012)
	Haematococcus pluvialis	Antitumoral	Rao et al. (2013)
	Fuji Chemical Industry Co. Ltd.	UV protection	Yoshihisa et al. (2014)
	Sigma-Aldrich	Anti-aging	Tominaga et al. (2017)
	Swiss Pharmaceutical Industries	UV protection	Lyons and O'Brien (2002)
	—	Moisturizer	Singh et al. (2020)
	—	Anti-aging	Chalyk et al. (2017)
	—	Anti-inflammatory	Tso and Lam (1996)
β-Carotene	*Dunaliella salina*	UV protection	Stahl et al. (2000)
	Purified, Sigma-Aldrich	High intensity light protection	Freitas et al. (2019b)
	—	IR protection	Darvin et al. (2011)
	—	Sun protection/Anti-aging	Gartner et al. (2003)
Canthaxanthin	*Nannochloropsis* spp.	Tanning pill	Mourelle et al. (2017)
Carotenoids mix	*Dunaliella salina*	UV protection	Sies and Stahl (2004)
	Betatene (Cognis, Germany), Lyc-o-mato, (Lycored, Israel), Covitol (Cognis, Germany)	UV protection	Césarini et al. (2003)
	Various microalgae	Antioxidant	Hettwer and Stebler (2019)
	Microalgae and vascular plants	UV protection	Balić and Mokos (2019)
	—	Tanning/sun allergies	Blime (1996)
	—	UV protection	Pradier and Fanchon (1995)

TABLE 1.1 *(Continued)*

Pigments	Source	Application	References
Carotenoids-rich extract	*Porphyridium cruentum*	Sun protection	In-ki et al. (2018)
Fucoxanthin	Diatoms and brown seaweed	UV protection	Joshi et al. (2018)
Lutein, β-carotene, and lycopene	–	Antioxidant	Yeum (2005)
Phytoene and Phytofluene	*Dunaliella salina*	Anti-aging	Oppen-Bezalel and Shaish (2009)
Zeaxanthin	–	Sun protection/tanning	Giehart and Fox (2012)

A complementary effect was found in *in vitro* studies devoted to the evaluation of the effect of β-carotene and flavonoids as photoprotectors, in special, regarding the blockage of melanogenesis induced by visible light. β-carotene showed to be effective in protecting cells from photoinduced death, being ideal for sunscreens formulations, as a topical antioxidant (Freitas et al., 2019b). The same effect was observed *in vivo*, in clinical trials (Stahl et al., 2006) with humans, where after 10–12 weeks of lycopene oral-intake, the sensitivity towards UV-induced erythema was decreases, as lycopene acted as protector against UV induced damage. In addition, Darvin et al. (2011) have explored the topical use of β-carotene as an antioxidant for skin protection against infra-red (IR) radiation, demonstrating that the pre-treatment with a superficial antioxidant was the most important defense against external stressors. Moreover, anti-inflammatory capacity has been also associated to β-carotene. This carotenoid is able to modulate biological targets such as cyclooxygenase-2 (COX-2), along with antioxidant capacity (radical quenching) (Berthon et al., 2017).

Regarding astaxanthin, Tominaga et al. (2012) showed the use of astaxanthin from *Haematococcus pluvialis* as supplement for skin health, including effects on the prevention of hyperpigmentation and photoaging. The treatment included an oral intake of astaxanthin supplement, in addition to topical application twice a day, in a group of 20 to 55 aged healthy individuals. The authors observed that after 4 weeks' treatment, an improvement in skin texture was observed and, after 8 weeks, there was improvement on wrinkles depth, age spot size, elasticity, and moister content.

Astaxanthin have been also associated with UV protection effect in fibroblast and melanocytes cells (Lyons and O'Brien, 2002). The oral treatment with astaxanthin-rich extract prevented UV-induced effect on superoxide dismutase (SOD) and cellular glutathione (GSH), which are usually reduced by this radiation. Astaxanthin was also able to decrease melanin production and skin thickening induced by UV radiation when applied in an astaxanthin-liposomal formulation, in a hairless mice model, before UV exposure (Hama et al., 2012). Thus, Yoshihisa et al. (2014) showed that astaxanthin decreased the protein content and the mRNA levels of inducible nitric oxide synthase (iNOS) and COX-2, plus decreasing the concentration of prostaglandins released from keratinocytes after UV irradiation. The authors also observed that astaxanthin could inhibit UV induced apoptosis in keratinocytes (*in vitro*). In clinical trials with healthy women between 35 and 60 years, astaxanthin increased skin moisture content and reduced wrinkles, being pointed out for anti-aging applications and maintenance of skin hydration and structure (Tominaga et al., 2017; Singh et al., 2020).

Astaxanthin has also demonstrated anticancer capacity in a skin cancer mouse model, reducing the UV induced tumor by up to 96% (Rao et al., 2013). The use of astaxanthin was also related to anti-age-related changes in skin in clinical trials. In trials performed with healthy adult Caucasian males and females over the age of 40, this pigment decreased levels of desquamation and microbial presence after 15 days treatment, and even more after 29 days treatment (Chalyk et al., 2017).

Ng et al. (2020) have summarized several clinical trials of oral intake of astaxanthin. In general, the studies last from 2 to 16 weeks, with a range of observed effects. Overall, astaxanthin intake had improved skin appearance, reduced, and/or reverse age-related changes, improving skin moisture, elasticity, and integrity, reducing wrinkles and age spots, and increasing protection against UV damage.

Other carotenoids, such as fucoxanthin, coming from diatoms and brown seaweeds, can reduce tyrosinase activity and, consequently, melanogenesis (Joshi et al., 2018). Thus canthaxanthin, from *Nannochloropsis* spp. was suggested for tanning pills, giving a yellow/orange color after regular intake (Mourelle et al., 2017). Moreover, two colorless carotenoids precursors (phytoene and phytofluene) from *Dunaliella salina*, can be used for the prevention of skin aging as a consequence of UV radiation (Oppen-Bezalel and Shaish, 2009). In addition, carotenoid mixtures have also been suggested for photoprotection against UV radiation, preventing skin aging and skin cancers (Césarini et al., 2003), sunburn reduction (Sies and Stahl, 2004), melanogenesis, and erythema formation (Balić and Mokos, 2019).

Furthermore, several patents have been made with carotenoids for cosmetical/cosmeceutical applications. Overall, the claims are related to skin protection against sun/UV radiation. A European patent from JBC Cosmetics claims the oral intake of carotenes for tanning and prevention of sun allergies (Blime, 1996). While other suggests the use of a formulation of β-carotene, lutein, and lycopene for sun protection and anti-aging (Gartner et al., 2003).

A French patent from the big cosmetic company L'Oréal, claim the use of carotenes (α-, β-, and γ-carotene and lycopene) together with vitamins for UV protection (Pradier and Fanchon, 1995). A similar claim was later found in a German patent, postulating the use of carotenoids from several microalgae - e.g., *Tetradesmus*, *Isochrysis*, *Chlorella*, and *Arthrospira* genera-for sun protection, as an antioxidant, reducing harmful effects of blue light (Hettwer and Stebler, 2019). With the microalgae *Porphyridium cruentum*, a Korean patent suggests the use of carotenoid-rich extract for sun protection and also for scalp irritation relief (In-Ki et al., 2018).

The patent claims are sometimes associated to microalgae and cyanobacteria; in other cases, are vaguely described and in some extent trying to cover all carotenoid-producer organisms, as in the case of an American patent that suggests the oral intake of zeaxanthin for tanning, and protection against sunburn (Giehart and Fox, 2012), and other that claims the use of astaxanthin as anti-inflammatory (Tso and Lam, 1996). A similar strategy was used in an international patent that claims the use of carotenoids mix (lutein, β-carotene, and lycopene) as an antioxidant ingredient, also decreasing DNA oxidation and lipid peroxidation (Yeum, 2005).

1.5   USE OF PHYCOBILIPROTEINS (PBPs) IN COSMETICS

PBPs are hydrophilic pigments, with a blue (PC, APC) or pink color (PE, phycoerythrocyanin). Due to their solubility in water, the use of these kinds of pigments in cosmetics for skincare, specially creams and lotions, is relatively easy. In addition, PBPs have a great range of heat and pH stability (Arad and Yaron, 1992), what can be an asset regarding the process associated with the development of the formulation, as well as its stability when regarding the application in different regions of the body with variable pH. PBPs are used in cosmetics as natural dyes, to reduce toxicity and skin damage, as well as allergies related to synthetic dyes (Balboa et al., 2015). As functional ingredients, PBPs have been correlated to several bioactivities, such as antioxidant, antiviral, antimicrobial, and antitumoral, among others (Pagels et al., 2019). The versatility of PBPs increases their biotechnological potential, since these molecules can be an active ingredient with different functions and claims.

The main source of PBPs for cosmetical use is *Arthrospira platensis*, although, in more fundamental studies, other cyanobacteria and microalgae have been suggested. Table 1.2 gather some examples of the use of PBPs for cosmetic applications.

Regarding skin benefits, some studies and patents relate PBPs mainly to an anti-inflammatory and antioxidant action. As anti-inflammatories, PBPs can act as inhibitors of COX-2 enzyme, reducing pain and inflammation; and also, as skin-soothing agents, in cases of rosacea and inflammatory acne (Pieloch, 2006; Leung et al., 2013). Also, these pigments can reduce autoimmune responses (Cervantes-Llanos et al., 2018), and epithelial apoptosis (Kim et al., 2018). As antioxidants, PBPs mechanisms involve mainly scavenge of free radicals (Zhou et al., 2005). Specifically, PC has been described as anti-aging due to antioxidant capacity responsible for delaying the effects of skin degradation caused by oxidative stress (Mobin et al., 2019).

TABLE 1.2 Cosmetic Applications of Phycobiliproteins from Microalgae and Cyanobacteria

Pigments	Source	Application	References
Phycocyanin	*Aphanizomenon flos-aquae*	Wound healing	Castangia et al. (2016)
	Arthrospira platensis	Dye (waterproof make-up)	Shimamatsu and Tanabe (1981)
	Arthrospira platensis	Anti-inflammatory	Pieloch (2006); Leung et al. (2013); Cervantes-Llanos et al. (2018); Kim et al. (2018)
	Arthrospira platensis	Skincare, reduction of serum lipids	Fujiwaka and Matsushima (2006)
	Arthrospira platensis	Skin hydration/UV protection	Haizhou et al. (2018)
Phycobiliprotein-rich extract	*Arthrospira platensis* and red microalgae	Dye (eye shadow and lipstick)	Arad and Yaron (1992)
	Aphanizomenon flos-aquae	Anti-inflammatory	Scoglio et al. (2012)
	Arthrospira platensis	Anti-aging	Mobin et al. (2019)
	Anacystis nidulans	Sun protection, anti-aging	Canovas et al. (2018)

PC has also been described as an inhibitor of skin lipases, with direct benefits to skin health, acting as a tranquilizer by reducing common skin undesirable symptoms, such as itching and pruritus (Fujiwaka and Matsushima, 2006; Spolaore et al., 2006). Moreover, this pigment can be used as wound healing, as it has an indirect effect on keratinocytes migration, leading to a faster and more efficient healing (Madhyastha et al., 2012).

Arad and Yaron (1992) described the use of PE from red microalgae in the composition of eye shadows. The application of PE reported in this study is comparable to those reported in a previous patent from the Japanese company Dainippon Ink and Chemicals, that used *Arthrospira platensis* PC in a blue-product line. The brand also included the use of the blue pigment in food products (ice-cream, gum, and candy) (Arad and Yaron, 1992; Spolaore et al., 2006).

Unfortunately, depending on the target epidermal layer, the topical administration of PC formulations may be limited, due to its high molecular weight. In order to overcome this, Castangia et al. (2016) have increase the bioavailability of PC as a wound healer, by encapsulating it in phospholipid vesicles, immobilized with hyaluronan sodium salt.

Another Japanese patent described a method for the obtention of an insoluble PC from *Arthrospira platensis* through a treatment with ethanol subsequent to its extraction. By giving an insoluble characteristic to the pigment, the dye becomes more resistant to sweat, being able to be applied in waterproof make-up (Shimamatsu and Tanabe, 1981; Sekar and Chandramohan, 2008). Moreover, a Chinese patent use PC from *Arthrospira platensis* as an active ingredient for cosmetical uses, together with microalgae and vascular plants extracts. The patent claims an improvement in skin hydration and UV protective effects (Haizhou et al., 2018).

Additionally, two American patents use PBPs as cosmetical ingredients, the first, a PC-rich extract from *Aphanizomenon flos-aquae*, claimed to act in prophylaxis or treatment of skin diseases, and decrease chronic inflammation (Scoglio et al., 2012). The second, claim the use of PBPs extracted from *Anacystis nidulans* to repair effects of sun exposure and premature aging on human skin (Canovas et al., 2018).

1.6 USE OF CHLOROPHYLLS AND DERIVATES IN COSMETICS

Chlorophyll is the most present compound in green natural dyes, and also abundant in microalgae and cyanobacteria. Industrially, it is common the use of vascular plants as a source of this pigment, with extremely low cost (e.g., grass). The industrial production of chlorophyll is estimated in ca. 1 billion tons per year (Mandal et al., 2020). Chlorophyll from microalgae and cyanobacteria can be an opportunity in a biorefinery process, by using co-products to cover the cost of production, allowing the use of chlorophyll as a potential product.

Besides chlorophyll, its derivatives have been described as bioactive molecules; it is the case of pheophytin α and β, described, and used for skin topical treatments in their purified forms (Zepka et al., 2019; Mandal et al., 2020). Chlorophyll and its derivates have been associated to several bioactivities, including antitumoral and anti-inflammatory, wound healer and photo-sensitizer in photodynamic therapy (Mandal et al., 2020). Their use with cosmetical potential is summarized in Table 1.3.

Chlorophyll use as colorant in dental powder has been patented by Colgate-Palmolive Company in a water-soluble form, including also several forms of CHL (Salzmann and Schiraldi, 1960). Regarding derivates, Lyapina et al. (2010) found that pheophorbide α extracted from *Arthrospira platensis*, together with Protoporphyrin IX (purified commercial reagent) can be used in skin for wound healing. The use of these pigments as photosensitizers in the skin, decreased the production of leukocytes and inhibited SOD enzyme activity. Moreover, Nakamura et al. (1996) have described the use of purified pheophorbide α as antitumoral in mouse skin tumors. The compound, applied topically, was able to inhibit tumor promotion mainly due to the anti-inflammatory activity by suppressing leukocytes activation.

TABLE 1.3   Cosmetic Applications of Chlorophylls from Microalgae and Cyanobacteria and its Derivatives

Pigments	Source	Application	References
Chlorophyll and chlorophyllins	–	Colorant in dental powder	Salzmann and Schiraldi (1960)
Chlorophyllin	Aldrich Chemical Co.	Chemopreventive	Park and Surh (1996); Chung et al. (1999)
Pheophorbide α	*Arthrospira platensis*	Wound healer/antioxidant	Lyapina et al. (2010)
	Toyo-Hakka Co. Ltd.	Antitumoral/anti-inflammatory	Nakamura et al. (1996)
Pheophytin α	*Cyanobium* sp., *Nodosilinea* sp. and *Arthrospira platensis*	Anti-obesogenic	Freitas et al. (2019a)
	Ulva prolifera (seaweed)	Anti-inflammatory	Okai and Higashi-Okai (1997)
Pheophytin α and β	*Camellia sinensis* (Green tea)	Antitumoral	Higashi-Okai et al. (1998)
Protoporphyrin IX	Aldrich Chemical Co.	Wound healer/antioxidant	Lyapina et al. (2010)

Other chlorophyll derivate, pheophytin α, showed anti-obesogenic capacity. The compounds, extracted from the marine cyanobacteria *Cyanobium* sp., *Nodosilinea* sp., and from commercial *Arthrospira platensis* biomass, exhibited lipid-reducing capacity in zebrafish models (Freitas et al., 2019a). Moreover, pheophytin α extracted from the green seaweed *Ulva prolifera*, has been also suggested as anti-inflammatory, as the extract showed a suppression on superoxide radical production and reduced the induction of edema formation in mouse skin (Okai and Higashi-Okai, 1997). Also, Higashi-Okai et al. (1998) has observed antitumoral activity of pheophytin α and β extracts from the vascular plant *Camellia sinensis* (green tea)-against mouse skin tumor, suppressing the skin tumorigenesis and inflammatory reactions induced by the tumor promoter.

In the case of CHL, its use has been associated with chemopreventive activity against mouse skin carcinogenesis and its administration can be performed orally, with a quick distribution through skin cells. Chemoprevention bases in the use of a drug before the formation or in the initial stages of cancer development (Park and Surh, 1996). Chemopreventive effect was also observed by Chung et al. (1999) in mouse skin cells, where the CHL activity was effective in the promotional and progression stage, but not effective if applied before the tumor induction. Moreover, CHL has been also related to antioxidant capacity, prevention of body odor and increasing wound healing (Nahata et al., 1983; Suryavanshi et al., 2015).

1.7 MICROALGAL AND CYANOBACTERIAL EXTRACTS IN COSMETICS

Microalgal and cyanobacterial pigments can be used as pure compounds or, more commonly, as raw extracts. Purification of compounds may represent up to 80% of the cost of production, and the use of extracts can be advantageous due to the lower production-associated costs (Acién et al., 2012) and greater stability. For example, carotenoids can be sold as oleoresin, containing ca. 20–30% of the specific carotenoid (Panis and Carreon, 2016). Also, it is possible to take advantage of compounds synergistic interaction, increasing the bioactive potential of the extract, plus, the presence of several classes of compounds represents the use of a single extract for different uses: polysaccharides for moisturizing, MAAs for UV absorption and PBPs as antioxidant and anti-inflammatories. Table 1.4 summarizes the main application of cyanobacterial and microalgal extracts as cosmetical bioactive ingredients. Microalgae and cyanobacteria extracts have already been used as cosmetic ingredients, in special in face, hands, and body creams and lotions (Mobin et al., 2019), but the extracts can also be used for hair care, as hair growth promoters (Nethravathy et al., 2019).

In the case of cyanobacteria, ethanolic extracts from *Arthrospira platensis* are suggested as potential tyrosinase inhibitors (Sahin, 2018) and as a source of carotenoids, increasing skin protection against age-related changes (Darvin et al., 2011). Furthermore, Morone et al. (2019) gather several studies with cyanobacteria extracts with antioxidant activity, in special, organic solvent extracts (methanolic and ethanolic). Singh et al. (2014) evaluated the antioxidant capacity of four cyanobacteria methanolic extracts (*Plectonema*

TABLE 1.4 Cosmetic Application of Microalgal and Cyanobacterial Extracts

Microalgae/ Cyanobacteria	Extraction Solvent	Application	References
Anabaena sp., *Nostoc* sp., *Calothrix* sp., *Oscillatoria* sp., and *Phormidium* sp.	Ethanol	Antioxidant	Babić et al. (2016)
Arthrospira platensis	Ethanol	Photoprotection and anti-aging	Darvin et al. (2011); Sahin (2018)
Cylindrotheca closterium, Odontella mobiliensis, Pseudo nitzschia and *Pseudo delicatissima*	Non-polar (acetone+resin)	Anti-inflammatory	Lauritano et al. (2016)
Dunaliella salina	Ethanol and methanol	Antioxidant and antimicrobial	Cakmak et al. (2014)
Leptolyngbya-like sp. and *Nodosilinea antarctica*	Ethanol and acetone	Antioxidant, anti-inflammatory, and antiproliferative	Lopes et al. (2020)
Nannochloropsis oculata	–	Photoprotection and elasticity enhancer	Mobin et al. (2019)
Plectonema boryanum, Hapalosiphon intricatus, Anabaena doliolum and *Oscillatoria acuta*	Methanol	Antioxidant	Singh et al. (2014)
Skeletonema marinoi	Non-polar (acetone+resin)	Anticancer	Lauritano et al. (2016)
Synechocystis spp., *Leptolyngbya* spp. and *Oscillatoria* sp.	Methanol	Antioxidant	Aydaş et al. (2013)
Tetraselmis tetrathele	Methanol	Antioxidant and anti-wrinkle	Farahin et al. (2019)

boryanum, Hapalosiphon intricatus, Anabaena doliolum and *Oscillatoria acuta*), being *Hapalosiphon intricatus* extract the one with highest capacity to scavenge DPPH (2,2-diphenyl-1-picrylhydrazyl) radical. Similar potential is suggested for the methanolic extracts of *Synechocystis* spp., *Leptolyngbya* spp., *Oscillatoria* sp. (Aydaş et al., 2013) and ethanolic extracts of *Anabaena* sp., *Nostoc* sp., *Calothrix* sp., *Oscillatoria* sp. and *Phormidium* sp. (Babić et al., 2016). In general, methanolic and ethanolic extracts from microalgae and cyanobacteria contain bioactive compounds with potential application in cosmetics such as polyphenols, flavonoids, and carotenoids (Morone et al.,

2019). Thus, Lopes et al. (2020) showed the potential of ethanolic and acetonic extracts (containing lutein, zeaxanthin, β-carotene, among others carotenoids) from cyanobacteria (*Leptolyngbya*-like sp. and *Nodosilinea antarctica*) with antioxidant, anti-inflammatory, and antiproliferative capacity, with potential application in psoriasis treatment.

In the case of microalgae, a *Nannochloropsis oculata* carotenoid-rich extract is suggested to increase skin elasticity and provide photoprotection (Mobin et al., 2019). In terms of bioactive potential, *Dunaliella salina* methanolic and ethanolic extracts have been suggested as containing antioxidant and antimicrobial (antibacterial and antifungal) capacity, including antibiotic capacity against food pathogenic *Escherichia coli* and *Salmonella enteritidis* (Cakmak et al., 2014). Furthermore, Lauritano et al. (2016) studied non-polar extracts (acetone extraction followed by a resin recovery) from 32 species of microalgae in terms of the bioactive potential, including antioxidant, anti-inflammatory, anticancer, antidiabetic, antibacterial, and anti-biofilm capacities. Six diatom extracts have shown bioactive potential without presenting cytotoxicity in HepG2 cells: *Cylindrotheca closterium*, *Odontella mobiliensis, Pseudo nitzschia,* and *Pseudo delicatissima* showed anti-inflammatory capacity, *Skeletonema marinoi* showed anticancer capacity and *Leptocylindrus danicus* and *Leptocylindrus aporus* showed anti-biofilm properties. Moreover, Farahin et al. (2019) observed the stability of *Tetraselmis tetrathele* methanolic extracts in cosmetical creams, as a source of phenolic compounds, carotenoids, FAs, and vitamins, and suggested the topical application of the formula for sunscreens and anti-wrinkles.

1.8 MARKET: CHALLENGES AND OPPORTUNITIES

Changes in the perception of consumers in relation to the cosmetic industry has brought changes in relation to products and regulatory measures and the so-called "green label" represents the greatest potential of microalgae and cyanobacteria in the industry. Consumers are increasingly aware of health care, and it is incredibly common to associate the idea of natural compounds with health benefits. The European Union (EU) participates, in a meaningful way, in encouraging natural products and, therefore, in the decrease in the use of synthetic pigments. Regarding cosmetics and natural cosmetics, the European companies L'Oréal and Unilever represent key players in industry, followed by the Americans Estée Lauder and Procter and Gamble. In terms of consumers, Asia represents 41% of the market, followed by North America (24%) and Europe (24%) (L'Oréal, 2019).

The global market for cosmetics was valued at ca. $220 billion (USD) in 2019, with the main players in sale being L´Oréal, with ca. $32 billion (USD) and Unilever with ca. $22 billion (USD) (L´Oréal, 2019). Besides, the global market for algal products (including cyanobacteria, microalgae, and seaweed) was evaluated in $3.8 billion (USD) in 2017 with an estimative of reaching $5.2 billion in 2023 (USD), being Asia, North America, and Europe the biggest consumers (Markets and Markets, 2018a). In terms of products, seaweeds are the most representative source of algal products, followed by cyanobacteria, in special, *Arthrospira platensis*.

When it comes to pigments, the market value is shared mainly by food and cosmetical industry. For food industry, the use of pigments (natural and synthetics) represents a value of $3.8 billion (USD) in 2018 (Markets and Markets, 2018b) and on cosmetical industry, the pigment market is estimated in ca. $700 million (USD) in 2019, with a predicted growth of 8% per year by 2024 (Markets and Markets, 2019). More specifically, the total market of carotenoids was evaluated in $1.5 billion (US) in 2019, with a prediction of $2.0 billion (US) by 2026; and β-carotene, lutein, and astaxanthin are responsible for 60% of the market (Gong and Bassi, 2016). Regarding PBPs, the market value is ca. $100 million (USD) and it is estimated to be over $200 million (USD) by 2028 (Pagels et al., 2019). Finally, the chlorophyll market was estimated at $280 million (USD) in 2019, possibly reaching $460 million (USD) by 2025 (Value Market Research, 2020).

The use of microalgae for cosmetics is regulated by the demand of the market in relation to more natural sources of ingredients in cosmetic products. Figure 1.2 shows a SWOT analysis of the use of microalgae and cyanobacteria pigments for cosmetical applications.

In general, appearance is one of the main factors for choosing a product, and the use of color to make the product more appealing is a widely used strategy for attracting consumers' attention that is not restricted to the cosmetic industry. Thus, the pigments obtained from microalgae and cyanobacteria, in addition to all the bioactive attributes discussed throughout this chapter, are part of an economic strategy to increase the visibility of the product, either by natural origin or by the color (Baker and Günther, 2004). Moreover, the presence of other compounds in microalgal and cyanobacterial extracts represents an opportunity for the use of different bioactivities in one single product.

The main constraints in the use of microalgae as a source of pigments for cosmetical industry is related to the production-associated costs, the extraction, processing, and further purification of pigments (Acién et al.,

2012). In addition, weather conditions are a huge problem on microalgal and cyanobacteria culture in open space, as productivity and pigment content can be affected by changes in temperature or light disponibility. Furthermore, there is still no standard procedure for the obtention of pigments, and the protocols must be adapted to each strain or target product, leading to an inconsistency in production that may vary the characteristics of the final product.

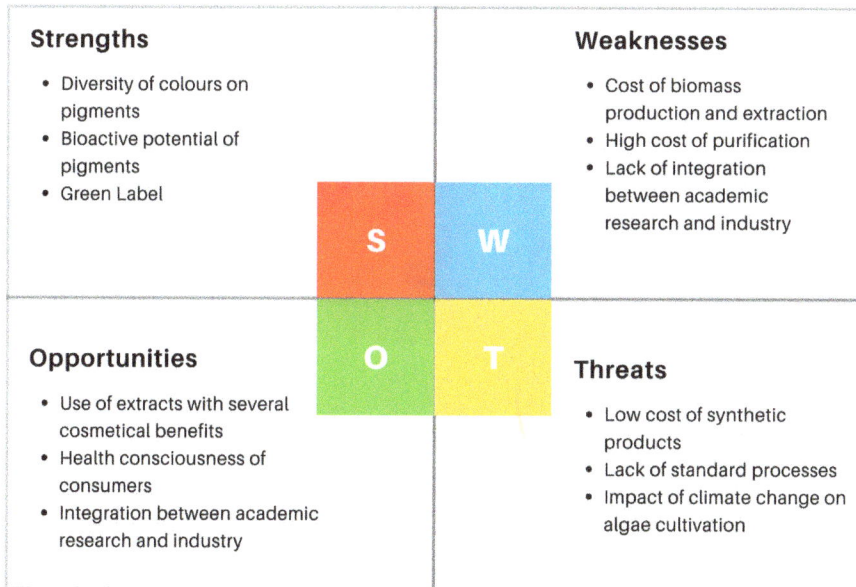

FIGURE 1.2 SWOT analysis of the use of microalgal and cyanobacterial pigments for cosmetical applications.

After all, a complex framework is needed between academic research and industry. Much of the advances and uses of microalgae and cyanobacterial pigments are protected by industrial protocols or patents. On the other hand, in academic research, the examples given usually take into account specific bioactivities and potential of specific compounds, as pointed in several articles and reviews that have been discussed herein. The joint work of the two R&D mechanisms is necessary for a sustainable advance, increasing the potential for the use of microalgae and cyanobacteria as sources of cosmetic ingredients with benefits for the well-being of consumers, whether pigments or not.

ACKNOWLEDGMENTS

A PhD fellowship (SFRH/BD/136767/2018) for author Fernando Pagels was granted by Fundação para a Ciência e Tecnologia (FCT, Portugal) under the auspices of Programa Operacional Capital Humano (POCH), supported by the European Social Fund and Portuguese Funds (MECTES). This work was financially co-supported by the strategical funding from FCT UIDB/04423/2020 and UIDP/04423/2020 and the BlueHuman-BLUE biotechnology as a road for innovation on human's health aiming smart growth in Atlantic Area (EAPA_151/2016).

KEYWORDS

- anti-aging
- anti-inflammatory
- antioxidant
- antitumoral
- astaxanthin
- carotenoids
- chlorophyll
- fucoxanthin
- pheophytin
- photoprotection
- phycobiliproteins
- phycocyanin
- phycoerythrin
- skincare
- β-carotene

REFERENCES

Acién, F. G., Fernández, J. M., Magán, J. J., & Molina, E., (2012). Production cost of a real microalgae production plant and strategies to reduce it. *Biotechnol. Adv., 30*, 1344–1353. https://doi.org/https://doi.org/10.1016/j.biotechadv.2012.02.005.

Anis, M., Ahmed, S., & Hasan, M. M., (2017). Algae as nutrition, medicine and cosmetic: The forgotten history, present status, and future trends. *World J. Pharm. Pharm. Sci., 6,* 1934–1959. https://doi.org/https://doi.org/10.20959/wjpps20176-9447.

Anunciato, T. P., & Da Rocha, F. P. A., (2012). Carotenoids and polyphenols in nutricosmetics, nutraceuticals, and cosmeceuticals. *J. Cosmet. Dermatol., 11*(1), 51–54. https://doi.org/10.1111/j.1473-2165.2011.00600.x.

Arad, S. M., & Yaron, A., (1992). Natural pigments from red microalgae for use in foods and cosmetics. *Trends Food Sci. Technol., 3,* 92–97. https://doi.org/10.1016/ 0924-2244(92) 90145-M.

Ashawat, M., Banchhor, M., Saraf, S., & Saraf, S., (2009). Herbal cosmetics: Trends in skincare formulation. *Pharmacogn. Rev., 3*(5), 82–89.

Babaoglu, A. S., Ozturk, S., & Aslim, B., (2013). Phenylalanine ammonia-lyase (PAL) enzyme activity and antioxidant properties of some cyanobacteria isolates. *Food Chem., 136*(1), 164–169. https://doi.org/10.1016/j.foodchem.2012.07.119.

Babić, O., Kovač, D., Rašeta, M., Šibul, F., Svirčev, Z., & Simeunović, J., (2016). Evaluation of antioxidant activity and phenolic profile of filamentous terrestrial cyanobacterial strains isolated from forest ecosystem. *J. Appl. Phycol., 28,* 2333–2342. https://doi.org/10.1007/ s10811-015-0773-4.

Baker, R., & Günther, C., (2004). The role of carotenoids in consumer choice and the likely benefits from their inclusion into products for human consumption. *Trends Food Sci. Technol., 15*(10), 484–488. https://doi.org/10.1016/j.tifs.2004.04.0094.

Balboa, E. M., Conde, E., Soto, M. L., Pérez-Armada, L., & Domínguez, H., (2015). Cosmetics from marine sources. In: Kim, S. K., (ed.), *Handbook of Marine Biotechnology* (pp. 1015–1042). Springer Berlin Heidelberg, Berlin. https://doi.org/10.1007/ 978-3-642-53971-8_44.

Balić, A., & Mokos, M., (2019). Do we utilize our knowledge of the skin protective effects of carotenoids enough? *Antioxidants, 8*(8), 259. https://doi.org/10.3390/antiox8080259.

Berthon, J. Y., Nachat-Kappes, R., Bey, M., Cadoret, J. P., Renimel, I., & Filaire, E., (2017). Marine algae as attractive source to skin care. *Free Radic. Res., 51,* 555–567. https://doi.org/10.1080/10715762.2017.1355550.

Blime, J. C., (1996). *Oral Composition for the Prevention of Sun Allergies Containing a Carotenoid, a Tocopherol, Ascorbic Acid, and Selenium.* European Patent, EP0712630A3.

Cakmak, Y. S., Kaya, M., & Asan-Ozusaglam, M., (2014). Biochemical composition and bioactivity screening of various extracts from *Dunaliella salina*, a green microalga. *Experimental and Clinical Sciences Journal, 13,* 679–690. https://doi.org/10.17877/DE290R-6669.

Canovas, E. R., Cerro, M. T. L., Ros, D. G. L., Turk, O. D., Yebra, J. S., & De Roca, J. M., (2018). *Compositions for Protecting Skin Comprising DNA Repair Enzymes and Phycobiliprotein.* US Patent, US20180207079A1.

Castangia, I., Manca, M. L., Catalán-Latorre, A., Maccioni, A. M., Fadda, A. M., & Manconi, M., (2016). Phycocyanin-encapsulating hyalurosomes as carrier for skin delivery and protection from oxidative stress damage. *J. Mater. Sci. Mater. Med., 27,* 75. https://doi.org/10.1007/s10856-016-5687-4.

Cervantes-Llanos, M., Lagumersindez-Denis, N., Marín-Prida, J., Pavón-Fuentes, N., Falcon-Cama, V., Piniella-Matamoros, B., Camacho-Rodríguez, H., et al., (2018). Beneficial effects of oral administration of C-phycocyanin and phycocyanobilin in rodent models of experimental autoimmune encephalomyelitis. *Life Sci., 194,* 130–138. https://doi.org/ https://doi.org/10.1016/j.lfs.2017.12.032.

Césarini, J. P., Michel, L., Maurette, J. M., Adhoute, H., & Béjot, M., (2003). Immediate effects of UV radiation on the skin: modification by an antioxidant complex containing carotenoids. *Photodermatol. Photoimmunol. Photomed., 19*, 182–189. https://doi.org/10.1034/j.1600-0781.2003.00044.x.

Chalyk, N. E., Klochkov, V. A., Bandaletova, T. Y., Kyle, N. H., & Petyaev, I. M., (2017). Continuous astaxanthin intake reduces oxidative stress and reverses age-related morphological changes of residual skin surface components in middle-aged volunteers. *Nutr. Res., 48*, 40–48. https://doi.org/10.1016/j.nutres.2017.10.006.

Chisvert, A., Benedé, J. L., & Salvador, A., (2018). Tanning and whitening agents in cosmetics: Regulatory aspects and analytical methods. In: Salvador, A., & Chisvert, A., (eds.), *Analysis of Cosmetic Products* (pp. 107–121). Elsevier, Boston. https://doi.org/https://doi.org/10.1016/B978-0-444-63508-2.00006-0.

Chiu, S. P., & Chuang, L. W., (2017). Analysis on the development trend of green cosmetics. In: *2017 IEEE International Conference on Consumer Electronics*. Taiwan. https://doi.org/10.1109/ICCE-China.2017.7991110.

Chung, W. Y., Lee, J. M., Park, M. Y., Yook, J. I., Kim, J., Chung, A. S., Surh, Y. J., & Park, K. K., (1999). Inhibitory effects of chlorophyllin on 7,12-dimethylbenz[α]anthracene-induced bacterial mutagenesis and mouse skin carcinogenesis. *Cancer Lett., 145*, 57–64. https://doi.org/https://doi.org/10.1016/S0304-3835(99)00229-3.

Couteau, C., & Coiffard, L., (2018). Microalgal application in cosmetics. In: Levine, I. A., & Fleurence, J., (eds.), *Microalgae in Health and Disease Prevention* (pp. 317–323). Academic Press, Boston. https://doi.org/10.1016/B978-0-12-811405-6.00015-3.

D'Errico, M., Lemma, T., Calcagnile, A., De Santis, L. P., & Dogliotti, E., (2007). Cell type and DNA damage specific response of human skin cells to environmental agents. *Mutation Research/Fundamental and Molecular Mechanisms of Mutagenesis, 614*(1, 2), 37–47. https://doi.org/10.1016/j.mrfmmm.2006.06.009.

Darvin, M. E., Fluhr, J. W., Meinke, M. C., Zastrow, L., Sterry, W., & Lademann, J., (2011). Topical beta-carotene protects against infra-red-light-induced free radicals. *Exp. Dermatol., 20*(2), 125–129. https://doi.org/10.1111/j.1600-0625.2010.01191.x.

Farahin, A. W., Yusoff, F. M., Basri, M., Nagao, N., & Shariff, M., (2019). Use of microalgae: *Tetraselmis tetrathele* extract in formulation of nanoemulsions for cosmeceutical application. *J. Appl. Phycol., 21*, 1743–1752. https://doi.org/10.1007/s10811-018-1694-9.

Freitas, J. V., Junqueira, H. C., Martins, W. K., Baptista, M. S., & Gaspar, L. R., (2019b). Antioxidant role on the protection of melanocytes against visible-light-induced photodamage. *Free Radic. Biol. Med., 121*, 399–407. https://doi.org/10.1016/j.freeradbiomed.2018.12.028.

Freitas, S., Silva, N. G., Sousa, M. L., Ribeiro, T., Rosa, F., Leão, P. N., Vasconcelos, V., Reis, M. A., & Urbatzka, R., (2019a). Chlorophyll derivatives from marine cyanobacteria with lipid-reducing activities. *Mar. Drugs, 17*(4), 229. https://doi.org/10.3390/md17040229.

Fujiwaka, N., & Matsushima, M., (2006). *Skin Function-Activating External Preparation Comprising Natural Blue Pigment as Active Ingredient*. Japanese Patent, JP2006036744A.

Gartner, C., Heinrich, U., & Stahl, W., (2003). *Sunscreen Agent for Oral Administration*. European Patent, EP1210073B1.

Giehart, D., & Fox, J., (2012). *Protection Against Sunburn and Skin Problems with Orally-Ingested High-Dosage Zeaxanthin*. US Patent, US8088363B2.

Gong, M., & Bassi, A., (2016). Carotenoids from microalgae: A review of recent developments. *Biotechnol. Adv., 34*, 1396–1412. https://doi.org/10.1016/j.biotechadv.2016.10.005.

Graham, H. K., Eckersley, A., Ozols, M., Mellody, K. T., & Sherratt, M. J., (2019). Human skin: Composition, structure, and visualization methods. In: Limbert, G., (ed.), *Skin Biophysics. Studies in Mechanobiology, Tissue Engineering and Biomaterials* (Vol. 22, pp. 1–18). Springer, Cham. https://doi.org/10.1007/978–3-030–13279–8_1.

Guedes, A. C., Amaro, H. M., & Malcata, F. X., (2011a). Microalgae as sources of carotenoids. *Mar. Drugs, 9*, 625–644. https://doi.org/10.3390/md9040625.

Guedes, A. C., Amaro, H. M., & Malcata, F. X., (2011b). Microalgae as sources of high added-value compounds: A brief review of recent work. *Biotechnol. Prog., 27*, 597–613. https://doi.org/10.1002/btpr.575.

Haizhou, L., Juanjuan, W., & Jie, G., (2018). *A Kind of Cosmetics Containing Phycocyanin and Preparation Method Thereof*. Chinese Patent, CN105055250B.

Hama, S., Takahashi, K., Inai, Y., Shiota, K., Sakamoto, R., Yamada, A., Tsuchiya, H., et al., (2012). Protective effects of topical application of a poorly soluble antioxidant astaxanthin liposomal formulation on ultraviolet-induced skin damage. *J. Pharm. Sci., 101*(8), 2909–2916. https://doi.org/10.1002/jps.23216.

Hettwer, S., & Stebler, S., (2019). *Cosmetic Compositions and Methods for Protection Against the Effects of Electromagnetic Radiation in the Gigahertz range*. German Patent, DE102017218642A1.

Higashi-Okai, K., Otani, S., & Okai, Y., (1998). Potent suppressive activity of pheophytin a and b from the non-polyphenolic fraction of green tea (*Camellia sinensis*) against tumor promotion in mouse skin. *Cancer Lett., 129*(2), 223–228. https://doi.org/10.1016/S0304-3835(98)00113-X.

Hosikian, A., Lim, S., Halim, R., & Danquah, M. K., (2010). Chlorophyll extraction from microalgae: A review on the process engineering aspects. *Int. J. Chem. Eng., 391632*. https://doi.org/10.1155/2010/391632.

In-Ki, H., Hill, K., Eun-Young, M., Hak-Soo, K., & Jeong-Gu, L., (2018). *Cosmetic Composition with Microalgae Extract for Anti-UV and Skin-Irritation Alleviation Effect*. South Korean Patent, KR101856480B1.

Jeffrey, S. W., (1974). Profiles of photosynthetic pigments in the ocean using thin-layer chromatography. *Mar. Biol., 26*, 101–110. https://doi.org/10.1007/BF00388879.

Jeffrey, S. W., Wright, S. W., & Zapata, M., (2012). Microalgal classes and their signature pigments. In: Roy, S., Llewellyn, C. A., Egeland, E. S., & Johnsen, G., (eds.), *Phytoplankton Pigments* (pp. 3–77). Cambridge University Press, Cambridge. https://doi.org/10.1017/cbo9780511732263.004.

Joshi, S., Kumari, R., & Upasani, V. N., (2018). Applications of algae in cosmetics: An overview. *Int. J. Innov. Res. Sci. Eng. Technol., 7*, 1269–1278. https://doi.org/10.15680/IJIRSET.2018.0702038.

Khavkin, J., & Ellis, D. A. F., (2011). Aging skin: Histology, physiology, and pathology. *Facial Plast. Surg. Clin. North Am., 19*, 229–234. https://doi.org/https://doi.org/10.1016/j.fsc.2011.04.003.

Kim, K. M., Lee, J. Y., Im, A. R., & Chae, S., (2018). Phycocyanin protects against UVB-induced apoptosis through the PKC α/βII-Nrf-2/HO-1 dependent pathway in human primary skin cells. *Molecules, 23*(2), 478. https://doi.org/10.3390/molecules23020478.

Kishore, N., Kumar, P., Shanker, K., & Verma, A. K., (2019). Human disorders associated with inflammation and the evolving role of natural products to overcome. *Eur. J. Med. Chem., 179*, 272–309. https://doi.org/https://doi.org/10.1016/j.ejmech.2019.06.034.

L'Oréal, (2019). *L'Oréal Annual Report*. https://www.loreal-finance.com/eng/annual-report (accessed on 2 July 2021).

Lauritano, C., Andersen, J. H., Hansen, E., Albrigtsen, M., Escalera, L., Esposito, F., Helland, K., et al., (2016). Bioactivity screening of microalgae for antioxidant, anti-inflammatory, anticancer, anti-diabetes, and antibacterial activities. *Front. Mar. Sci., 3*, 68. https://doi.org/10.3389/fmars.2016.00068.

Leung, P. O., Lee, H. H., Kung, Y. C., Tsai, M. F., & Chou, T. C., (2013). Therapeutic effect of C-phycocyanin extracted from blue-green algae in a rat model of acute lung injury induced by lipopolysaccharide. *Evidence-based Complement. Altern. Med.,* 916590. https://doi.org/10.1155/2013/916590.

Lopes, G., Clarinha, D., & Vasconcelos, V., (2020). Carotenoids from cyanobacteria: A biotechnological approach for the topical treatment of psoriasis. *Microorganisms, 8*(2), 302. https://doi.org/10.3390/microorganisms8020302.

Lyapina, E. A., Machneva, T. V., Larkina, E. A., Tkachevskaya, E. P., Osipov, A. N., & Mironov, A. F., (2010). Effect of photosensitizers pheophorbide a and protoporphyrin IX on skin wound healing upon low-intensity laser irradiation. *Biophysics, 55*, 296–300. https://doi.org/10.1134/S0006350910020223.

Lyons, A. B., Trullas, C., Kohli, I., Hamzavi, I. H., & Lim, H. W., (2020). Photoprotection beyond ultraviolet radiation: A review of tinted sunscreens. *J. Am. Acad. Dermatol.* https://doi.org/10.1016/j.jaad.2020.04.079.

Lyons, N. M., & O'Brien, N. M., (2002). Modulatory effects of an algal extract containing astaxanthin on UVA-irradiated cells in culture. *J. Dermatol. Sci., 30*(1), 73–84. https://doi.org/10.1016/S0923-1811(02)00063-4.

Madhyastha, H., Madhyastha, R., Nakajima, Y., Omura, S., & Maruyama, M., (2012). Regulation of growth factors-associated cell migration by C-phycocyanin scaffold in dermal wound healing. *Clin. Exp. Pharmacol. Physiol., 39*(1), 13–19. https://doi.org/10.1111/j.1440-1681.2011.05627.x.

Mandal, M. K., Chanu, N. K., & Chaurasia, N., (2020). Cyanobacterial pigments and their fluorescence characteristics: Applications in research and industry. In: Singh, P., Kumar, A., Singh, V. L., & Shrivistava, A., (eds.), *Advances in Cyanobacterial Biology* (pp. 55–72). Elsevier, Boston. https://doi.org/10.1016/b978-0-12-819311-2.00005-x.

Markets and Markets, (2018a). *Algae Products Market by Type (Lipids, Carrageenan, Carotenoids, Alginate, and Algal Protein), Application (Food & Beverages, Nutraceuticals & Dietary Supplements, Personal Care, Feed, and Pharmaceuticals), Source, form, and Region-Global Forecast to 2023.* https://www.marketsandmarkets.com/Market-Reports/algae-product-market-250538721.html (accessed on 2 July 2021).

Markets and Markets, (2018b). *Food Colors Market by Type (Natural, Synthetic, Nature-Identical), Application (Beverages, Processed Food, Bakery & Confectionery Products, Oils & Fats, Dairy Products, Meat, Poultry, Seafood), Form, Solubility, and Region-Global Forecast to 2023.* https://www.marketsandmarkets.com/Market-Reports/food-colors-market-36725323.html (accessed on 2 July 2021).

Markets and Markets, (2019). *Cosmetic Pigments Market by Composition (Organic, Inorganic), Type (Special Effect, Surface Treated, Nano), Application (Facial Makeup, Eye Makeup, Lip Products, Nail Products, Hair Color Products), and Region-Global Forecast to 2024.* https://www.marketsandmarkets.com/Market-Reports/cosmetic-pigment-market-179525453.html (accessed on 2 July 2021).

Masojídek, J., Torzillo, G., Koblížek, M., Masojídek, J., Torzillo, G., & Koblížek, M., (2013). Photosynthesis in microalgae. In: Richmond, A., & Hu, Q., (eds.), *Handbook of Microalgal Culture* (pp. 21–36). John Wiley & Sons, New Jersey. https://doi.org/10.1002/9781118567166.ch2.

Meléndez-Martínez, A. J., Stinco, C. M., & Mapelli-Brahm, P., (2019). Skin carotenoids in public health and nutricosmetics: The emerging roles and applications of the UV radiation-absorbing colorless carotenoids phytoene and phytofluene. *Nutrients, 11*(5), 1093. https://doi.org/10.3390/nu11051093.

Mobin, S. M. A., Chowdhury, H., & Alam, F., (2019). Commercially important bioproducts from microalgae and their current applications: A review. *Energy Procedia, 160*, 752–760. https://doi.org/https://doi.org/10.1016/j.egypro.2019.02.183.

Morone, J., Alfeus, A., Vasconcelos, V., & Martins, R., (2019). Revealing the potential of cyanobacteria in cosmetics and cosmeceuticals: A new bioactive approach. *Algal Res., 41*, 101541. https://doi.org/10.1016/j.algal.2019.101541.

Mourelle, M. L., Gómez, C. P., & Legido, J. L., (2017). The potential use of marine microalgae and cyanobacteria in cosmetics and thalassotherapy. *Cosmetics, 4*(4), 46. https://doi.org/10.3390/cosmetics4040046.

Nahata, M. C., Slencsak, C. A., & Kamp, J., (1983). Effect of chlorophyllin on urinary odor in incontinent geriatric patients. *Drug Intell. Clin. Pharm., 17*(10), 732–734. https://doi.org/10.1177/106002808301701006.

Nakamura, Y., Murakami, A., Koshimizu, K., & Ohigashi, H., (1996). Inhibitory effect of pheophorbide α, a chlorophyll-related compound, on skin tumor promotion in ICR mouse. *Cancer Lett., 108*(2), 247–255. https://doi.org/10.1016/S0304-3835(96)04422-9.

Nethravathy, M. U., Mehar, J. G., Mudliar, S. N., & Shekh, A. Y., (2019). Recent advances in microalgal bioactives for food, feed, and healthcare products: Commercial potential, market space, and sustainability. *Compr. Rev. Food Sci. Food Saf., 18*(6), 1882–1897. https://doi.org/10.1111/1541-4337.12500.

Ng, Q. X., De Deyn, M. L. Z. Q., Loke, W., Foo, N. X., Chan, H. W., & Yeo, W. S., (2020). Effects of astaxanthin supplementation on skin health: A systematic review of clinical studies. *J. Diet. Suppl.* https://doi.org/10.1080/19390211.2020.1739187.

Odjadjare, E. C., Mutanda, T., & Olaniran, A. O., (2017). Potential biotechnological application of microalgae: A critical review. *Crit. Rev. Biotechnol., 37*, 37–52. https://doi.org/10.3109/07388551.2015.1108956.

Okai, Y., & Higashi-Okai, K., (1997). Potent anti-inflammatory activity of pheophytin a derived from edible green alga, *Enteromorpha prolifera* (Sujiao-nori). *Int. J. Immunopharmacol., 19*(6), 355–358. https://doi.org/10.1016/S0192-0561(97)00070-2.

Oppen-Bezalel, L., & Shaish, A., (2009). Application of the colorless carotenoids, phytoene, and phytofluene in cosmetics, wellness, nutrition, and therapeutics. In: Ben-Amortz, A., Polle, J. E. W., & Rao, D. V. S., (eds.), *The Alga Dunaliella* (pp. 423–444). CRC Press, Florida. https://doi.org/10.1201/b10300-19.

Pagels, F., Guedes, A. C., Amaro, H. M., Kijjoa, A., & Vasconcelos, V., (2019). Phycobiliproteins from cyanobacteria: Chemistry and biotechnological applications. *Biotechnol. Adv., 37*, 422–443. https://doi.org/https://doi.org/10.1016/j.biotechadv.2019.02.010.

Panis, G., & Carreon, J. R., (2016). Commercial astaxanthin production derived by green alga *Haematococcus pluvialis*: A microalgae process model and a techno-economic assessment all through production line. *Algal Res., 18*, 175–190. https://doi.org/https://doi.org/10.1016/j.algal.2016.06.007.

Pareek, S., Sagar, N. A., Sharma, S., Kumar, V., Agarwal, T., González-Aguilar, G. A., & Yahia, E. M., (2017). Chlorophylls: Chemistry and biological functions, In: Yahia, E. M., (ed.), *Fruit and Vegetable Phytochemicals: Chemistry and Human Health* (pp. 269–284). Wiley, UK. https://doi.org/doi:10.1002/9781119158042.ch14.

Park, K. K., & Surh, Y. J., (1996). Chemopreventive activity of chlorophyllin against mouse skin carcinogenesis by benzo[α]pyrene and benzo[α]pyrene-7,8-dihydrodiol-9,10-epoxide. *Cancer Lett., 102*, 143–149. https://doi.org/https://doi.org/10.1016/0304-3835(96)04173-0.

Peng, J., Yuan, J. P., Wu, C. F., & Wang, J. H., (2011). Fucoxanthin, a marine carotenoid present in brown seaweeds and diatoms: Metabolism and bioactivities relevant to human health. *Mar. Drugs, 9*(10), 1806–1828. https://doi.org/10.3390/md9101806.

Pieloch, M., (2006). *Method of Use and Dosage Composition of Blue-Green Algae Extract for Inflammation in Animals*. US Patent, US7025965B1.

Pradier, F., & Fanchon, C., (1995). *Photoprotective Composition Administered Orally*. French Patent, FR2698268B1.

Queiroz, Z. L., Jacob-Lopes, E., & Roca, M., (2019). Catabolism and bioactive properties of chlorophylls. *Curr. Opin. Food Sci., 26*, 94–100. https://doi.org/10.1016/j.cofs.2019.04.004.

Rao, A. R., Sindhuja, H. N., Dharmesh, S. M., Sankar, K. U., Sarada, R., & Ravishankar, G. A., (2013). Effective inhibition of skin cancer, tyrosinase, and antioxidative properties by astaxanthin and astaxanthin esters from the green alga *Haematococcus pluvialis. J. Agric. Food Chem., 61*(16), 3842–3851. https://doi.org/10.1021/jf304609j.

Sahin, S. C., (2018). The potential of *Arthrospira platensis* extract as a tyrosinase inhibitor for pharmaceutical or cosmetic applications. *South African J. Bot., 119*, 236–243. https://doi.org/10.1016/j.sajb.2018.09.004.

Sahota, A., (2013). Introduction to sustainability. In: Sahota, A., (ed.), *Sustainability: How the Cosmetics Industry is Greening up* (pp. 1–15). John Wiley and Sons, New Jersey. https://doi.org/10.1002/9781118676516.ch1.

Salzmann, G. M., & Schiraldi, R. J., (1960). *Chlorophyll Dental Powder*. US Patent, US2941926A.

Sánchez, M., Bernal-Castillo, J., Rozo, C., & Rodríguez, I., (2003). Spirulina (*Arthrospira*): An edible microorganism: A review. *Univ. Sci., 8*, 7–24.

Schluchter, W. M., Shen, G., Alvey, R. M., Biswas, A., Saunée, N. A., Williams, S. R., Mille, C. A., & Bryant, D. A., (2010). Phycobiliprotein biosynthesis in cyanobacteria: Structure and function of enzymes involved in post-translational modification. In: Hallenbeck, P. C., (ed.), *Advances in Experimental Medicine and Biology* (pp. 211–228). Springer, New York. https://doi.org/10.1007/978-1-4419-1528-3_12.

Scoglio, S., Canestrari, F., Benedetti, S., & Zolla, L., (2012). *Extracts of Aphanizomenon flosaquae and Nutritional, Cosmetic and Pharmaceutical Compositions Containing the Same*. US Patent, US8337858B2.

Scott, D. A., (2016). A review of ancient Egyptian pigments and cosmetics. *Stud. Conserv., 61*(4), 185–202. https://doi.org/10.1179/2047058414Y.0000000162.

Sekar, S., & Chandramohan, M., (2008). Phycobiliproteins as a commodity: Trends in applied research, patents, and commercialization. *J. Appl. Phycol., 20*, 113–136. https://doi.org/10.1007/s10811-007-9188-1.

Shegokar, R., & Mitri, K., (2012). Carotenoid lutein: A promising candidate for pharmaceutical and nutraceutical applications. *J. Diet. Suppl., 9*(3), 183–210. https://doi.org/10.3109/19390211.2012.708716.

Shimamatsu, S., & Tanabe, Y., (1981). *Cosmetic Containing Hard-Soluble Phycocyanin*. Japanese Patent, JPS5663911A.

Sies, H., & Stahl, W., (2004). Carotenoids and UV protection. *Photochem. Photobiol. Sci.,* 3(8), 749–752. https://doi.org/10.1039/b316082c.

Singh, D. P., Prabha, R., Meena, K. K., Sharma, L., & Sharma, A. K., (2014). Induced accumulation of polyphenolics and flavonoids in cyanobacteria under salt stress protects organisms through enhanced antioxidant activity. *Am. J. Plant Sci., 5*, 43916. https://doi.org/10.4236/ajps.2014.55087.

Singh, K. N., Patil, S., & Barkate, H., (2020). Protective effects of astaxanthin on skin: Recent scientific evidence, possible mechanisms, and potential indications. *J. Cosmet. Dermatol., 19*(1), 22–27. https://doi.org/10.1111/jocd.13019.

Spolaore, P., Joannis-Cassan, C., Duran, E., & Isambert, A., (2006). Commercial applications of microalgae. *J. Biosci. Bioeng., 101*(2), 87–96. https://doi.org/10.1263/jbb.101.87.

Spritz, R. A., & Hearing, V. J., (2013). Abnormalities of pigmentation. In: Rimoin, D., Pyeritz, R., & Korf, B., (eds.), *Emery and Rimoin's Principles and Practice of Medical Genetics* (pp. 1–44). Academic Press, Oxford. https://doi.org/https://doi.org/10.1016/B978-0-12-383834-6.00154-3.

Stahl, W., Heinrich, U., Aust, O., Tronnier, H., & Sies, H., (2006). Lycopene-rich products and dietary photoprotection. *Photochem. Photobiol. Sci., 5*(2), 238–242. https://doi.org/10.1039/b505312a.

Stahl, W., Heinrich, U., Jungmann, H., Sies, H., & Tronnier, H., (2000). Carotenoids and carotenoids plus vitamin E protect against ultraviolet light-induced erythema in humans. *Am. J. Clin. Nutr., 71*(3), 795–798. https://doi.org/10.1093/ajcn/71.3.795.

Stewart, S., (2017). *Painted Faces: A Colorful History of Cosmetics* (p. 288). Amberley Publishing Limited, United Kingdom.

Stoicescu, D. M., (2020). Skin lesions. In: Stoicescu, D. M., (ed.), *General Medical Semiology Guide* (Part II, pp. 1–178). Academic Press. https://doi.org/https://doi.org/10.1016/B978-0-12-819640-3.00001-2.

Suryavanshi, S., Sharma, D., Checker, R., Thoh, M., Gota, V., Sandur, S. K., & Sainis, K. B., (2015). Amelioration of radiation-induced hematopoietic syndrome by an antioxidant chlorophyllin through increased stem cell activity and modulation of hematopoiesis. *Free Radic. Biol. Med., 85*, 56–70. https://doi.org/https://doi.org/10.1016/j.freeradbiomed.2015.04.007.

Tobin, D. J., (2017). Introduction to skin aging. *J. Tissue Viability, 26*, 37–46. https://doi.org/https://doi.org/10.1016/j.jtv.2016.03.002.

Tominaga, K., Hongo, N., Fujishita, M., Takahashi, Y., & Adachi, Y., (2017). Protective effect of astaxanthin on skin deterioration. *J. Clin. Biochem. Nutr., 61*(1), 33–39. https://doi.org/10.3164/jcbn.17-35.

Tominaga, K., Hongo, N., Karato, M., & Yamashita, E., (2012). Cosmetic benefits of astaxanthin on humans' subjects. *Acta Biochim. Pol., 59*(1), 43–47. https://doi.org/10.18388/abp.2012_2168.

Tso, M., & Lam, T. T., (1996). *Method of Retarding and Ameliorating Central Nervous System and Eye Damage.* US Patent, US5527533A.

Value Market Research, (2020). *Global Chlorophyll Extract Market Report by Type (Liquid, Tablet and powder), by Application (Food Additive, Cosmetics and Dietary Supplement) and by Regions-Industry Trends, Size, Share, Growth, Estimation and Forecast, 2019–2026.* https://www.valuemarketresearch.com/report/chlorophyll-extract-market (accessed on 2 July 2021).

Yaar, M., & Gilchrest, B. A., (2007). Photo ageing: Mechanism, prevention, and therapy. *Br. J. Dermatol., 157*, 874–887. https://doi.org/10.1111/j.1365-2133.2007.08108.x.

Yeum, K. J., (2005). *Synergistic Effect of Compositions Comprising Carotenoids Selected from Lutein, Beta-carotene, and Lycopene.* International Patent, WO2005087208A3.

Yoshihisa, Y., Rehman, M. U., & Shimizu, T., (2014). Astaxanthin, a xanthophyll carotenoid, inhibits ultraviolet-induced apoptosis in keratinocytes. *Exp. Dermatol., 23*(3), 178–183. https://doi.org/10.1111/exd.12347.

Zakar, T., Laczko-Dobos, H., Toth, T. N., & Gombos, Z., (2016). Carotenoids assist in cyanobacterial photosystem II assembly and function. *Front. Plant Sci., 7*, 295. https://doi.org/10.3389/fpls.2016.00295.

Zhou, Z. P., Liu, L. N., Chen, X. L., Wang, J. X., Chen, M., Zhang, Y. Z., & Zhou, B. C., (2005). Factors that affect antioxidant activity of C-phycocyanins from *Spirulina platensis. J. Food Biochem., 29*, 313–322. https://doi.org/10.1111/j.1745-4514.2005.00035.x.

CHAPTER 2

ALGAE AS NUTRACEUTICAL, FUNCTIONAL FOOD, AND FOOD INGREDIENTS

UMESH PAWAR,[1] NIVAS DESAI,[1] UTTAM DETHE,[1] VISHAL APARADH,[1] and DATTATRAY GAIKWAD[2]

[1]Department of Botany, S.P.K. College, Sawantwadi–416510, Maharashtra, India

[2]Sub-Center, Babasaheb Ambedkar Marathwada University, Aurangabad, Osmanabad–413501, Maharashtra, India

ABSTRACT

The increasing demand of algal food supplement is growing tremendously. Algae are being consumed for functional benefits beyond the traditional considerations of nutrition and health. There is considerable evidence for the health benefits of algal-derived food products, but there remain considerable challenges in quantifying these benefits, as well as possible adverse effects. Algae are primary producers presenting a remarkable source of different nutrients. The high protein content of algal species is one of the main reasons to consider them as an important source of proteins, oils from microalgae rich in some PUFAs seem particularly suitable for children, pregnant women, vegetarians, and patients with fish allergies. Algae also represent an important source of vitamins, minerals, antioxidants, and natural colorants, the incorporation of the whole biomass in food and feed could be used to provide the color, increment nutritional value, and improve texture or resistance to oxidation.

2.1   INTRODUCTION

Algae are the most abundant primary producers on the global surface. The algae are the lower plants which contain chlorophyll in cells and are typical inhabiting in aquatic environment (freshwater and saline ocean waters) (Blazencic, 2007), and on the basis of dimensions they are divided into macroalgae and microalgae. There are three main classes, or phyla, of seaweed: Phaeophyceae (brown algae), Rhodophyta (red algae), and Chlorophyta (green algae). Thousands of species comprise each phylum (Rindi et al., 2012; Guiry and Guiry, 2019). A special group of microalgae are blue-green algae, also called cyanobacteria because of their prokaryotic cell type, group of organisms for the isolation of novel and biochemically active natural products (Singh et al., 2005). Algae have been consumed by coastal localities worldwide since the beginning of human civilization (Dillehay et al., 2008). Microalgae are also utilized in the cosmetics industry or as animal feed (Huntley et al., 2015; Ariede et al., 2017).

2.1.1   NUTRITIONAL VALUE

Algae are a rich and varied source of pharmacologically active natural products and nutraceuticals. While nutraceutical and pharmaceutical content in the baseline algae strain is very small, current market values for these products are extremely high. The major products currently being commercialized or under consideration for commercial extraction include carotenoids, phycobilins, fatty acids (FAs), polysaccharides, vitamins, sterols, and biologically active molecules for use in human and animal health. The upcoming sections will bring into focus the use of algae as a potential source of pharmaceutical and nutraceutical ingredients. The growing use of algae biomass for nutraceutical purposes is expected to provide an attractive revenue stream for algae producers. While nutraceutical content in the baseline algae strain is very small, current market values for these products are extremely high. Physiologically-active nutraceuticals from algae include food supplements, dietary supplements, value-added processed foods as well as non-food supplements such as tablets, soft gels, capsules, etc.

Some of the noteworthy products that can be derived from algae: omega-3 polyunsaturated fatty acids (PUFA), carotenoids, astaxanthin, and β-carotene.

The markets for both pharmaceuticals and nutraceuticals are growing quickly worldwide, and it is this global scope that particularly attracts marketers. A growing proportion of today's promising pharmaceutical and nutraceutical research focuses on the production of promising compounds from algae. Thus, the untapped potential of algae in the field of pharmaceuticals and nutraceuticals has to be still explored to grow and capitalize on tremendous global marketing opportunities.

2.1.2 FUNCTIONAL FOOD AND INGREDIENTS

At the beginning of the 1980s, the functional food concept first time came in Japan. The meaning of that concept was protecting the health of people and to reduce the high health costs (Arai, 1996). Functional food is defined as food that contains more functional ingredients that provide additional health benefits (Lordan et al., 2011). Nowadays, a wide variety of compounds such as polyphenols, PUFA, or phytosterols obtained, for example, from wine, fish byproducts, or plants are utilized to prepare new functional foods (Plaza et al., 2009). Algae play an important role in human diet in recent years (Turner, 2003; Craigie, 2010). The algal metabolite compounds are a vital source of ingredients for the development of food products they are also good source of proteins, minerals, vitamins, amino acids, lipids, fatty acids, polysaccharides, nucleic acid, and carotenoids, etc., which have huge value as nutraceuticals (Tokuşoglu and Una, 2003; Priyadarshani and Rath, 2012).

2.2 ALGAE AS NUTRACEUTICAL

In the last few decades, interest has grown in seaweeds as nutraceuticals, or functional foods, which gave dietary benefits beyond their macronutrient content. In addition, seaweed has been explored for metabolites with biological activity, to produce therapeutic products (Davis and Vasanthi, 2011; Zerrifi et al., 2018). Throughout human history, seaweeds have been used as food, folk remedies, dyes, and mineral-rich fertilizers (Shannon and Abu-Ghannam, 2019). Marine algae present a good antioxidant activity (Biriş-Dorhoi et al., 2019). They are an important source of vitamins, minerals, proteins, PUFA, antioxidants, etc., (Blazencic, 2007; Gouveia et al., 2008).

Seaweeds are also rich in bioactive compounds, including polyphenols, polysaccharides (Teng et al., 2013), peptides, sterols, flavonoids, alkaloids, proteins, and other bioactive compounds (Jmel et al., 2019; Randhir et al., 2020).

Algae contain enormous bioactive compounds having wide biological activities such as antioxidant, anti-bacterial, anti-viral, anti-carcinogenic, etc. Some algae contain high fiber content in the form of sulfated galactans or carragenates in red algae, fucans, alginates, and laminarin, etc. The intake of high fiber diet has a progressive effect on the human body that curtails the risks of cancer, diabetes, obesity, and hypercholesterolemia. The high fiber content diet showed a good immunological activity (German et al., 2013). Alginate in *Undaria pinnatifida* showed a positive effect on cardiovascular disease, and alginic acid demonstrated to reduce hypertension in hypertensive rats (Ikeda et al., 2006). Alginic acid, xyloglucans in *Sargassum vulgare* and sulfated fucans in *Undaria pinnatifida* have anti-viral activity against herpes type-1 and cytomegalovirus in humans (Thompson and Dragar, 2004; Hemmingson et al., 2006). The fucoidans from algae acts as anti-coagulant, anti-thrombotic agents and also showed anti-tumoral properties in rat model studies (Lee et al., 2013; Maruyama et al., 2013; De Jesus Raposo et al., 2015). A sulfated polysaccharide from the *Porphyra* sp. has apoptotic property in carcinogenic cells (Maruyama et al., 2013; Raposo et al., 2015).

2.2.1 ANTIOXIDANTS

The in vitro antioxidant activities were proved by different studies. Marine algae are rich sources of antioxidant compounds. Microalgae and macroalgae contain antioxidant organic compounds and enzymes that control the oxidative damage (Cornish and Garbary, 2010). There are two broad categories of antioxidant activity of algae, and they are decreasing oxidative stress on the gut microbiome by limiting reactive oxygen species (ROS) within the digestive tract and transportation of epithelial cells into the blood for distribution throughout the body (Gobler et al., 2011). Natural antioxidants, found in many algae, are important bioactive compounds that play an important role against various diseases and aging processes through the protection of cells from oxidative damage. The detected antioxidant compounds in algae from these genera and others have potential anti-aging, dietary, anti-inflammatory, anti-bacterial, antifungal, cytotoxic, anti-malarial, anti-proliferative, and anticancer properties (Zubia et al., 2007).

2.2.2 PROTEINS

Protein constitutes 5–47% of seaweed dry mass. Red seaweeds have the greatest protein content, while green has less, and brown the least (Cerna, 2011). Of the total amino acids (aa) in seaweeds, approximately 42% to 48% are essential aa (Wong and Cheung, 2000). In terms of a score (on a scale of 0.0–1.0), where egg protein has a score of 1.0, most seaweeds have a higher score than all plant-based proteins, with the exception of soy, which has a score of 1.0. *Undaria pinnatifida* has an amino acid score of 1.0, equal to that of egg and soy, *Pyropia/Porphyra* 0.91, and *Laminaria saccharina* 0.82 (Murata and Nakazoe, 2001). However, the high poly-phenolic content of seaweeds can reduce the digestibility of algal proteins, giving a slightly lower score on the protein digestibility-corrected aa scale (Wong and Cheung, 2001). Despite this, seaweeds still represent a viable alternative to animal-derived protein; if other high-aa scoring vegan foods, such as soy or mycoprotein, are included in the diet. Nori, a dried sheet product of *Porphyra yezoensis* is known to contain an exceptionally high (12.5–51.5% w/w) protein content among seaweeds and can be expected to produce a high quantity of free amino acids after degradation (Uchida et al., 2017). *Spirulina* as a plant-based source of complete protein, makes it an ideal dietary supplement choice for vegetarians (De Marco et al., 2014).

2.2.3 LIPIDS

Lipids play a vital role for all living organism as it is component of cell membrane, actively engaged in cell signaling (Eyster, 2007). Algal lipids consist of phospholipids, glycolipids, and non-polar glycerolipids. Phospho-lipids and glycolipids are important for membrane function. Lipid membranes contain sterols such as fucosterol and β-sitosterol (Fahy et al., 2005) and it has health benefits (Arul et al., 2012). Algae contain many of the major lipids of plants, such as the glycosylglycerides and the usual phosphoglycerides. In addition, more unusual compounds such as the betaine lipids, chlorosulfoli-pids or various other sulfolipids may be major components of some species or orders. These acyl lipids have characteristic fatty acid compositions and are often highly enriched in PUFAs. PUFA are vital components in human nutrition and are known to have several beneficial effects for human health. Betaine lipids are widely distributed in algae, where they display different functions as donors of diacylglycerols and fatty acids (FAs) to be used in the biosynthesis of other lipid classes (Rey et al., 2019).

2.2.4 CARBOHYDRATES

The photosynthate produced during photosynthesis is the major carbohydrates in the algae. Mostly these carbohydrates are used for building structural element and also energy storage. The main polysaccharides obtained from marine algae are: alginate, agar, and carrageenan. Those hydrocolloids are commonly used in food, pharmaceutical industries, and in biotechnology, among others, due to their ability to form highly viscous solutions and gels. Agar and carrageenan form thermoreversible gels. However, agar melts at a higher temperature than carrageenan (Hernández-Carmona et al., 2013).

2.2.5 ALKALOIDS

Alkaloids have been isolated from macroalgae. Marine algae contain 44 alkaloids, consisting of 1 phenylethylamine, 41 indole, and 1 naphthyridine derivates. In the halogenated alkaloid group, there can be found 25 bromine-containing compounds, among which 7 have chlorine and 5 have sulfur, additionally. In brown algae are rarely found, but in red algae, alkaloids are more abundant than in green algae. Some alkaloids were produced by host organisms on algae. Like communes in which was isolated from the mycelium of a strain of *Penicillium* sp. on the *Enteromorpha intestinalis*, Ascosalipyr-rolidinone was isolated from fungus *Ascochyta salicornia* on green alga *Ulva* sp. and citrinadin A was isolated from *Penicillium citrinum* separated from a marine red alga (Guven et al., 2013).

2.2.6 MINERALS AND VITAMINS

Marine algae in particular often contain a wide range of minerals (magnesium, selenium, chromium, zinc) (Sugimura et al., 1976; Thomson et al., 1996; Mouritsen, 2013). Processed seaweeds are widely used as mineral and metal nutritional supplements (Kay, 1991). Some minerals may be excessively concentrated by algae. Iodine is often found in algae at levels greater than recommended for human consumption, causing thyroid problems in regular consumers (Phaneuf et al., 1999). Cesium is also concentrated in seaweed, which is of concern only when the water is contaminated with radioactive Caesium. Algae are also rich in many vitamins, such as A, C, B1, B2, B3, and B6, as well as minerals, such as iodine, calcium, potassium, magnesium, and iron. They can be consumed from cooked to dried or raw

(Siva et al., 2015). Algae are used for some products such as medicines, vitamins, vaccines, nutraceuticals, and other nutrients. Many types of algae and the products derived from them have shown medicinal values and nutritional applications.

2.3   FUNCTIONAL FOOD INGREDIENTS

The increased interest in functional food ingredients and other natural food products has been recognized to promote excellent health, decrease the risk of mainly non-communicable diseases, and enhance cost effective care by promoting a quality of life (Shahidi, 2008). The positive impact of nutritious foods on human health has been long realized, which has led to the development of several innovative functional ingredients and functional food products. Functional foods have been linked with health improvement, quality of life, and decrease illness risk (Gouveia et al., 2010). Algae-derived nutrients and bioactive compounds are being investigated for their potential biological activities (Batista et al., 2013; Nuno et al., 2013). Edible algae can be directly consumed or used as a raw material for preparing food. Nowadays, there are diverse species of edible seaweeds be cultured as large-scale including *Porphyra yezoensis*, *Saccharina japonica*, and *Undaria pinnatifida*. Seaweeds contain a large amount of nutrient contents such as proteins, dietary fiber, vitamins, and minerals. Regular consumption of seaweeds has many benefits. As an addition to these benefits, the medicinal properties of seaweed bioactive have been recognized. Seaweeds are used for treatment and or for prevention of goiter, which is caused by the lack of iodine in the diet. Several studies have shown various remedial effects of algal species against non-communicable diseases such as inflammation, obesity, diabetes, hypertension, and viral infections (Shao et al., 2017). A clinical study showed that regular consumption of *Undaria* seaweed can effectively minimize the risk of breast cancer in women, while an oral administration of seaweed extracts (*Fucus vesiculosus*, *Macrocystis pyrifera* and *Laminaria japonica*) with zinc, manganese, and vitamin B6, potentially decreased osteoarthritis symptoms in a mixed population. Seaweeds are well-known for their antioxidant capacities and bioactive polyphenolic compounds (Ganesan et al., 2019). *Gracilaria canaliculata* can be used up in the form of powder added to food items and ready to use (Madhu et al., 2011). In coastal areas of all continents, seaweeds are used in human and animal nutrition, so that they are widely cultivated algal crops.

Species such as *Porphyra* sp., *Chondrus crispus*, *Himanthalia elongata*, and *Undaria pinnatifida* are very interesting to consumers and the food industry due to their low content in calories and high content in vitamins, minerals, and dietetic fiber (Plaza et al., 2008). On the other hand, microalgal biomass is usually available in the form of powder, tablets, capsules, liquids, and, also, it can be incorporated into different food products. However, the consumption of microalgal biomass is restricted to very few taxa, and the most important in human nutrition are *Spirulina* and *Chlorella* genera. *Spirulina* as a superfood is a plant that can nourish the body by providing most of the protein require by the body. It helps to prevent the annoying sniffling and sneezing of allergies, reinforces the immune system, helps to control high blood pressure and cholesterol, and helps to protect against cancer. The recommended daily dose is typically 3–5 grams, which can be spread out twice to thrice a day (Oluwakemi and Sharma, 2017). *Spirulina* comes in capsules, tablets, powders, and flakes. *Spirulina* is a safe source of protein, nutrients, vitamins, and minerals that has been used for centuries, though there are no known side effects associated with spirulina, the body may react to it based on individual current state of health. The reactions can be reduced by increasing the water intake, reducing stress levels, eating according to nutritional type, and getting plenty of rest.

2.4 CONCLUSION

Algae are a sustainable source of bioactive compounds for human health and functional food applications. The global burden of diseases such as type 2 diabetes, hypertension, obesity, cancer, antibiotic resistance, and heart disease places a huge strain on the finances and resources of health services in affected countries. This may be improved by the inclusion of algal-based supplements in the diet, as part of overall lifestyle improvement. 'Algal Food as Pharma' could be promoted in terms of the natural health and nutritional benefits of dietary macroalgae based on epidemiological studies. Having reviewed the literature on the benefits of algae consumption, data from scientific medical studies may inform public health systems in the design of dietary intervention plans and may be beneficial for policymakers, educators, practitioners, researchers, and academics who contribute to the promotion of public health. The functional properties of seaweed can be incorporated into food, from fat replacers to antioxidant, fiber, and antimicrobial enhancers. We should focus on the marine agronomy, mariculture, and bioengineering of algae for future prospecting and challenges.

KEYWORDS

- algae
- alkaloids
- antioxidants
- carbohydrates
- food ingredients
- functional food
- lipids
- minerals
- nutraceuticals
- nutrients
- proteins
- seaweeds
- vitamins

REFERENCES

Arai, S., (1996). Studies of functional foods in Japan; State of the art. *Biosci. Biotechnol. Biochem., 60*, 9–15.

Ariede, M. B., Candido, T. M., Jacome, A. L. M., Velasco, M. V. R., De Carvalho, J. C. M., & Baby, A. R., (2017). Cosmetic attributes of algae: A review. *Algal Res., 25*, 483–487.

Arul, A. B., Al Numair, K., Al Saif, M., & Savarimuthu, I., (2012). Effect of dietary beta-sitosterol on fecal bacterial and colonic biotransformation enzymes in 1,2-dimethylhydrazine induced colon carcinogenesis. *Turk. J. Med. Sci., 42*, 1307–1313.

Batista, A. P., Gouveia, L., Bandarra, N. M., Franco, J. M., & Raymundo, A., (2013). Comparison of microalgal biomass profiles as novel functional ingredients for food products. *Algal Research, 2*(2), 164–173.

Biriş-Dorhoi, E., Michiu, D., Taloş, I., & Tofa, M., (2019). Algae as functional food: Review. *Hop and Medicinal Plants, 28*, 1–2.

Blazencic, J., (2007). *Sistematika Algi*. Beograd: NNK International.

Cerna, M., (2011). Seaweed proteins and amino acids as nutraceuticals. In: Kim, S. K., (ed.), *Advances in Food and Nutrition Research* (pp. 297–312). Academic Press, San Diego, California, USA.

Cornish, M. L., & Garbary, D. J., (2010). Antioxidants from macroalgae: Potential applications to human health and nutrition. *Algae, 25*, 155–171.

Davis, G. D. J., & Vasanthi, A. H. R., (2011). Seaweed metabolite database: A database of natural compounds from marine algae. *Bioinformation, 5*, 361–364.

De Jesus, R. M. F., De Morais, A. M., & De Morais, R. M., (2015). Marine polysaccharides from algae with potential biomedical applications. *Marine Drugs, 13*(5), 2967–3028.

De Marco, E. R., Steffolani, M. E., Martinez, C. S., & Leon, A. E., (2014). Effects of spirulina biomass on the technological and nutritional quality of bread wheat pasta. *LWT-Food Sci. Technol., 58*, 102–108.

Dillehay, T. D., Ramirez, C., Pino, M., Collins, M. B., Rossen, J., & Pino, N. J., (2008). Monte Verde: Seaweed, food, medicine, and the peopling of South America. *Science, 320*, 784–786.

Eyster, K. M., (2007). The membrane and lipids as integral participants in signal transduction: Lipid signal transduction for the non-lipid biochemist. *Adv. Physiol. Educ., 31*, 5–16.

Fahy, E., Subramaniam, S., Brown, H. A., Glass, C. K., Merrill, A. H., & Murphy, R. C., (2005). A comprehensive classification system for lipids. *J. Lipid Res., 46*, 839–861.

Ganesan, A. R., Uma, T., & Gaurav, R., (2019). Seaweed nutraceuticals and their therapeutic role in disease prevention. *Food Science and Human Wellness, 8*, 252–263.

German, J. B., Zivkovic, A. M., Dallas, D. C., & Smilowitz, J. T., (2013). Nutrigenomics and personalized diets: What will they mean for food? *Annual Review of Food Science Technology, 2*, 97–123.

Gobler, C. J., Berry, D. L., Dyhrman, S. T., Wilhelm, S. W., Salamov, A., & Lobanov, A. V., (2011). Niche of the harmful alga *Aureococcus anophagefferens* revealed through ecogenomics. *Proc Natl. Acad. Sci. USA., 108*, 4352–4357.

Gouveia, L., Batista, A. P., Sousa, I., Raymundo, A., & Bandarra, N. M., (2008). Microalgae in novel food products. In: Papadoupoulos, K., (ed.), *Food Chemistry Research Developments* (pp. 75–112). New York: Nova Science Publishers.

Gouveia, L., Marques, A. E., Sousa, J. M., Moura, P., & Bandarra, N. M., (2010). Microalgae-source of natural bioactive molecules as functional ingredients. *Food Science and Technology Bulletin, 7*(2), 21–37.

Guiry, M. D., & Guiry, G. M., (2019). *Algae Base*. National University of Ireland, Galway.

Hemmingson, J. A., Falshaw, R., Furneaux, R. H., & Thompson, K., (2006). Structure and anti-viral activity of the galactofucan sulfates extracted from *Undaria pinnatifida* (Phaeophyta). *Journal of Applied Phycology, 18*, 185–193.

Hernandez-Carmona, G., Freile-Pelegrin, Y., & Garibay, E., (2013). Conventional and alternative technologies for the extraction of algal polysaccharides. *Functional Ingredients from Algae for Foods and Nutraceuticals*, 475–516.

Huntley, M., Johnson, Z., Brown, S., Sills, D., Gerber, L., Archibald, I., Machesky, S., et al., (2015). Demonstrated large-scale production of marine microalgae for fuels and feed. *Algal Res., 10*, 249–265.

Ikeda, K., Kitamura, A., Machida, H., Watanabe, M., & Negishi, H., (2006). Effect of *Undaria pinnatifida* (Wakame) on the development of cerebrovascular diseases in stroke-prone spontaneously hypertensive rats. *Clinical and Experimental Pharmacology and Physiology, 30*(1, 2), 44–48.

Kay, R. A., (1991). Microalgae as food and supplement. *Crit. Rev. Food Sci. Nutr., 30*, 555–573.

Kwon, M. J., & Nam, T. J., (2006). Porphyran induces apoptosis related signal pathway in AGS gastric cancer cell lines. *Life Sciences, 79*(20), 1956–1962.

Lee, J. C., Hou, M. F., Huang, H. W., Chang, F. R., & Yeh, C. C., (2013). Marine algal natural products with anti-oxidative, anti-inflammatory, and anticancer properties. *Cancer Cell International, 13*(1), 55–62.

Lordan, S., Paul, R. R., & Stanton, C., (2011). Marine bioactives as functional food ingredients: Potential to reduce the incidence of chronic diseases. *Marine Drugs, 9*(6), 1056–1100.

Madhu, B. K., Bereket, T., Bikshal, B. K., Phani, R. S., & Ravi, A., (2011). Protein rich marine red algae-*Gracilaria canaliculata* as an additive for diet. *Journal of Pharmacy Research, 4*(11), 4306–4307.

Maruyama, H., Tamauchi, H., Hashimoto, M., & Nakano, T., (2013). Antitumor activity and immune response of Mekabu fucoidan extracted from sporophyll of *Undaria pinnatifida. In Vivo, 17*(3), 245–249.

Mouritsen, O. G., (2013). *Seaweeds, Edible, Available and Sustainable.* Chicago: University of Chicago Press.

Murata, M., & Nakazoe, J., (2001). Production and use of marine algae in Japan. *Japan Agricultural Research Quarterly, 35,* 281–290.

Nuno, K., Villarruel-Lopez, A., Puebla-Perez, A. M., Romero-Velarde, E., & Puebla-Mora, A. G., (2013). Effects of the marine microalgae *Isochrysis galbana* and *Nannochloropsis oculata* in diabetic rats. *Journal of Functional Foods, 5*(1), 106–115.

Phaneuf, D., Cote, I., Dumas, P., Ferron, L. A., & LeBlanc, A., (1999). Evaluation of the contamination of marine algae (seaweed) from the St. Lawrence River and likely to be consumed by humans. *Environmental Research, 80,* 175–182.

Plaza, M., Cifuentes, A., & Ibanez, E., (2008). In the search of new functional food ingredients from algae. *Trends in Food Science and Technology, 19*(1), 31–39.

Plaza, M., Herrero, M., Cifuentes, A., & Ibanez, E., (2009). Innovative natural functional ingredients from microalgae. *J. Agric. Food Chem., 57,* 7159–7170.

Priyadarshani, I., & Rath, B., (2012). Commercial and industrial applications of microalgae: A review. *Journal of Algal Biomass Utilization, 3*(4), 89–100.

Randhira, A., Laird, D. W., Maker, G., Trengove, R., & Moheimania, N. R., (2020). Microalgae: A potential sustainable commercial source of sterols. *Algal Research, 46,* 101772.

Rindi, F., Soler-Vila, A., & Guiry, M. D., (2012). Taxonomy of marine macroalgae used as sources of bioactive compounds. In: Hayes, M., (ed.), *Marine Bioactive Compounds* (pp. 1–53). Springer, New York, USA.

Shahidi, F., (2008). Nutraceuticals and functional foods: Whole versus processed foods. *Trends in Food Science and Technology, 20*(9), 376–387.

Shannon, E., & Abu-Ghannam, N., (2019). Seaweeds as nutraceuticals for health and nutrition. *Phycologia, 58*(5), 563–577.

Shao, L. L., Xu, J., Shi, M. J., Wang, X., Li, Y. T., Kong, L. M., & Zhou, T., (2017). Preparation, antioxidant, and antimicrobial evaluation of hydroxamated degraded polysaccharides from *Enteromorpha prolifera. Food Chem., 237,* 481–487.

Singh, S., Kate, B. N., & Banerjee, U. C., (2005). Bioactive compounds from cyanobacteria and microalgae: An overview. *Critical Reviews in Biotechnology, 25*(3), 73–95.

Slavin, J., (2013). Fiber and prebiotics: Mechanisms and health benefits. *Nutrients, 5*(4), 1417–1435.

Sugimura, Y., Suzuki, Y., & Miyake, Y., (1976). The content of selenium and its chemical form in seawater. *Journal of the Oceanographical Society of Japan, 32,* 235–241.

Thompson, K. D., & Dragar, C., (2004). Anti-viral activity of *Undaria pinnatifida* against herpes simplex virus. *Phytotherapy Research, 18*(7), 551–555.

Thomson, C. D., Smith, T. E., Butler, K. A., & Packer, M. A., (1996). An evaluation of urinary measures of iodine and selenium status. *Journal of Trace Elements in Medicine and Biology, 10,* 214–222.

Tokuşoglu, O., & Una, M. K., (2003). Biomass nutrient profiles of three microalgae: *Spirulina platensis, Chlorella vulgaris,* and *Isochrisis galbana. Journal of Food Science, 68*(4), 1144–1148.

Turner, N. J., (2003). The ethnobotany of edible seaweed (*Porphyra abbottae* and related species; Rhodophyta: Bangiales) and its use by first nations on the Pacific Coast of Canada. *Can. J. Bot., 81,* 283–293.

Uchida, M., Kurushima, H., Ishihara, K., Murata, Y., Touhata, K., Ishida, N., Niwa, K., & Araki, T., (2017). Characterization of fermented seaweed sauce prepared from nori (*Pyropia yezoensis*). *J. Biosci. Bioeng., 123*(3), 327–332.

Wong, K. H., & Cheung, P. C. K., (2001). Nutritional evaluation of some subtropical red and green seaweeds. Part II: *In vitro* protein digestibility and amino acid profiles of protein concentrates. *Food Chemistry, 72*, 11–17.

Wong, K., & Cheung, P. C., (2000). Nutritional evaluation of some subtropical red and green seaweeds: Part I: proximate composition, amino acid profiles and some physicochemical properties. *Food Chemistry, 71*, 475–482.

Zerrifi, S., El Khalloufi, F., Oudra, B., & Vasconcelos, V., (2018). Seaweed bioactive compounds against pathogens and microalgae: Potential uses on pharmacology and harmful algae bloom control. *Marine Drugs, 16*, 55.

Zubia, M., Robledo, D., & Freile-Pelegrin, Y., (2007). Antioxidant activities in tropical marine macroalgae from the Yucatan Peninsula, Mexico. *J. Appl. Phycol., 19*, 449–458.

CHAPTER 3

FOOD BIOACTIVE COMPOUNDS FROM MICROALGAE

PRICILA P. NASS, TATIELE C. DO NASCIMENTO,
ANDRÊSSA S. FERNANDES, LEILA Q. ZEPKA, and
EDUARDO JACOB-LOPES

*Department of Food Science and Technology,
Federal University of Santa Maria (UFSM), Roraima Avenue,
1000, 97105-900, Santa Maria, RS, Brazil*

ABSTRACT

The market trends, and consequently, industrial, and commercial interest, have turned to foods with natural ingredients. Thus, the richness of bioactive food biocompounds from microalgae has long inspired their exploitation, fundamentally supported by balanced composition, containing multiple molecules with myriad biological functionalities. To validate this assumption, several microalgae-based bioproducts are already commercially available, either as protein, fatty acids (FAs), and pigments. Also, diverse emerging bioproducts (polysaccharides, vitamins, phycoerythrin (PE), and sterols) are yet to be fully explored. In general, these molecules are less competitive commercially than traditional sources. However, some microalgae biomolecules have advantages over synthetic molecules, due to its chemical conformation, making their use commercially attractive for the food sector, despite the higher production costs. In that context, encouraged by the growing interest in food biocompounds, this chapter provides an overview of the bioactive food compounds microalgae-based, the current status of the market of such compounds as well as to regulatory issues in this field.

3.1   INTRODUCTION

The growing demand for natural products has strengthened microalgae a promising alternative for obtaining natural compounds. Because a consequence of this trend, the worldwide market estimation of microalgae is assessed to be around US \$5.7 billion, out of which about US \$2.5 billion is generated by the health food ingredients (Koyande et al., 2019; Jacob-Lopes et al., 2019; Sathasivam et al., 2019; Tang et al., 2020; Katiyar and Arora, 2020).

Regarding the collection of microorganisms, we refer to as microalgae is a polyphyletic group, with an estimated number of 72,500 species consistently cataloged. However, the species considered safe and commercially consolidated as food supplements, ingredients, or additives, comprise *Arthrospira (Spirulina), Chlorella* sp., *Porphyridium cruentum, Crypthecodinium cohnii, Haematococcus pluvialis, Phaeodactylum tricornutum, Dunaliella* sp., *Nannochloropsis* sp., *Nitzschia* sp. and *Schizochytrium* sp. (Jacob-Lopes et al., 2019; Torres-Tiji et al., 2020).

In this sense, due to the metabolic and taxonomic diversity, combined with the high biotechnological potential, microalgae are excellent producers of bioactive molecules, such as proteins with essential amino acids (EAAs), fatty acids (FAs), carbohydrates, vitamins, and pigments related to many human biological activities (Guedes et al., 2011; Borowitzka, 2013; Patias et al., 2017; Fernandes et al., 2017, 2020; Niccolai et al., 2019; Nascimento et al., 2020). The main bioactive properties associated with these compounds are related to the anti-inflammatory, antioxidant, anti-aging, immunosuppressive, antihypertensive, hypocholesterolemic, photoprotective, and neurotransmitting activities (Fernandes et al., 2017; Patias et al., 2017; Sathasivam et al., 2019; Jacob-Lopes et al., 2019; Nascimento et al., 2019, 2020).

Thus, this chapter provides a comprehensive overview of the potentially valuable bioactive food compounds from the microalgae, including the biological properties of these compounds. Moreover, it is revised the current status of the market of such biocompounds as well as to regulatory milestones for application.

3.2   MICROALGAE-BASED FOOD BIOACTIVE COMPOUNDS

The difference in morphological, physiological, and genetic characteristics between the different microalgae species provides these microorganisms with the ability to synthesize a wide range of bioactive molecules in different proportions (Sathasivam et al., 2019). Thus, in recent decades, greater attention has been paid to studies on microalgae biochemicals with potential for application as functional food ingredients (Vaz et al., 2016; Bernaerts et al., 2018; Caporgno and Mathys, 2018).

It is known that the constitution of biomass is rich mainly in proteins, with values that can reach up over 50% of dry weight in some species (Singh et al., 2020). Thus, since the beginning of studies with microalgae, this class of biomolecules was the most explored bioactive food compounds in these matrices (Jacob-Lopes et al., 2019). However, currently, emerging bioactive compounds such as polysaccharides, FAs, pigments, and vitamins are changing their status in the field of microalgae research and gaining space in relevant research (Camacho et al., 2019; Levasseur et al., 2020).

Among the class of macromolecules, microalgae are recognized as potent and promising sources of proteins (Khanra et al., 2018). The great advantage of microalgal proteins is in the constitution of amino acids in their structures, as they present a complete profile of EAAs that are often not found in higher plants (Koyande et al., 2019). These essential compounds are histidine, isoleucine, leucine, lysine, methionine, phenylalanine, threonine, tryptophan, and valine. Other amino acids such as cysteine, proline, arginine, glutamine, glycine, proline, tyrosine, and aspartic acid also appear in the protein fraction of some species (Mobin et al., 2019; Torres-Tiji et al., 2020).

In particular, the presence of EAAs is very significant for the population that follows a vegetarian and vegan diet since microalgae biomass becomes an alternative to obtain these essential molecules (Koyande et al., 2019). Likewise, other significant factors such as favorable nutritional and functional properties, low allergenicity, and high protein content (~50%) drive the application of these biopolymers as functional, nutritional, and therapeutic commodities (Soto-Sierra et al., 2018). An example of this is the application of some microalgae proteins in formulations of anti-obesity functional foods, since they stimulate the production of the cholecystokinin hormone, responsible for regulating appetite suppression (Patias et al., 2018).

From the fraction of carbohydrates in microalgae, polysaccharides have become a frequent target of scientific research as compounds with bioactive properties. These compounds are complex and heterogeneous macromolecules, composed of different monosaccharide units, and some may contain some degree of neutral acids or sulfate groups (called sulfated polysaccharides) (Raposo et al., 2015).

According to the microalgae group, a diversity of structures is produced. Some species synthesize homopolysaccharides (units of monosaccharides of the same type), and most of them have in their cellular constitution heteropolysaccharides (units of different monosaccharides). Some of these compounds are components of the cell walls, while others are secreted out of cells and are denominated exopolysaccharides (EPSs) (Raposo et al., 2015; Bernaerts et al., 2018; Li et al., 2020).

The high structural diversity reflects the bioactivity diversity of these molecules. Some polysaccharides when supplemented in diets have shown positive effects on human health. β-glucans, for example, in the human organism, act as soluble fiber and are considered great allies in the reduction of LDL cholesterol and cardiovascular diseases (Khoury et al., 2011). On the other hand, sulfated polysaccharides derived from algae have already proven to have several important properties such as cancer preventive properties, antioxidant activity, anticoagulant, immunomodulatory, antiviral activities, anti-inflammatory, antinociceptive, antihyperlipidemic, and antihepatotoxic activities (Wang et al., 2014).

Among the sulfated structures, spirulan, an algae-specific polysaccharide (*Arthrospira platensis*), also has been shown to have antiviral activity and capacity of reduction in plasma cholesterol levels (Lee et al., 2001; Samuels et al., 2002). In particular, a study realized by Guzmán et al. (2003) showed that crude polysaccharide extracts from *Chlorella stigmatophora* and *Phaeodactylum tricornutum* exhibited a high anti-inflammatory effect. Antiviral effects have been observed for polysaccharides in *Porphyridium* sp. microalgae (Huleihel et al., 2001). As well as antioxidant properties have been related to the fraction of extracellular polysaccharide from microalgae *Rhodella reticulata* (Chen et al., 2010). EPSs produced by microalgae *Chlorella pyrenoidosa*, *Chlorococcum* sp. and *Scenedesmus* sp. exhibited free radicals scavenging abilities and exhibited significant antitumor effects for colon cancer cell (Zhang et al., 2019). Considering these different biological activities, microalgae polysaccharides are suitable for application in many distinct areas, such as for food purposes as additives that improve the quality and texture of food or to obtain healthy, nutraceutical, and functional foods (Raposo et al., 2014, 2015; Bernaerts et al., 2018).

Microalgae are recognized as an important source of lipids. The most strains present levels from 20% to 50% of the dry biomass (w/w), depending on the species and cultivation conditions (Barkia et al., 2019). In particular, *Botryococcus braunii* (green microalgae), under optimal growth conditions can reach lipid concentrations of up to 86% (Tandon and Jin, 2017). The lipid profile of microalgae is basically consisting of triacylglycerols (TAG) molecules, monounsaturated fatty acids (MUFA), and polyunsaturated fatty acids (PUFA) (Tang et al., 2020). Especially, PUFA, i.e., omega fatty acids (FAs), receive greater attention since they are valuable as health food supplements (Barkia et al., 2019). The nutritional importance of this class of biocompounds is associated with vital health functions, as they are considered essential for biological processes. Furthermore, the relevance of obtaining these compounds exogenously is based on the fact that human beings are unable to synthesize some of these FAs (Kaur et al., 2014).

In certain species, PUFAs are present in the constitution of biomass in concentrations between 25 and 60% of the total lipids (Vaz et al., 2016). The PUFAs profile in microalgae include ω-6 FAs as linoleic acid (LNA, 18:2n-6), γ-linolenic acid (GLA, 18:3n-6), and arachidonic acid (ARA, 20:4n-6), as well as ω-3 FAs which include α-linolenic acid (ALA, 18:3n-3), docosapentaenoic acid (DPA, 22:5n-3), docosahexaenoic acid (DHA, 22:6n-3), and eicosapentaenoic acid (EPA, 20:5n-3) (Morais et al., 2015; Katiyar and Arora, 2020).

Among the ω-3 long-chain FAs, EPA, and DHA are well accepted as being essential components of a healthy diet, having beneficial effects on cardiovascular disease (Delgado-Lista et al., 2012), development especially of the neural system (Innis, 2007; Campoy et al., 2012), prevention, and treatment of cancer (Berquin et al., 2008), lowers the incidence of diabetes (Woodman et al., 2003) arteriosclerosis, and thrombosis (Liu et al., 2016) and arthritis (Lee et al., 2012). The cardioprotective effect of EPA and DHA FAs is related to the mechanisms of reduction of triglyceride (TG) levels, attenuation of atherosclerotic plaques, the exertion of antidysrhythmic, anti-thrombotic, and anti-inflammatory effects, lowering of systolic and diastolic blood pressures, and improvement in endothelial function (Bradberry et al., 2013). Also, DHA is considered a vital supplementary component in feeding during pregnancy and breastfeeding, as it plays an important role in the development of infants, especially the brain and retina (Echeverría et al., 2017; Sun et al., 2018).

The ω-6 fatty acid ARA, as well as EPA and DHA acids, play important roles in regulating body homeostasis. The human body is converted into eicosanoids and can regulate diverse sets of homeostatic and inflammatory processes linked to in numerous diseases, including inflammation, infection, cancer, and cardiovascular diseases (Saini and Keum, 2018). GLA, in addition to contributing to prostaglandins biosynthesis, has beneficial effects against tumor cells, dermatitis, diabetes, schizophrenia, multiple sclerosis, and rheumatoid arthritis (Raja et al., 2018). Linoleic and α-linolenic FAs are essential nutrients for the immune system and other processes related to the tissue regeneration, as well as or the synthesis of the cell membrane prosta-glandins (Raposo et al., 2013). At the same time are associated with positive effects against some pathologies. Some evidence indicates that ALA has cardiovascular-protective, neuroprotective, anti-cancer, anti-osteoporotic, antioxidative, and anti-inflammatory effects (Kim et al., 2014).

Also, epidemiological studies indicate a positive association between the intake of LNA with lower levels of plasma low-density lipoprotein-cholesterol (LDL-C) (Harris, 2008), as well as long-term glycemic control (Salas-Salvadó et al., 2011), and cardiovascular protection. Another relevant factor associated with these FAs is the ability to become more biologically more active substances. At the metabolic level, LNA is metabolized by the enzymes elongases and desaturases to generate mainly ARA (Marangoni et al., 2020). In contrast, ALA is the precursor of EPA and DHA (Kim et al., 2014).

Another class of biocompounds that can be obtained from the lipid frac-tion of microalgae are sterols. These compounds, considered membrane lipids, are present in higher concentrations in eukaryotic microalgae than in prokaryotic ones. However, microalgae with prokaryotic cell structure, although they do not have large concentrations of sterols, can synthesize different compositions of these molecules (Volkman, 2016). Thus, micro-algae produce uncommon sterols, also known as phytosterols (C28 and C29 sterols), such as brassicasterol, clionasterol, squalene stigmasterol, campes-terol, β-sitosterol, and ergosterol (Randhir et al., 2020).

Due to their bioactive properties, these uncommon compounds are being increasingly incorporated into functional foods (Luo et al., 2015). This class of compounds is associated with biological benefits, particularly due to the hypocholesterolemic properties (Devaraj et al., 2004). Also, sterols from microalgae presented potent antioxidant properties that provide protection from stroke, heart, and coronary diseases, anticarcinogenic properties, and anti-inflammatory activity (Sanjeewa et al., 2016; Matos et al., 2017).

Furthermore, sterols molecules, particularly cholesterol, are precursors to a variety of compounds important for human metabolisms, such as steroid hormones (estrogens and progesterone), vitamin D, and bile salts (Wollam and Antebi, 2011). Likewise, Blaga et al. (2018) described ergosterol as an ergocalciferol (Vitamin D_2) precursor. Another application of sterols in the nutritional area is the use of biomass from some species with significant concentrations of these compounds in formulating rations to promote the growth of juveniles, especially oysters (Raposo et al., 2013). Noteworthy, the β-sitosterol is a phytosterol that has already been marketed as a nutraceutical component due to its ability to reduce serum cholesterol in hypercholesterol-emic individuals (Luo et al., 2015).

In terms of micro-nutrients, microalgae contain high levels of essential vitamins compared to staple foods (Chew et al., 2017). They are able to synthesize hydrophilic vitamins including B-complex (thiamine-B_1, riboflavin-B_2, niacin-B_3, pantothenic acid-B_5, pyridoxine-B_6, biotin-B_7, folic acid-B_9, and cobalamin-B_{12}) and vitamin C; and lipophilic vitamins as pro-vitamin A, vitamin E and K (Nazih and Bard, 2018; Mobin et al., 2019). Although little documented, it is known that some microalgae species are also capable of synthesizing vitamin D (Ljubic et al., 2020).

These biocompounds have consolidated applications in metabolic functions essential for health, as they are precursors of important enzymatic cofactors (Galasso et al., 2019). Furthermore, vitamins play a vital role in the immune system with participation in the cell formation and blood clotting mechanism and to display strong antioxidant activity (Raposo et al., 2013). Vitamin B_{12} supplementation is of particular interest to people who follow strict vegetarian or vegan diets. This is because the main source of this vitamin is foods of animal origin, such as meat, milk, and eggs. Thus, vitamin B_{12} is often used as a food supplement (Galasso et al., 2019).

Among the biochemical components, microalgae can synthesize large quantities of pigments that act as colorant and food supplements. These compounds are classified into three classes: carotenoids, chlorophylls, and phycobiliproteins (PBPs) (Rodrigues et al., 2015; D'Alessandro and Filho, 2016). Microalgae synthesize numerous carotenoid structures, including all known xanthophylls from conventional sources (e.g., lutein, zeaxanthin, antheraxanthin). Also, are able to synthesize specific pigments such as ketocarotenoids, glycosylated carotenoids, allenic, and acetylenic carotenoids (e.g., astaxanthin, echinenone, fucoxanthin, neoxanthin, diadinoxanthin, canthaxanthin), many found only in these microorganisms (Takaichi and Mochimaru, 2007; Guedes et al., 2011; Takaichi, 2011).

These phytochemicals are associated with several important biological functions, including pro-vitamin A activities, well documented for β-carotene and other carotenoids containing β rings. However, its bioactive values go beyond the pro-vitamin activity. Many studies have reviewed the properties of these compounds and a number of other bioactivities are associated with these pigments, which include potent antioxidant activities, neurological disorders, immune system functions, prevention of degenerative chronic diseases as diabetes, obesity, certain types of cancer, macular degeneration, and cardiovascular diseases (Rodriguez-Concepcion et al., 2018; Sathasivam and Ki, 2018; Khalid et al., 2019).

In addition, due to the antioxidant properties of carotenoids can also act as protective agents of lipid peroxidation. A recent study carried out on rats demonstrated that the carotenoids of the microalgae *Scenedesmus obliquus* was able to protect against oxidative body tissues stress by their capacity to improve the activity of some antioxidant enzymes and to reduce lipid peroxidation (Nascimento et al., 2019, 2020). It is worth mentioning that these molecules represent a successful model for food components from microalgae since β-carotene and astaxanthin are produced and marketed from microalgae *Dunaliella salina* and *Haematococcus pluvialis*, respectively (Gong and Bassi, 2016). At the same time, other less established pigments as lutein and zeaxanthin are also gaining momentum in the market of natural pigments (Novoveská et al., 2019).

All photosynthetic beings, such as microalgae, have chlorophyll α in their composition. On the other hand, chlorophyll *b* is exclusively found in the *Chlorophyta* and their descendants. The chlorophyll fraction in microalgae may show some chlorophyll derivatives as pheophorbides, chlorophyllides, pheophytins, oxidized, and allomerized compounds (Mulders et al., 2014). In addition, other series of chlorophylls can be found in these microorganisms such as chlorophyll *c*, which is exclusively found in the descendants of the *Rhodophyta* (Jeffrey and Wright, 2005), chlorophyll *d* and *f* synthesized by cyanobacteria (Airs et al., 2014). However, knowledge about its bioactivities is lacking, since most studies only explore the series of chlorophyll α and β and their derivative compounds.

In fact, microalgae have high chlorophyll content per unit of biomass, sometimes more abundant than in higher plants (Galasso et al., 2019). Furthermore, in green microalgae (*Chlorophyta*), this class of pigments (specifically chlorophylls α and β) is predominant compared to carotenoids (Fernandes et al., 2020). Even so, considering the number of microalgae species in existence, few studies address the complete profile of chlorophylls in microalgae (Fernandes et al., 2017, 2020; Maroneze et al., 2019).

Although some microalgae biomass is commercialized with the appeal of the high content of bioactive chlorophylls, the real bioactivity of these compounds in the biomass is still unclear. Nevertheless, chlorophylls, and their derivatives compounds from other food sources or in isolated extract show a significant degree of absorption and subsequent bioavailability (Ferruzzi and Blakeslee, 2007; Gallardo-Guerrero et al., 2008; Gandul-Rojas et al., 2009; Chen and Roca, 2018) and are related to prominent biological activities as antimutagenic effect, anti-inflammatory, antigenotoxic properties and potent antioxidant capacity (Lanfer-Marquez et al., 2005; Pérez-Gálvez et al., 2017). Given these characteristics, these compounds are currently promising candidates as food bioactive compounds.

PBPs are bioactive substances formed by a proportion of protein with an associated chromophore (called phycobilin) commonly present in cyanobacteria and red algae (D'Alessandro and Filho, 2016). Phycoerythrin (PE) and phycocyanin (PC) stemming the metabolism of *Porphyridium cruentum* and *Arthrospira platensis* (*Spirulina*), respectively, are the most well-known and studied PBPs (Raposo et al., 2013). Besides, phycobilin has antioxidant, hypolipidemic, and anti-inflammatory properties, for which it is showing great interest in the application in functional food (Fernández-Rojas et al., 2014; Baky et al., 2015; Pagels et al., 2019). In contrast, PE presents applications in the biomedical field as a fluorescent agent, tool for research and diagnosis (Pan-Utai and Lamtham, 2019; Sathasivam et al., 2019).

The important capacity for the synthesis of enzymes by microalgae has also been reported. Protease, galactosidases, amylases, phytases, lipase, laccases, cellulases, carbonic anhydrase and antioxidant enzymes are some examples used in the food industry (Brasil et al., 2017). Furthermore, it is worth mentioning the fraction of antifreeze proteins (AFPs) found in some microalgae species (Jung et al., 2014). These compounds, despite having their unknown bioactive capacity, are considered important in the food industry due to their capacity for frozen food preservation. Examples of its application are in meat products, where they were able to reduce the damage caused by freezing, loss of drip and loss of proteins and also improve the juiciness of the meat after thawed (Xiang et al., 2020).

Finally, due to the richness of microalgal biodiversity, these microorganisms have the potential to serve as a natural pool of biochemical for use in the food industry, and some have a consolidated share in the market.

3.3 CURRENT MARKET OF MICROALGAE-BASED BIOACTIVE FOOD COMPOUNDS

A few years ago, the market for bioactive compounds was dominated by synthetic molecules or animal and vegetable sources. Nowadays, the demand for natural products and the viability of industrial production of microalgae-based products appear as an opportunity for commercial expansion (Jacob-Lopes et al., 2019). The global market for microalgae products is estimated to reach approximately US $53.43 billion by 2026 (Rahman, 2020). Today, most of the microalgae biocompound market is related to the healthy food segment, and large companies such as BASF, Unilever, and Dow Chemical are involved in projects related to the production of microalgae (Fernandez et al., 2017). In 2017, Taiwan, Japan, the USA, China, Brazil, Spain, Israel, Germany, and Myanmar were the leading producers of microalgae biomass and derived products (Ramirez-Merida et al., 2017). Currently, the market is dominated by the United States, Asia, and Oceania. It is believed that shortly, Europe will become one of the leaders (Rahman, 2020).

As shown in Table 3.1, microalgae-based products are widely exploited, unicellular protein (total dry biomass), and carotenoid (β-carotene, astaxanthin) PC, and polyunsaturated fatty acid have their share in the market for bioactive compounds consolidated. In contrast, other molecules such as PE, chlorophylls, sterols, vitamins, and polysaccharides are considered emerging proposals that have required efforts from the research and development sector to expand their presence and expand the portfolios of commercialized products (Jacob-Lopes et al., 2019). Additionally, the varied applications of microalgae-based products are made possible due to the plurality of their bioactive properties (Hu, 2019).

The main classes of microalgae used in the commercial production of compounds include Cyanophyceae (*Arthrospira* (*Spirulina*), *Oscillatoria*, *Nostoc*), Chlorophyceae (*Chlamydomonas*, *Dunaliella salina*, *Chlorella*), Prymnesiophyceae (*Isochrysis galbana*) and Bacillariophyceae (*Thalassiosira weissflogii* and *Cyclotella cryptica*) (Singh et al., 2020). However, *Chlorella* and *Spirulina* dominate the world market (Koyande et al., 2019).

Among microalgae-based products, dry microalgae biomass was the first product aimed at the industry. The biomass utilization as food was a traditional practice of many ancient peoples, and even today, it is consolidated worldwide as a protein supply (Ramirez-Merida et al., 2017). As a result of discoveries about its bioactivity, it has currently been frequently associated with the prophylaxis of several pathologies, such as hypolipidemic, hypoglycemic, and anti-obesity (Patias et al., 2018; Soto-Sierra et al., 2018).

Microalgae dry biomass production is equivalent to approximately 19,000 tons/year, generating an estimated annual value of 5.7 billion dollars (Jacob-Lopes et al., 2019). Of this total produced, about 12,000 and 5,000 ton/year are equivalent to Spirulina and *Chlorella* biomass, respectively (Mudliar and Shekh, 2019). According to estimates by Mudliar and Shekh (2019), Spirulina dry biomass market volume could reach US $380 million by 2027.

Because of the global trend towards healthy eating habits, the natural color of foods is considered the largest segment of the food coloring market,

TABLE 3.1 Microalgae-based Products Market, Sources, Applications, and Bioactivity

Market	Product	Microalgae Source	Applications	Bioactivity	References
Consolidated	Protein (dry biomass)	*Chlorella*, Spirulina	Food supplements, nutraceuticals, functional foods	Hypolipidemic, hypoglycemic, anti-obesity	Fernandez et al. (2017); Ramirez-Merida et al. (2017); Patias et al. (2018); Soto-Sierra et al. (2018)
	β-carotene	*D. salina*	Natural food pigments, food supplements, feed additives, pharmaceuticals, cosmetics	Antioxidant, anti-inflammatory, immunological modulation	Barkia et al. (2019); Hu (2019); Jacob-Lopes et al. (2019); Sathasivam et al. (2019); Tang et al. (2020)
	Astaxanthin	*H. pluvialis*	Natural food pigments, food supplements, pharmaceuticals, cosmetics	Antioxidant, anti-inflammatory, antitumor	Hu (2019); Jacob-Lopes et al. (2019); Tang et al. (2020)
	Phycocyanin	Spirulina	Natural food pigments, pharmaceuticals, cosmetics	Antioxidant, anti-inflammatory, neuroprotective effects	Romay et al. (2005); Hu (2019); Singh et al. (2020)
	DHA and EPA	*Schizochytrium*, *P. tricornutum*, *Ulkenia* sp., *C. cohnii*	Food supplements, feed additive	Antioxidant, anti-inflammatory, neuroprotective, hepatoprotective effects, anti-bacterial, antitumor	Echeverria et al. (2017); Jacob-Lopes et al. (2019); Singh et al. (2020); Tang et al. (2020)

TABLE 3.1 *(Continued)*

Market	Product	Microalgae Source	Applications	Bioactivity	References
Emerging	Chlorophylls	*P. autumnale, C. sorokiniana, S. bijuga, C. thermophila*	Natural food pigments, feed additive, nutraceuticals	Antioxidant, anti-inflammatory	Rodrigues et al. (2015); García and Galán (2017); Fernandes et al. (2017, 2020); Sarkar et al. (2020)
	Phycoerythrin	*Porphyridium, Rhodella*	Chemicals, natural food pigments	Antioxidant, anticarcinogenic, antigenotoxic, antimutagenic	Román et al. (2002); Borowitzka (2013); Hu (2019)
	Sterols	*H. pluvialis, A. solitaria, N. carneum*	Nutraceuticals	Anti-hypocholesterolemic, anti-inflammatory activity; antitumor	Randhir et al. (2020); Singh et al. (2020)
	Vitamins	*Chlamydomonas, Chlorella, Scenedesmus, Dunaliella, Tetraselmis*	Nutraceuticals, food supplements, functional food	Antioxidant; cell formation, blood coagulation, immunological modulation	Becker (2013); Koyande et al. (2019); Sathasivam et al. (2019); Singh et al. (2020)
	Polysaccharides	*Porphyridium*	Functional food, cosmetics, tissue engineering	Prebiotics, hypolipemic, hypoglycemic, antioxidants, antivirals	Khoury et al. (2011); Raposo et al. (2013–2015); Gaignard et al. (2019); Hu (2019)

representing more than 80% of this sector's total revenue (Fernandes et al., 2020). According to an industry report, it is estimated that the global food color market will arrive in 2025, with revenues of the US $2.97 billion (Global Natural Food Colors, 2018). The production of microalgal carotenoids arose initially through the cultivation of *Dunaliella* and *Haematococcus*, among which β-carotene and astaxanthin have consolidated economic participation in the market (Hu, 2019; Jacob-Lopes et al., 2019; Tang et al., 2020).

β-carotene is widely marketed for food supplements, natural pigment, feed additives, pharmaceuticals, and cosmetics (Hu, 2019). Along with *D. salina*, the *D. tertiolecta*, *D. bardawil*, *B. braunii*, *C. nivalis*, *C. acidophila*, *Chlorococcum* sp., *Chlamydocapsa* sp., *Tetraselmis* sp., *C. sorokiniana* and *C. striolata* dominate the production of β-carotene (Sathasivam et al., 2019). The β-carotene market is estimated to range between US $224 and 285 million in 2019 (Mudliar and Shekh, 2019).

Astaxanthin is recognized for its bioactive abilities, as well as β-carotene, its applications include food supplements, natural food pigments, pharmaceuticals, and cosmetics (Hu, 2019). The current production of microalgal astaxanthin is dominated by *H. pluvialis*, followed by *C. zofingiensis, C. nivalis, B. braunii, C. vulgaris, C. striolata, Monoraphidium* sp., *Chlamydocapsa* sp., *Neospongiococcum* sp., *Chlorococcum* sp. and *S. obliquus* (Khoo et al., 2019).

The prices of nutraceutical grade astaxanthin originating from *H. pluvialis* can reach the US $6,000/kg depending on the cultivation configuration (Dawidziuk et al., 2017), it is estimated that the market of this ketocarotenoid reaches 770 million dollars up to 2024 (Mudliar and Shekh, 2019). The global market of microalgal astaxanthin still faces some problems of competitiveness due to production costs compared to its synthetic form. While the cost of producing *H. pluvialis* astaxanthin can reach US $3600/kg (Li et al., 2011), the cost of synthetic production is the US $1000/kg (Olaizola, 2003). However, synthetic astaxanthin has its use restricted to aquaculture, making natural production necessary for human consumption and animal feed (Li et al., 2011).

Another pigment with a consolidated place in the market is PC, which is obtained from Spirulina and marketed as a natural blue pigment, being used as a natural dye for health food (beverage, dairy products, confectionery, jellies, etc.), and nutritional ingredient. The selling price of PC ranges from US $500–100,000/kg, depending on purity (Borowitzka,

2013). In 2019, the global market for this PBP was valued at US $60 million (Tang et al., 2020).

DHA and EPA are the main sources of microalgae omega-3 FAs, their market prices vary in the range of US $80–160/kg, with an estimated 2025 reaching US $898.7 million (Hu, 2019). Among the microalgal oils rich in PUFAs, authorized, and marketed for human consumption are oils extracted from *Schizochytrium* (EPA and DHA), *Ulkenia* sp. (DHA), *Crypthecodinium cohnii* (DHA), and *Phaeodactylum tricornutum* (EPA) (Echeverría et al., 2017).

While the aforementioned microalgal products include a portfolio of concrete applications from a commercial point of view, others are potentially indicated or are in the process of development (Fernandez et al., 2017; Jacob-Lopes et al., 2019). In recent years, emerging health issues, such as hypersensitivity, have emerged from synthetic compounds in food (Tang et al., 2020). Consequently, efforts were directed towards the commercial development of new microalgae products (Singh et al., 2020).

Chlorophyll is among these emerging compounds, in addition to pigmentation properties the bioactive activity has been boosting the scientific community in order to make production feasible and increase the number of microalgal pigments available for commercialization (Rodrigues et al., 2015; Fernandes et al., 2017, 2020; Sarkar et al., 2020). Although the production potential has been demonstrated for *Phormidium autumnale, Chlorella sorokiniana, Scenedesmus bijuga,* and *Chlorella thermophila* (Rodrigues et al., 2015; Fernandes et al., 2017, 2020; Sarkar et al., 2020) the commercial chlorophylls available they are produced only from plant sources, mainly spinach (Clark, 2016).

Another pigment in commercial development is PE, produced from red microalgae (*Porphyridium* and *Rhodella*) (Borowitzka, 2013; Hu, 2019). This PBP is widely used as a fluorescent probe and analytical reagent (Román et al., 2002), and to a lesser extent, in the food industry as a natural red-pink pigment (Rahman, 2020). In general, the selling price of PE ranges from US $500 to 50,000/kg (Hu, 2019).

Steroids such as β-sitosterol, campesterol, brassicasterol, stigmasterol, and ergosterol are lipid biocomposites of commercial importance due to their bioactive properties (Singh et al., 2020). The global market for these phytosterols is estimated to reach US $935 million by 2022. In this context, microalgae represent a potential commercial source, as it has been suggested that microalgae phytosterols have higher productivity than those of rapeseed plants, for example (Randhir et al., 2020). The species with the most significant commercial relevance are *Haematococcus pluvialis*, *Anabaena solitaria* and *Nostoc carneum* (Randhir et al., 2020).

Microalgae are also considered potential sources of vitamins; from these microorganisms, it is possible to obtain substantial amounts of vitamins A, C, E, K, and B complex (Becker, 2013). The most-reported species are those of the genus *Chlamydomonas*, *Chlorella*, *Scenedesmus*, *Dunaliella*, and *Tetraselmis* (Koyande et al., 2019; Sathasivam et al., 2019). In terms of prices, the global forecast for the vitamin market by 2023 is US $7.35 billion. According to a research report, the increased demand for functional and enriched processed food products, the widespread vitamin deficiencies, and the fortification of meat and dairy products are factors that currently drive the market for vitamins.

Microalgal polysaccharides are promising, and the interest of the scientific community in these biopolymers is recent (Gaignard et al., 2019). EPSs obtained from *Porphyridium* species have been one of the most explored (Raposo et al., 2013–2015). Additionally, β-glucans, produced by many green algae, have been identified as exceptional prebiotics acting in the reduction of LDL cholesterol and the risk of cardiovascular diseases (Khoury et al., 2011). Also, the synthesis of water-soluble lubricants and thickening agents is made possible from microalgae polysaccharides (Sathasivam et al., 2019).

The bioactive properties of polysaccharides have aroused interest in several industrial sectors, such as pharmaceutical, nutraceutical, cosmetic, and food production (Gaignard et al., 2019). In particular, sulfated polysaccharides found abundantly in different microalgae species (Raposo et al., 2013–2015) have few equivalents in terrestrial plants and have chemical and biological characteristics that allow their application in the development of innovative systems for tissue engineering (Silva et al., 2012). Despite this, the commercialization of this microalgal biopolymer is still in the initial stage and therefore, the market values are not clear (Hu, 2019).

Finally, microalgal bioactive compounds have specific advantages over established alternatives, mainly due to a chemical conformation that is more effective than those of conventional competitors, making their use commercially attractive for multiple market segments (Jacob-Lopes et al., 2019). However, in order to expand the global market for microalgae products, commercial authorization is required in regional markets through various regulations that will be demonstrated in the next section.

3.4 SAFETY AND REGULATORY ISSUES OF MICROALGAE-BASED FOOD COMPOUNDS

There is a high interest in the application of microalgae, or their bioproducts, as food ingredients; therefore, it is fundamental to know their safety. However, the regulatory frameworks that control the use of microalgae like foods or food ingredients differ substantially in different regions of the world (Table 3.2) (Jacob-Lopes et al., 2019; Matos, 2019; Torres-Tiji et al., 2020; Zanella and Vianello, 2020).

TABLE 3.2 Some Quality Standards for Microalgae

Parameter	EU	USA	China	India	Japan	Brazil
Moisture (%)	<10.0	<7.0	–	<9.0	<7.0	<10.0
Ash (%)	<10.0	–	–	<9.0	–	–
Protein (%)	20–55	6–71	6–71	>55	–	–
Lipid (%)	12–60	2–40	2–18	–	–	–
Carbohydrates (%)	25–60	4–64	8–64	–	–	–
Arsenic (ppm)	<0.1	–	–	–	–	<1.0
Lead (ppm)	<3.0	<0.2	–	–	–	<0.3
Cadmium (ppm)	<3.0	<0.2	–	–	–	<0.1
Mercury (ppm)	<0.1	<0.025	–	–	–	<0.5
Microorganism	–	–	–	–	–	–
Standard plate count ($\times 10^6$ g^{-1})	<0.1	<0.2	–	–	<0.005	–
Mold (number/g)	<100	<100	–	–	<100	–
Coliforms (number/g)	Absent	Absent	–	–	<10	<10
Salmonella sp.	Absent	Absent	–	–	–	Absent
Staphylococcus sp. (number/g)	<100	Absent	–	–	–	<500

Source: Becker (1994); AOAC (2002); ISO (2002, 2009a, b, 2015); EN (2009); Matos (2019); Zanella and Vianello (2020); FAO (1997); Chacon-Lee and Gonzalez-Marino (2010); IS (1990); Jassby (1988); Anvisa (2001, 2013).

Specifically, in European Union (EU), the European Food and Safety Authority (EFSA) recognizes the application of *Arthrospira platensis* (Spirulina), *Chlorella pyrenoidosa*, *Chlorella luteoviridis*, and *Chlorella vulgaris* as food, because of its long history of use, can be commercialized in the EU without the need to comply with Regulation EU 2015/2283 on novel foods.

While *Odontella aurita* and *Tetraselmis chui* must follow current legislation of novel foods. As for your bioproducts from microalgae, are authorized for use β-carotene from *Dunaliella salina* (Commission Directive 2008/128/EC), DHA, and EPA extracted from *Schizochytrium* sp. (Commission Decision 2003/427/EC), oil-rich in PUFA obtained from *Ulkenia* sp. (Commission Decision 2009/777/EC), astaxanthin-rich oleoresin extracted from *Haematococcus pluvialis*, and oil-rich in EPA obtained from *Phaeodactylum tricornutum* are approved extracts and labeled as "novel food" (Matos, 2019; EFSA, 2020; Torres-Tiji et al., 2020).

Consequently, a similar standard is held in Canada, where Health Canada is the organization that supervise food safety and determine that any food that is new or has changed compared to existing food products is classified as novel foods, and its safety must be assessed by Canadian Food Inspection Agency (CFIA, 2020). In addition to these, Canada considered safe to consume *A. platensis*, *C. vulgaris*, *C. sorokiniana*, *C. regularis*, *D. salina*, and *Euglena gracilis* (CFIA, 2020; Zanella and Vianello, 2020).

In the United States (USA) jurisdiction, the Food and Drug Administration (FDA) grants GRAS status (generally recognized as safe) (US FDA, 2018). Similarly, like the EU, in the USA, *A. platensis*, *C. vulgaris*, *D. Salina*, *H. pluvialis*, and *P. tricornutum*, authorized as foods and extracts. Moreover, *Chlamydomonas reinhardtii*, *Auxenochlorella prototothecoides*, *Dunaliella bardawil* and *E. gracilis*, also have GRAS status (US FDA, 2018; Jacob-Lopes et al., 2019; Matos, 2019; Torres-Tiji et al., 2020).

Regulations in India, Japan, China, and Brazil have been consulted, and the findings regarding the safety of microalgae as food are summarized, *A. platensis*, *D. salina*, and *H. pluvialis* have been found to be considered safe to consume in India, Japan, China, and Brazil. *Chlorella* is also widely assumed as safe for human consumption, but the approved species of *Chlorella* varies among countries: *C. prototothecoides* is approved in Japan, *C. pyrenoidesa* is approved in China, and *C. vulgaris* is approved in Japan and Brazil. Furthermore, *Ulkenia* sp. has been found to be considered safe in Brazil. Finally, *E. gracilis* is approved by China and Japan (FSSAI, 2016; Anvisa, 2018; Torres-Tiji et al., 2020).

3.5 CONCLUSION

This chapter investigated the potential of microalgae to be exploited in the bioactive ingredients in the food industries. Various renewable bioactive natural compounds with beneficial properties can be extracted from microalgae biomass and incorporated into industrial products to replace synthetic materials. However, new processing perspectives should be explored in order to emerge bioactive compounds not yet marketed.

KEYWORDS

- anti-inflammatory
- antioxidant
- biological properties
- biomass
- fatty acids
- food ingredients
- food safety
- health-food markets
- microalgae
- phycoerythrin
- pigments
- polysaccharides
- protein
- sterols
- vitamins

REFERENCES

Airs, R. L., Temperton, B., Sambles, C., Farnham, G., Skill, S. C., & Llewellyn, C. A., (2014). Chlorophyll *f* and chlorophyll *d* are produced in the cyanobacterium *Chlorogloeopsis fritschii* when cultured under natural light and near-infrared radiation. *FEBS Lett., 588*(20), 3770–3777.

Anvisa, (2001). http://portal.anvisa.gov.br/documents/33880/2568070/RDC_12_2001.pdf/ 15ffddf6-3767-4527-bfac-740a0400829b (accessed on 2 July 2021).

Anvisa, (2013). http://portal.anvisa.gov.br/documents/33880/2568070/rdc0042_29_08_2013. pdf/c5a17d2d-a415-4330-90db-66b3f35d9fbd (accessed on 2 July 2021).

Anvisa, (2018). http://portal.anvisa.gov.br/documents/10181/3898888/IN282018COMP.pdf/ db9c7460-ae66-4f78-8576-dfd019bc9fa1 (accessed on 2 July 2021).

AOAC, (2002). http://www.aoacofficialmethod.org/index.php?mainpage=productinfo&prod uctsid=533 (accessed on 2 July 2021).

Baky, H. H. A. E., Baroty, G. S. E., & Ibrahe, E. A., (2015). Functional characters evaluation of biscuits sublimated with pure phycocyanin isolated from Spirulina and Spirulina biomass. *Nutr. Hosp., 32*(1), 231–241.

Barkia, I., Saari, N., & Manning, S. R., (2019). Microalgae for high-value products towards human health and nutrition. *Mar. Drugs, 17*(5), 1–29.

Becker, E. W., (1994). *Microalgae: Biotechnology and Microbiology* (Vol. 10, p. 293). Cambridge University Press, Cambridge.

Becker, E. W., (2013). Microalgae for human and animal nutrition. In: Richmond, A., & Hu, Q., (eds.), *Handbook of Microalgal Culture* (pp. 461–503). John Wiley and Sons, New York.

Bernaerts, T. M., Gheysen, L., Kyomugasho, C., Kermani, Z. J., Vandionant, S., Foubert, I., & Van-Loey, A. M., (2018). Comparison of microalgal biomasses as functional food ingredients: Focus on the composition of cell wall related polysaccharides. *Algal Res., 32*, 150–161.

Berquin, I. M., Edwards, I. J., & Chen, Y. Q., (2008). Multi-targeted therapy of cancer by omega-3 fatty acids. *Cancer Lett., 269*(2), 363–377.

Blaga, A. C., Ciobanu, C., Caşcaval, D., & Galaction, A. I., (2010). Enhancement of ergosterol production by *Saccharomyces cerevisiae* in batch and fed-batch fermentation processes using n-dodecane as oxygen-vector. *Biochem. Eng. J., 131*, 70–76.

Borowitzka, M. A., (2013). High-value products from microalgae-their development and commercialization. *J. Appl. Phycol., 25*, 743–756.

Bradberry, J. C., & Hilleman, D. E., (2013). Overview of omega-3 fatty acid therapies. *Pharmacol. Ther., 38*(11), 680–691.

Brasil, B. D. S. A. F., De Siqueira, F. G., Salum, T. F. C., Zanette, C. M., & Spier, M. R., (2017). Microalgae and cyanobacteria as enzyme biofactories. *Algal Res., 25*, 76–89.

Camacho, F., Macedo, A., & Malcata, F., (2019). Potential industrial applications and commercialization of microalgae in the functional food and feed industries: A short review. *Mar. Drugs, 17*(312), 1–25.

Campoy, C., Escolano-Margarit, M. V., Anjos, T., Szajewska, H., & Uauy, R., (2012). Omega 3 fatty acids on child growth, visual acuity, and neurodevelopment. *Br. J. Nutr., 107*(2), 85–106.

Caporgno, M. P., & Mathys, A., (2018). Trends in microalgae incorporation into innovative food products with potential health benefits. *Front. Nutr., 58*(5), 1–10.

CFIA, (2020). https://www.canada.ca/en/health-canada/services/food-nutrition/genetically-modified-foods-other-novel-foods.html (accessed on 2 July 2021).

Chacon-Lee, L. T., & Gonzalez-Marino, E. G., (2010). Microalgae for "healthy" foods-possibilities and challenges. *Compr. Rev. Food. Sci. F., 9*, 655–675.

Chen, B., You, W., Huang, J., Yu, Y., & Chen, W., (2010). Isolation and antioxidant property of the extracellular polysaccharide from *Rhodella reticulata*. *World J. Microb. Biot., 26*(5), 833–840.

Chen, K., & Roca, M., (2018). *In vitro* bioavailability of chlorophyll pigments from edible seaweeds. *J. Funct. Foods, 41*, 25–33.

Chew, K. W., Yap, J. Y., Show, P. L., Suan, N. H., Juan, J. C., Ling, T. C., Lee, D. J., & Chang, J. S., (2017). Microalgae biorefinery: High value products perspectives. *Bioresour. Technol., 229*, 53–62.

Clark, M., (2016). *Handbook of Textile and Industrial Dyeing Principles, Processes and Types of Dyes* (Vol. 1, p. 680). Woodhead Publishing, Cambridge.

D'Alessandro, E. B., & Filho, N. R. A., (2016). Concepts and studies on lipid and pigments of microalgae: A review. *Renew Sust. Energ. Rev., 58*, 832–841.

Dawidziuk, A., Popiel, D., Luboinska, M., Grzebyk, M., Wisniewski, M., & Koczyk, G., (2017). Assessing contamination of microalgal astaxanthin producer *Haematococcus* cultures with high-resolution melting curve analysis. *J. Appl. Genet., 58*(2), 277–285.

Delgado-Lista, J., Perez-Martinez, P., Lopez-Miranda, J., & Perez-Jimenez, F., (2012). Long chain omega-3 fatty acids and cardiovascular disease: A systematic review. *Br. J. Nutr., 107*(2), 201–213.

Devaraj, S., Jialal, I., & Vega-López, S., (2004). Plant sterol-fortified orange juice effectively lowers cholesterol levels in mildly hypercholesterolemic healthy individuals. *Arterioscler Thromb. Vasc. Biol., 24*(3), 25–28.

Echeverría, F., Valenzuela, R., Hernandez-Rodas, M. C., & Valenzuela, A., (2017). Docosahexaenoic acid (DHA), a fundamental fatty acid for the brain: New dietary sources. *Prostaglandins Leukot. Essent. Fatty Acids, 124*, 1–10.

EFSA, (2020). https://ec.europa.eu/food/safety/novel_food/catalogue/search/public/index. cfm?ascii=N (accessed on 2 July 2021).

EN, (2009). https://cdn.standards.iteh.ai/samples/29302/799c23fe6a6a4e8cb2dd72ae1050 e81f/SIST-EN-15763-2010.pdf (accessed on 2 July 2021).

FAO, (1997). http://www.fao.org/3/w7241e/w7241e0h.htm#chapter%206%20%20%20oil%20 production (accessed on 2 July 2021).

Fernandes, A. S., Nogara, G. P., Menezes, C. R., Cichoski, A. J., Mercadante, A. Z., Jacob-Lopes, E., & Zepka, L. Q., (2017). Identification of chlorophyll molecules with peroxyl radical scavenger capacity in microalgae *Phormidium autumnale* using ultrasound-assisted extraction. *Food Res. Int., 99*, 1036–1041.

Fernandes, A. S., Petry, F. C., Mercadante, A. Z., Jacob-Lopes, E., & Zepka, L. Q., (2020). HPLC-PDA-MS/MS as a strategy to characterize and quantify natural pigments from microalgae. *Curr. Res. Nutr. Food Sci., 3*, 100–112.

Fernandez, F. G. A., Sevilla, J. M. F., & Grima, E. M., (2017). Microalgae: The basis of mankind sustainability. In: Moya, B. L., De Gracia, D. S., & Mazadiego, L. F., (eds.), *Case Study of Innovative Projects-Successful Real Cases* (pp. 123–140). IntechOpen, London.

Fernández-Rojas, B., Hernández-Juárez, J., & Pedraza-Chaverri, J., (2014). Nutraceutical properties of phycocyanin. *J. Funct. Foods, 11*, 375–392.

Ferruzzi, M. G., & Blakeslee, J., (2007). Digestion, absorption, and cancer preventative activity of dietary chlorophyll derivatives. *Nutr. Res. Rev., 27*(1), 1–12.

FSSAI, (2016). https://archive.fssai.gov.in/home/fss-legislation/notifications/gazette-notification.html (accessed on 2 July 2021).

Gaignard, C., Gargouch, N., Dubessay, P., Delattre, C., Pierre, G., Laroche, C., Fendri, I., et al., (2019). New horizons in culture and valorization of red microalgae. *Biotechnol. Adv., 37*(1), 193–222.

Galasso, C., Gentile, A., Orefice, I., Ianora, A., Bruno, A., Noonan, D. M., Sansone, C., Albini, A., & Brunet, C., (2019). Microalgal derivatives as potential nutraceutical and food supplements for human health: A focus on cancer prevention and interception. *Nutr., 11*(6), 1–22.

Gallardo-Guerrero, L., Gandul-Rojas, B., & Mínguez-Mosquera, M. I., (2008). Digestive stability, micellarization, and uptake by Caco-2 human intestinal cell of chlorophyll derivatives from different preparations of pea (*Pisum sativum* L.). *J. Agric. Food Chem., 56*(18), 8379–8386.

Gandul-Rojas, B., Gallardo-Guerrero, L., & Mínguez-Mosquera, M. I., (2009). Influence of the chlorophyll pigment structure on its transfer from an oily food matrix to intestinal epithelium cells. *J. Agric. Food Chem., 57*(12), 5306–5314.

García, J. L., Vicente, M., & Galán, B., (2017). Microalgae, old sustainable food, and fashion nutraceuticals. *Microb. Biotechnol., 10*(5), 1017–1024.

Global Natural Food Colors Market Size & Share Industry Report 2018–2025, (2018). https://www.grandviewresearch.com/industry-analysis/natural-food-colors-market (accessed on 2 July 2021).

Gong, M., & Bassi, A., (2016). Carotenoids from microalgae: A review of recent developments. *Biotechnol. Adv., 34*(8), 1396–1412.

Guedes, A. C., Amaro, H. M., & Malcata, F. X., (2011). Microalgae as sources of carotenoids. *Mar. Drugs, 9*(4), 625–644.

Guzmán, S., Gato, A., Lamela, M., Freire-Garabal, M., & Calleja, J. M., (2003). Anti-inflammatory and immunomodulatory activities of polysaccharide from *Chlorella stigmatophora* and *Phaeodactylum tricornutum*. *Phytother. Res., 17*(6), 665–670.

Harris, W. S., (2008). Linoleic acid and coronary heart disease. *Prostaglandins Leukot. Essent. Fatty Acids, 79*(35), 169–171.

Hu, I. C., (2019). Production of potential coproducts from microalgae. In: Pandey, A., Chang, J. S., Soccol, C. R., Lee, D. J., & Chisti, Y., (eds.), *Biofuels from Algae* (pp. 345–358). Elsevier BV, Netherlands.

Huleihel, M., Ishanu, V., Tal, J., & Arad, S. M., (2001). Antiviral effect of red microalgal polysaccharides on *Herpes simplex* and *Varicella zoster* viruses. *J. Appl. Phycol., 13*(2), 127–134.

Innis, S. M., (2007). Dietary (n-3) fatty acids and brain development. *J. Nutr., 137*(4), 855–859.

IS, (1990). https://archive.org/details/gov.in.is.12895.1990 (accessed on 2 July 2021).

ISO, (2002). https://www.iso.org/standard/37272.html (accessed on 2 July 2021).

ISO, (2009a). https://www.iso.org/standard/44807.html (accessed on 2 July 2021).

ISO, (2009b). https://www.iso.org/obp/ui/#iso:std:iso:1871:ed-2:v1:en (accessed on 2 July 2021).

ISO, (2015). https://www.iso.org/standard/63542.html (accessed on 2 July 2021).

Jacob-Lopes, E., Maroneze, M. M., Deprá, M. C., Sartori, R. B., Dias, R. R., & Zepka, L. Q., (2019). Bioactive food compounds from microalgae: An innovative framework on industrial biorefineries. *Curr. Opin. Food Sci., 25*, 1–7.

Jassby, A., (1988). Some public health aspects of microalgal products. In: Lembi, C. A., & Waaland, J. R., (eds.), *Algae and Human Affair* (pp. 181–201). Cambridge University Press, Cambridge.

Jeffrey, S. W., & Wright, S. W., (2005). Photosynthetic pigments in marine microalgae: Insights from cultures and the sea. *Environ. Sci., 1*, 33–90.

Jung, W., Gwak, Y., Davies, P. L., Kim, H. J., & Jin, E., (2014). Isolation and characterization of antifreeze proteins from the Antarctic marine microalga *Pyramimonas gelidicola*. *Mar. Biotechnol., 16*(5), 502–512.

Katiyar, R., & Arora, A., (2020). Health-promoting functional lipids from microalgae pool: A review. *Algal Res., 46*, 101800.

Kaur, N., Chugh, V., & Gupta, A. K., (2014). Essential fatty acids as functional components of foods: A review. *J. Food Sci. Technol., 51*(10), 2289–2303.

Khalid, M., Bilal, M., Iqbal, H. M., & Huang, D., (2019). Biosynthesis and biomedical perspectives of carotenoids with special reference to human health-related applications. *Biocatal. Agric. Biotechnol., 17*, 399–407.

Khanra, S., Mondal, M., Halder, G., Tiwari, O. N., Gayen, K., & Bhowmick, T. K., (2018). Downstream processing of microalgae for pigments, protein, and carbohydrate in industrial application: A review. *Food Bioprod. Process., 110*, 60–84.

Khoo, K. S., Chew, K. W., Ooi, C. W., Ong, H. C., Ling, T. C., & Show, P. L., (2019). Extraction of natural astaxanthin from *Haematococcus pluvialis* using liquid biphasic flotation system. *Bioresour. Technol., 290.*

Khoury, D., Cuda, C., Luhovyy, B. L., & Anderson, G. H., (2012). Beta-glucan: Health benefits in obesity and metabolic syndrome. *J. Nutr. Metab., 2012*, 1–28.

Kim, K. B., Nam, Y. A., Kim, H. S., Hayes, A. W., & Lee, B. M., (2014). α-Linolenic acid: Nutraceutical, pharmacological and toxicological evaluation. *Food Chem. Toxicol., 70*, 163–178.

Koyande, A. K., Chew, K. W., Rambabu, K., Tao, Y., Chu, D. T., & Show, P. L., (2019). Microalgae: A potential alternative to health supplementation for humans. *FSHW, 8*(1), 16–24.

Lanfer-Marquez, U. M., Barros, R. M., & Sinnecker, P., (2005). Antioxidant activity of chlorophylls and their derivatives. *Food Res. Int., 38*(9), 885–891.

Lee, J. B., Srisomporn, P., Hayashi, K., Tanaka, T., Sankawa, U., & Hayashi, T., (2001). Effects of structural modification of calcium spirulan, a sulfated polysaccharide from *Spirulina platensis*, on antiviral activity. *Chem. Pharm. Bull., 49*(1), 108–110.

Lee, Y. H., Bae, S. C., & Song, G. G., (2012). Omega-3 polyunsaturated fatty acids and the treatment of rheumatoid arthritis: A meta-analysis. *Arch. Med. Res., 43*(5), 356–362.

Levasseur, W., Perré, P., & Pozzobon, V., (2020). A review of high value-added molecules production by microalgae in light of the classification. *Biotechnol. Adv., 107*545.

Li, J., Zhu, D., Niu, J., Shen, S., & Wang, G., (2011). An economic assessment of astaxanthin production by large scale cultivation of *Haematococcus pluvialis*. *Biotechnol. Adv., 29*, 568–574.

Li, Y., Wang, C., Liu, H., Su, J., Lan, C. Q., Zhong, M., & Hu, X., (2020). Production, isolation, and bioactive estimation of extracellular polysaccharides of green microalga *Neochloris oleoabundans*. *Algal Res., 48*, 101883.

Liu, L., Hu, Q., Wu, H., Xue, Y., Cai, L., Fang, M., Liu, Z., et al., (2016). Protective role of n6/n3 PUFA supplementation with varying DHA/EPA ratios against atherosclerosis in mice. *J. Nutr. Biochem., 32*, 171–180.

Ljubic, A., Jacobsen, C., Holdt, S. L., & Jakobsen, J., (2020). Microalgae *Nannochloropsis oceanica* as a future new natural source of vitamin D₃. *Food Chem., 320*, 126627.

Luo, X., Su, P., & Zhang, W., (2015). Advances in microalgae-derived phytosterols for functional food and pharmaceutical applications. *Mar. Drugs, 13*(7), 4231–4254.

Marangoni, F., Agostoni, C., Borghi, C., Catapano, A. L., Cena, H., Ghiselli, A., & Riccardi, G., (2020). Dietary linoleic acid and human health: Focus on cardiovascular and cardiometabolic effects. *Atherosclerosis, 292*, 90–98.

Maroneze, M. M., Zepka, L. Q., Jacob-Lopes, E., Pérez-Gálvez, A., & Roca, M., (2019). Chlorophyll oxidative metabolism during the phototrophic and heterotrophic growth of *Scenedesmus obliquus*. *Antioxidants, 8*(12), 600.

Matos, A. P., (2019). Microalgae as a potential source of proteins. In: Galanakis, C., (ed.), *Proteins: Sustainable Source, Processing and Applications* (pp. 63–93). Academic Press, Cambridge.

Matos, J., Cardoso, C., Bandarra, N. M., & Afonso, C., (2017). Microalgae as healthy ingredients for functional food: A review. *Food Funct., 8*(8), 2672–2685.

Mobin, S. M., Chowdhury, H., & Alam, F., (2019). Commercially important bioproducts from microalgae and their current applications: A review. *Energy Procedia, 160*, 752–760.

Morais, M. G., Vaz, B. D. S., Morais, E. G., & Costa, J. A. V., (2015). Biologically active metabolites synthesized by microalgae. *Biomed Res. Int. 1,* 15.

Mudliar, S. N., & Shekh, A. Y., (2019). Recent advances in microalgal bioactives for food, feed, and healthcare products: Commercial potential, market space, and sustainability. *Compr. Rev. Food Sci. Food Saf., 18*(6), 1882–1897.

Mulders, K. J., Lamers, P. P., Martens, D. E., & Wijffels, R. H., (2014). Phototrophic pigment production with microalgae: Biological constraints and opportunities. *J. Phycol., 50*(2), 229–242.

Nascimento, T. C. D., Cazarin, C. B. B., Maróstica, Jr. M. R., Mercadante, A. Z., Jacob-Lopes, E., & Zepka, L. Q., (2020). Microalgae carotenoids intake: Influence on cholesterol levels, lipid peroxidation and antioxidant enzymes. *Food Res. Int., 128*, 108770.

Nascimento, T. C., Cazarin, C. B. B., Roberto, M. M., Risso, É. M., Amaya-Farfan, J., Grimaldi, R., Mercadante, A. Z., et al., (2019). Microalgae biomass intake positively modulates serum lipid profile and antioxidant status. *J. Funct. Foods, 58*, 11–20.

Nazih, H., & Bard, J. M., (2018). Microalgae in human health: Interest as a functional food. In: Levine, I., & Fleurence, J., (eds.), *Microalgae in Health and Disease Prevention* (pp. 211–226). Academic Press, Cambridge.

Niccolai, A., Chini, G., Rodol, L., Biondi, N., & Tredici, M. R., (2019). Microalgae of interest as food source: Biochemical composition and digestibility. *Algal Res., 42*, 101617.

Novoveská, L., Ross, M. E., Stanley, M. S., Pradelles, R., Wasiolek, V., & Sassi, J. F., (2019). Microalgal carotenoids: A review of production, current markets, regulations, and future direction. *Mar. Drugs, 17*(11), 640.

Olaizola, M., (2003). Commercial development of microalgal biotechnology: From the test tube to the marketplace. *Biomol. Eng., 20*(4–6), 459–466.

Pagels, F., Guedes, A. C., Amaro, H. M., Kijjoa, A., & Vasconcelos, V., (2019). Phycobiliproteins from cyanobacteria: Chemistry and biotechnological applications. *Biotechnol. Adv., 37*, 422–443.

Pan-utai, W., & Lamtham, S., (2019). Extraction, purification, and antioxidant activity of phycobiliprotein from *Arthrospira platensis. Process Biochem., 82*, 189–198.

Patias, L. D., Fernandes, A. S., Petry, F. C., Mercadante, A. Z., Jacob-Lopes, E., & Zepka, L. Q., (2017). Carotenoid profile of three microalgae/cyanobacteria species with peroxyl radical scavenger capacity. *Food Res. Int., 100*, 260–266.

Patias, L. D., Maroneze, M. M., Siqueira, S. F., Menezes, C. R., Zepka, L. Q., & Jacob-Lopes, E., (2018). Single-cell protein as a source of biologically active ingredients for the formulation of anti-obesity foods. In: Holban, A. M., & Grumezescu, A. M., (eds.), *Alternative and Replacement Foods* (pp. 317–353). Academic Press, Cambridge.

Pérez-Galvez, A., Viera, I., & Roca, M., (2017). Chemistry in the bioactivity of chlorophylls: An overview. *Curr. Med., 24*(40), 4515–4536.

Rahman, K. M., (2020). Food and high value products from microalgae: Market opportunities and challenges. In: Asrafu, A. M., Xu, J. L., & Wang, Z., (eds.), *Microalgae Biotechnology for Food, Health and High Value Products* (pp. 3–27). Springer, Singapore.

Raja, R., Coelho, A., Hemaiswarya, S., Kumar, P., Carvalho, I. S., & Alagarsamy, A., (2018). Applications of microalgal paste and powder as food and feed: An update using text mining tool. *BJBAS, 7*(4), 740–747.

Ramirez-Mérida, L. G., Zepka, L. Q., & Jacob-Lopes, E., (2017). Current production of microalgae at industrial scale. In: Pires, J. C. M., (ed.), *Recent Advances in Renewable Energy* (pp. 242–260). Bentham Science Publishers, Sharjah.

Randhir, A., Laird, D. W., Maker, G., Trengove, R., & Moheimani, N. R., (2020). Microalgae: A potential sustainable commercial source of sterols. *Algal Res., 46*, 101772.

Raposo, M. F. D. J., Morais, A. M. M. B., & Morais, R. M. S. C., (2014). Influence of sulphate on the composition and antibacterial and antiviral properties of the exopolysaccharide from *Porphyridium cruentum. Life Sci., 101*(1, 2), 56–63.

Raposo, M. F. J., Morais, A. M. B., & Morais, R. M. S. C., (2015). Marine polysaccharides from algae with potential biomedical applications. *Mar. Drugs, 13*(5), 2967–3028.

Raposo, M. F., Morais, R. M. S. C., & Morais, A. M. M. B., (2013). Bioactivity and applications of sulphated polysaccharides from marine microalgae. *Mar. Drugs, 11*, 233–252.

Rodrigues, D. B., Menezes, C. R., Mercadante, A. Z., Jacob-Lopes, E., & Zepka, L. Q., (2015). Bioactive pigments from microalgae *Phormidium autumnale. Food Res. Int., 77*, 273–279.

Rodriguez-Concepcion, M., Avalos, J., Bonet, M. L., Boronat, A., Gomez-Gomez, L., Hornero-Mendez, L. M. C., Meléndez-Martínez, A. J., et al., (2018). A global perspective on carotenoids: Metabolism, biotechnology, and benefits for nutrition and health. *Prog Lipid Res., 70*, 62–93.

Román, R. B., Alvárez-Pez, J. M., Fernández, F. G. A., & Grima, E. M., (2002). Recovery of pure B-phycoerythrin from the microalga *Porphyridium cruentum. J. Biotechnol., 93*, 73–85.

Romay, C., Gonzalez, R., Ledon, N., Remirez, D., & Rimbau, V., (2005). C-Phycocyanin: A biliprotein with antioxidant, anti-inflammatory, and neuroprotective effects. *Curr. Protein Pept. Sc., 4*(3), 207–216.

Saini, R. K., & Keum, Y. S., (2018). Omega-3 and omega-6 polyunsaturated fatty acids: Dietary sources, metabolism, and significance: A review. *Life Sci., 203*, 255–267.

Salas-Salvadó, J., Bulló, M., Babio, N., Martínez-González, M. Á., Ibarrola-Jurado, N., Basora, J., Estruch, R., Covas, M. I., et al., (2011). Reduction in the incidence of type 2 diabetes with the Mediterranean diet: Results of the predimed-reus nutrition intervention randomized trial. *Diabetes Care, 34*(1), 14–19.

Samuels, R., Mani, U. V., Iyer, U. M., & Nayak, U. S., (2002). Hypocholesterolemic effect of spirulina in patients with hyperlipidemic nephrotic syndrome. *J. Med. Food, 5*(2), 91–96.

Sanjeewa, K. K. A., Fernando, I. P. S., Samarakoon, K. W., Lakmal, H. H. C., Kim, E. A., Kwon, O., Dilshara, M. G., et al., (2016). Anti-inflammatory and anti-cancer activities of sterol rich fraction of cultured marine microalga *Nannochloropsis oculata. Algae, 31*(3), 277–287.

Sarkar, S., Manna, M. S., Bhowmick, T. K., & Gayen, K., (2020). Extraction of chlorophylls and carotenoids from dry and wet biomass of isolated *Chlorella thermophila*: Optimization of process parameters and modeling by artificial neural network. *Process Biochem., 96*, 58–72.

Sathasivam, R., & Ki, J. S., (2018). A review of the biological activities of microalgal carotenoids and their potential use in healthcare and cosmetic industries. *Mar. Drugs, 16*(1), 26.

Sathasivam, R., Radhakrishnan, R., Hashem, A., & Abd-Allah, E. F., (2019). Microalgae metabolites: A rich source for food and medicine. *Saudi J. Biol. Sci., 26*(4), 709–722.

Silva, T. H., Alves, A., Popa, E. G., Reys, L. L., Gomes, M. E., Sousa, S. S. S., Mano, J. F., & Reis, R. L., (2012). Marine algae sulfated polysaccharides for tissue engineering and drug delivery approaches. *Biomatter, 2*(4), 278–289.

Singh, S. K., Kaur, R., Bansal, A., Kapur, S., & Sundaram, S., (2020). Biotechnological exploitation of cyanobacteria and microalgae for bioactive compounds. In: Verma, M. L., & Chandel, A. K., (eds.), *Biotechnological Production of Bioactive Compounds* (pp. 107–137). Elsevier, Amsterdam.

Soto-Sierra, L., Stoykova, P., & Nikolov, Z. L., (2018). Extraction and fractionation of microalgae-based protein products. *Algal Res., 36*, 175–192.

Sun, G. Y., Simonyi, A., Fritsche, K. L., Chuang, D. Y., Hannink, M., Gu, Z., Greenlief, C. M., et al., (2018). Docosahexaenoic acid (DHA): An essential nutrient and a nutraceutical for brain health and diseases. *Prostaglandins Leukot. Essent. Fatty Acids, 136*, 3–13.

Takaichi, S., & Mochimaru, M., (2007). Carotenoids and carotenogenesis in cyanobacteria: Unique ketocarotenoids and carotenoid glycosides. *Cell. Mol. Life Sci., 64*(19, 20), 2607.

Takaichi, S., (2011). Carotenoids in algae: Distributions, biosynthesis, and functions. *Mar. Drugs, 9*(6), 1101–1118.

Tandon, P., & Jin, Q., (2017). Microalgae culture enhancement through key microbial approaches. *Renew Sust. Energ Rev., 80*, 1089–1099.

Tang, D. Y. Y., Khoo, K. S., Chew, K. W., Tao, Y., Ho, S. H., & Show, P. L., (2020). Potential utilization of bioproducts from microalgae for the quality enhancement of natural products. *Bioresour. Technol., 304*, 122997.

Torres-Tiji, Y., Fields, F. J., & Mayfield, S. P., (2020). Microalgae as a future food source. *Biotechnol. Adv., 41*, 107536.

US FDA, (2018). https://www.govinfo.gov/content/pkg/FR-2018-05-07/pdf/2018-09636.pdf (accessed on 2 July 2021).

Vaz, B., Moreira, J. B., Morais, M. G., & Costa, J. A. V., (2016). Microalgae as a new source of bioactive compounds in food supplements. *Curr. Opin. Food Sci., 7*, 73–77.

Volkman, J. K., (2016). Sterols in microalgae. In: Borowitzka, M. A., Beardall, J., & Raven, J. A., (eds.), *The Physiology of Microalgae* (pp. 485–505). Springer Nature, Switzerland.

Wang, L., Wang, X., Wu, H., & Liu, R., (2014). Overview on biological activities and molecular characteristics of sulfated polysaccharides from marine green algae in recent years. *Mar. Drugs, 12*(9), 4984–5020.

Wollam, J., & Antebi, A., (2011). Sterol regulation of metabolism, homeostasis, and development. *Annu. Rev. Biochem., 80*, 885–916.

Woodman, R. J., Mori, T. A., Burke, V., Puddey, I. B., Barden, A., Watts, G. F., & Beilin, L. J., (2003). Effects of purified eicosapentaenoic acid and docosahexaenoic acid on platelet, fibrinolytic and vascular function in hypertensive type 2 diabetic patients. *Atherosclerosis, 166*(1), 85–93.

Xiang, H., Yang, X., & Hu, Y., (2020). The properties, biotechnologies, and applications of antifreeze proteins. *Int. J. Biol. Macromol., 153*, 661–675.

Zanella, L., & Vianello, F., (2020). Microalgae of the genus *Nannochloropsis*: Chemical composition and functional implications for human nutrition. *J. Funct. Foods, 6*8, 103919.

Zhang, J., Liu, L., Ren, Y., & Chen, F., (2019). Characterization of exopolysaccharides produced by microalgae with antitumor activity on human colon cancer cells. *Int. J. Biol. Macromol., 128*, 761–767.

CHAPTER 4

SPIRULINA-DERIVED NUTRACEUTICALS AND THEIR APPLICATIONS IN THE FOOD INDUSTRY

MANONMANI KUMARAGURUPARASWAMI,[1]
DEEPAK SUBRAMANI,[1] SANGEETHA ARUNACHALAM,[1]
SENTHILKUMAR KANDASAMY,[2]
SANGEETHA GANDHI SIVASUBRAMANIYAN,[3] and
DHIVYA NALLAMUTHU[4]

[1]Department of Food Technology, Kongu Engineering College, Perundurai, Erode–638060, Tamil Nadu, India

[2]Department of Chemical Engineering, Kongu Engineering College, Perundurai, Erod–638060, Tamil Nadu, India

[3]Department of Food Technology, JCT College of Engineering and Technology, Coimbatore–641105, Tamil Nadu, India

[4]Department of Food Technology, Kalasalingam Academy of Research and Education, Virudhunagar–626126, Tamil Nadu, India

ABSTRACT

The major contributor for various diseases like cardiac failures, diabetes, obesity, cancer, etc., is unhealthy diets followed in frequently changing lifestyle of consumers. Nowadays, they need food not only as a source of energy but to treat and cure diseases. Those substances or compounds said to provide energy, nutritional, and health benefits are known as nutraceuticals and functional foods. This opened up the need to recognize spirulina as nutraceuticals because of its promising health promoting nutrients and bioactive compounds. It is a filamentous photosynthetic blue-green algal biomass that can be industrially cultivated, harvested, and processed. Some of the nutraceutical properties of spirulina are promoting growth, immune booster, antimicrobial, antioxidant, anticancer, antidiabetic property. Also, spirulina can be incorporated into various food products to enhance its nutraceutical value.

4.1 INTRODUCTION

Nowadays, most of the consumers are conscious about the dietary requirements to get rid of various illnesses such as cardiac malfunction, cancer, diabetes, pulmonary disorders, etc. They require the foodstuffs, capable of supplying a significant quantity of essential nutrients and bioactive compounds. This involves the production and development of nutraceuticals and functional foods. Nutraceuticals are substances or compounds obtained from food and its products that provide nutritional benefits and also promote the prevention or treatment of illness. The demand for nutraceutical products has been increasing in the global market. This made the manufactures to look for nutrient-rich edible sources that are cost-effective and easily available (Bishop and Zubeck, 2012).

Algae being photosynthetic, autotropic, and eukaryotic organisms can grow rapidly to produce huge biomass. Algae were classified into many categories based on the morphological characteristics, pigments, reserve food material, etc. Marine and freshwater habitat microalgae have been recognized as a source to generate food and bioactive compounds since ancient times, as well as being a biofuel feedstock. Spirulina (*Arthrospira platensis*), a cyanobacterium, belongs to the community of photosynthetic organisms, which is usually distributed in tropical and subtropical areas. They have been provided with various primary and secondary metabolites, regarded as 'superfood.' The Food and Agriculture Organization of the United Nations proposed that spirulina could be the perfect dietary aid for the 21st century, because of its potential nutrient profile than plant-based food products (Udayan et al., 2017). Also, Food and Drug Administration accepted spirulina as generally regarded as safe (GRAS). Their fascinating nutraceutical properties and balanced nutritional composition can be benefited in the form of nutraceuticals, functional foods, and dietary supplements (Morsy et al., 2014). This chapter discusses the various aspects of the spirulina to be used as nutraceuticals and its application in various food products.

4.2   FEATURES OF SPIRULINA

Spirulina is blue-green planktonic algae found in alkaline water made up of delicate cell walls formed from complex protein and sugars. It contains amino acids, vitamin B_{12}, phycobilisomes (PBS), and phytopigments like carotene and xanthophyll (Estrada et al., 2001). It is a filamentous non-nitrogen-fixing cyanobacterium but neglects any distinction for the heterocyst, akinete or hormogonium seen in any other filamentous nitrogen-fixing cyanobacteria (Fujisawa et al., 2010). It is generally characterized as normal helical coiling's or spirals. During laboratory cultivation or commercial development, genetic modification (GM) and morphological transformation (linear to spiral or vice versa) has been reported (Ma and Gao, 2009). It grows naturally in warm climates, in the alkaline water bodies (Usharani et al., 2012). This has the multicellular tubular trichomes that were structured over the total length. The non-heterocyst blue-green filaments, comprises of vegetative fungal cells initiates binary fission in a single plane that exhibits identifiable transverse walls. Filaments are swirling and exhibit whirling motility. Apical cells might be narrowly circular or pointing that can be capitate and calyptrate. The shorter cylindrical cells were present around the width of the trichomes. The pores are present in the concave area of the coil in hemi circular rows. Just below the cell wall is the plasma membrane encompassed by cytoplasm, and the cytoplasm consists of vacuoles, granules, thylakoid membranes, ribosomes, and DNA (Vonshak, 1997).

4.3   PRODUCTION PROCESS OF SPIRULINA BIOMASS

4.3.1   CULTIVATION OF SPIRULINA BIOMASS

Spirulina has been industrially produced because of its high nutritive value and biomass safety. The mass production of spirulina biomass can be obtained using cultivation techniques such as open pond system, photo-bioreactor (PBR), and hybrid system (a combination of both). Microalgae require selective medium and the factors influencing the cultivation in any of the above-mentioned systems were listed to be nutrient medium, salinity, light intensity, temperature, pH, inoculum size, agitation, etc. (Usharani et al., 2012).

4.3.1.1   OPEN POND SYSTEM

Open ponds were the most commonly used techniques to cultivate microalgae in large scale, also known as raceway ponds. They can be either natural (lakes, ponds, lagoons) or artificial (containers) waters (Ugwu et al., 2008). These raceway ponds can be characterized as open ponds supported with paddle system to circulate the algae and microbes (Christenson and Sims, 2011). Though open ponds construction was easy and economical, certain limitations hinder the growth of algae. The major limitations of these open ponds were improper utilization of light, large area requirement, higher evaporation rate, poor mixing, contamination by other microbes (Ugwu et al., 2008). These limitations opened up a venue for mass cultivation in a closed system.

4.3.1.2   PHOTOBIOREACTOR (PBR)

A PBR utilizes a light source to cultivate phototrophic microorganisms such as plants, mosses, macroalgae, microalgae, cyanobacteria, and purple bacteria in a confined area. The major advantages of PBRs (closed system) over open systems were hygienic cultivation, minimum contamination, pH, and temperature control, improved light intensity, uniform light illumination, reduced evaporation loss, enhance biomass nutrition, increased growth rate, rapid mass transfer (Soni et al., 2017). Commonly used PBRs are classified into flat plate, bubble column, and tubular reactors.

4.3.1.3   HYBRID REACTORS

These types of PBR were designed to get the benefits of two different reactors, to counter the drawback of one system over the other. For example, Acién-Fernández et al. (2001) reported an airlift-driven tubular PBR that has been proved to produce a continuous culture of microalgae. Improved production of biomass was achieved in the combined tubular air-lift reactors under limited lighting and laminar flow conditions (Converti et al., 2006). Combination of open pond and PBR was also possible in the case of a hybrid system.

4.3.2　HARVESTING AND PROCESSING OF SPIRULINA

The algal mass is retrieved from culture by a sequence of filtering steps (Shimamatsu, 2004). Harvesting techniques are majorly classified into, chemical, mechanical, biological-based method. In chemical methods, typically flocculation, wherein, synthetic polymers like ferric cation, ferric chloride, aluminum sulfate have been added to flocculate the biomass by charge neutralization (Christenson and Sims, 2011). Generally, it would be used to increase the particle size of biomass to facilitate easy removal in further steps. Centrifugation, sedimentation, filtration, dissolved air flotation, electrophoresis were the physical techniques (Soni et al., 2017). The biobased technique includes bioflocculation and microbial flocculation. Bioflocculation refers to the secretion of extracellular biopolymers, whereas microbial flocculation involves the addition of other microbes to flocculate the algal biomass, which was further removed by sedimentation process (Christenson and Sims, 2011).

The harvested spirulina shall be consumed in a fresh form as well as processed form. The shelf life of spirulina would be increased by several drying methods like sun drying, freeze-drying, drum drying, solar drying, etc. The dried forms were either in the shape of thin-film or rods. Further, it can be size reduced in the form of powders or pressed to obtain tablets. Spirulina may be incorporated into various food products to avail its nutraceutical property (Shimamatsu, 2004).

4.4　NUTRITIONAL COMPOSITION OF SPIRULINA

Spirulina contains both macronutrient and micronutrient in the significant quantity that makes it a wholesome food supplement. The food products incorporated with spirulina tends to show improved nutritional profile invariantly. In general, they possess nutrients like carbohydrates (15–25%), proteins (50–70%), lipids (6–13%), vitamins, and minerals (2.2–4.8%) and contains some natural pigments (Khan et al., 2005; Hoseini et al., 2013). The following paragraph briefly discusses the nutritional composition of spirulina:

1. **Carbohydrates:** Its composition varies depending on the growth conditions (temperature, pH, nutrient availability, etc.), by which the biomass has been cultivated and harvested (Braga et al., 2018). Total carbohydrate content solely comprises of glucose along with other

sugars like galactose, mannose, xylose, and rhamnose (Shekharam et al., 1987). It also consists of branched polysaccharides having antiviral and anti-immunomodulation activities.

2. **Proteins:** The protein content of spirulina has been comparable to proteins obtained from animals and plants. Spirulina is the richest source of microbial protein comprises of 460–360/kg (dry matter basis). Rather than the quantity of protein, the quality of protein was also not compromised in spirulina (Lupatini et al., 2017). It has been proved by many researchers that they contain both essential and non-essential amino acids (EAAs) in significant amount. EAAs like lysine, leucine, isoleucine, valine, threonine, histidine, tryptophan, phenylalanine, tyrosine, methionine, and cysteine were found, whereas non-essential acids like glutamic acid, serine, proline, aspartic acid, arginine, alanine, and glycine were present (Campanella et al., 1999; Safi et al., 2013). Biopeptides obtained from spirulina has potential health benefits because of its anti-coagulant, antioxidant, anti-hypertensive, anti-proliferative properties, etc., (Betoret et al., 2011). Besides this, protein digestibility of spirulina is higher compared to other protein sources based on the values of protein efficiency ratio and net protein utilization (Hoseini et al., 2013).

3. **Lipids:** The fatty acid profile of spirulina comprises of palmitic acid, palmitoleic acid, linoleic acid (LNA), γ-linolenic acid (GLA), stearic acid and oleic acid. Among these, ranges of palmitic acid, LNA and GLA found to be 42.3 to 47.6%, 13.1 to 31.5% and 12.9 to 29.4% respectively (Mühling et al., 2005). The fatty acid content of spirulina makes it as a substantial source of microbial lipids. Similarly, Chaiklahan et al. (2008) in their study reported that among the total fatty acids (FAs) extracted by solvent extraction method contains linoleic, and GLA of 21% and 18%, respectively.

4. **Vitamins and Minerals:** Spirulina contains a certain amount of vitamin B and C (thiamine, riboflavin, niacin, folate, choline, pantothenic acid, biotin, vitamin K1 and K2), trace amounts of vitamin A, E, K, and minerals like phosphorous, calcium, iron, sodium, magnesium, potassium, zinc, manganese, copper, selenium (Capelli and Cysewski, 2010; Ohmori and Ehira, 2013). Capelli and Cysewski (2010) reported that beta carotene level of spirulina was higher than carrots. Similarly, it exceeds the iron content in spinach.

5. **Pigments:** Spirulina consists of various pigments like carotenoids, xanthophyll, zeaxanthin, chlorophyll, and phycocyanin (PC) which contributes to the antioxidant activity (Morsy et al., 2014).

Due to their high value, they found various applications such as dietary supplements, nutraceuticals, cosmeceuticals, functional foods (Nuhu, 2013).

4.5 NUTRACEUTICAL PROPERTIES OF SPIRULINA

4.5.1 GROWTH AND DEVELOPMENT

Being a concentrated source of protein and micronutrients, intake of spirulina in daily diet would significantly stimulate the growth of children. As spirulina is devoid of the cell wall, it can be easily digested by infants and malnourished people having intestinal absorption issues (Sharoba, 2014). A study involving undernourished children from Kisantu, Congo found that the supplementation of spirulina (10 g per day) for 30 days significantly increased weight and height for age Z score. Oedema in population got reduced from 64% to 4% (Matondo et al., 2016). Abed et al. (2016) reported a significant increase in ferritin, iron, hemoglobin, mean corpuscule volume in blood on the introduction of spirulina in the daily diet of children from the Gaza strip. A study conducted reported that male rats fed with spirulina stimulated growth hormone and parathyroid hormone resulting in increased bone growth and bone strength (Cho et al., 2020). Kalafati et al. (2010) stated that the supplementation of spirulina for a period of 4 weeks enhanced cognitive functions and exercise performance by fat oxidation and an increase in the growth-stimulating hormone. A 68% reduction in undernutrition status and weight gain of 7 g/kg/day was found among children on feeding just 2 g of spirulina for 180 days (Fehmida et al., 2018). Voltarelli et al. (2011) found that, rate of protein synthesis has been increased whereas protein degradation decreased, resulting in excellent muscle development in the rat model fed with spirulina. In support of this animal study, Azabji-Kenfack et al. (2011) described the useful effect of spirulina on a fat-free mass of malnourished HIV infected adults.

4.5.2 IMMUNOMODULATORY POTENTIAL

Spirulina consumption in the form of pure biomass, hot water extract or its isolated bioactive compounds was found to have a promising effect in resisting deadly infections by stimulating innate and adaptive immune action. The oral intake of spirulina increased natural killer cells, interferon-gamma

(IFN-γ), and interleukin (IL) by activating immune cells through toll-like receptor pathway (Hirahashi et al., 2002). The immune reaction is mediated by a high molecular weight polysaccharide in spirulina named immulina (Pugh et al., 2001). Apart from polysaccharide-based immune modulation, a water-soluble pigment in spirulina named PC also found to exhibit immune stimulatory activity (Nemoto-Kawamura et al., 2004). A human clinical study by Løbner et al. (2008) confirmed an enhancement in blood immune markers with spirulina supplementation for 56 days. The daily supplementation of spirulina to HIV-1 patients for 12 months significantly increased CD4 (cluster of differentiation 4) cell and decreased viral loads, which resulted in fast rehabilitation of patients (Ngo-Matip et al., 2015).

4.5.3   ANTIMICROBIAL PROPERTY

A sulfated polysaccharide from spirulina named calcium spirulan was known to inhibit the proliferation of enveloped virus by preventing the entry of the virus into the host. Spirulina has its inhibition towards HIV-1, herpes simplex type 1, human cytomegalovirus, pseudorabies virus, influenza, measles, and mumps (Hayashi et al., 1996; Hernández-Corona et al., 2002). Immulina extracted from spirulina has antimicrobial activity against tetanus toxoid and *Candida albicans* (Løbner et al., 2008). Cyanovirin, a protein isolated from spirulina was found active against HIV-1 (Balzarini, 2007). Besides polysaccharides and proteins, spirulina has a wide range and high concentration of phenolic and flavonoid compounds exhibiting antimicrobial activity (Chakraborty et al., 2019). The ethanolic extract of spirulina inhibited adenovirus type 7, coxsackievirus B4, adenovirus type 40, astrovirus type 1 and rotavirus Wa strain by 53.3, 66.7, 50, 76.7, and 56.7% respectively (El-Baz et al., 2013). The bacterial species such as *E. coli, Pseudomonas aeruginosa, Salmonella typhi, Salmonella senftenberg, Listeria monocytogenes* and *Enterococcus faecalis* were inhibited by spirulina extracts (Hetta et al., 2014). Though there prevail numerous studies on in-vitro antimicrobial activity of spirulina, in vivo studies are needed to validate its effectiveness.

4.5.4   CARDIOPROTECTIVE AND HYPOCHOLESTEROLEMIC POTENTIAL

The main reasons behind cardiovascular dysfunctions are an increase in low-density lipoprotein (LDL) and decrease in very-LDL (Norata et al., 2009).

Spirulina being rich in gamma-linolenic acid can inhibit the accumulation of cholesterol in the body (Samuels et al., 2002). Studies have found that oral supplementation of spirulina decreased systolic and diastolic blood pressure, plasma triacylglycerols (TAG) and LDL, whereas it increased high-density lipoprotein, thereby enhancing cardiovascular functions (Torres-Duran et al., 2007; Huang et al., 2018; Hernández-Lepe et al., 2019). It inhibited pancreatic lipase activity and cholesterol absorption in the intestine. The rat model shown that excretion of bile acids and cholesterol were quite significant on the supplementation of spirulina (Nagaoka et al., 2005; Han et al., 2006). Spirulina consumption (250 mg per kg body weight) shown to have a defensive and progressive effect on drug-induced cardiac arrest (Khan et al., 2005).

4.5.5   CANCER TREATMENT

Spirulina can prevent cancer and shrink the tumor cells by activating natural killer cells (Akao et al., 2009). Calcium spirulan isolated from spirulina was found active against cancer types such as fibrosarcoma, carcinoma, and melanoma (Mishima et al., 1998). The water extract of spirulina reduced the viability of adenocarcinomic alveolar basal epithelial cells and reduced its spread due to inhibition of the cell cycle (G1 phase). Moreover, cytotoxic activity against normal cells was found absent (Czerwonka et al., 2018). The consumption of spirulina biomass of about 250–550 mg/kg/body weight inhibited tumor cells of skin and stomach (Dasgupta et al., 2001). A study involving supplementation of 1 g of spirulina to tobacco chewers from Kerala, India for a year resulted in complete regression of leukoplakia (Mathew et al., 1995). Fayyad et al. (2019) found that methanolic extract of spirulina exhibited a cytotoxic effect on breast cancer adenocarcinoma and intestinal carcinoma cell lines of mice by 78% and 35.5%, respectively. These studies suggest the promising application of spirulina as a sole anticancer drug or an adjunct together with other antitumor agents without causing side effects.

4.5.6   COMBATING OXIDATIVE STRESS

Oxidative stress is becoming the dominant cause for arising disorders like cancer, atherosclerosis, arthritis, diabetes, hypertension, cancer, cellular damage, neural dysfunctions, etc. (Pizzino et al., 2017). Improper functioning of innate antioxidant defense system in the body causes lipid peroxidation and free radical damage that subsequently leads to metabolic disorders (Nasirian et al., 2018). Spirulina is a good source of free radical scavenging antioxidants like β-carotene, xanthophyll, phycobiliproteins (PBP), tocopherol, superoxide dismutase (SOD) and phenolic compounds (Al-Dhabi and Arasu, 2016). The aqueous extract of spirulina was found to reduce the free radical-induced apoptosis of epithelial cells in mouse fibroblast cell-based assay (Chu et al., 2010). Iron-induced oxidative stress of in vitro neuroblastoma cells was reduced by spirulina due to its chelating property (Asghari et al., 2016). In a recent study, proven that spirulina inhibited nitrous oxide, hydroxyl, and DPPH free radicals, and lipid peroxidation. Moreover, rats supplemented with spirulina of about 300 mg/kg bodyweight for 12 weeks resulted in decreased plasma lipid peroxidation, which is reasoned by increased SOD and catalase enzymes (Okechukwu et al., 2019). Similarly, Nasirian et al. (2018) fed spirulina (20–30 mg /kg body weight) to rat for 35 days and found a substantial increase in the synthesis of plasma antioxidant enzymes leading to reduction of malondialdehyde (MDA), an oxidative stress marker.

4.5.7 ANTIDIABETIC ACTIVITY

The expensive and undesirable effects of synthetic antidiabetic drugs like biguanides and sulfonylureas necessitate the need to explore alternative antidiabetic agents. Spirulina was found to exhibit an antihyperglycemic effect by retarding intestinal glucose absorption (Fayzunnessa et al., 2011). The high fiber, flavonoids, chromium, and PC profile makes spirulina to exhibit antidiabetic property (El-Sayed et al., 2018). A human clinical study supplementing 7 g of spirulina twice a day to 40 diabetic patients found improvement from polydipsia, polyuria, polyphagia, and tiredness (Alam et al., 2016). Many animal studies proved the reduction of blood glucose level and lipid profile, and enhanced synthesis of insulin on introducing spirulina or its isolated PC in the diet (El-Sayed et al., 2018). A significant increase in hexokinase activity in the liver and glucose-6-phosphate activity in muscles was noticed on feeding spirulina (15 mg/kg body weight) to diabetic rats for 45 days, which in turn indicates enhancement of glucose metabolism (Layam and Reddy, 2007). PC from spirulina increased the liver and muscle glycogen level and suppressed the abnormally enlarged islets in the pancreas of diabetic mice (Ou et al., 2013). Though animal studies proved the effectiveness of spirulina as an antidiabetic agent, human studies are further needed to adequately prove its outstanding ability.

4.5.8 DETOXIFICATION AND RENAL PROTECTION

The overload of drugs and heavy metals, and accumulation of salts would temporarily or permanently damage the renal function leading to the concentration of toxins in the blood. The presence of antioxidants and chelating compounds in spirulina prevents glomerular and tubular renal dysfunctions (Rojas-Franco et al., 2018). Spirulina supplemented rat recovered from ethylene glycol induced nephrolithiasis as evident from lower blood urea nitrogen and alanine aminotransferase (ALT), and higher serum calcium levels compared to rat group not fed with spirulina (Al-Attar, 2010). The absence of abnormal morphological changes in kidney elevated plasma nitrite concentration and decreased blood urea and creatine levels were observed on feeding spirulina to rat having gentamicin induced nephrotoxicity (Avdagić et al., 2008). Similarly, Memije-Lazaro et al. (2018) stated that the progression of chronic kidney disease was prevented on spirulina supplementation by partially inhibiting the alteration of serum creatinine, uric acid, urinary protein, and creatine clearance. Moreover, it enhanced kidney function by preventing disturbances in electrolytic balance and urinary excretion (Rojas-Franco et al., 2018). The consumption of spirulina (250 mg) and zinc (2 mg) by 41 individuals for two times a day for a period of 16 weeks found to be an effective remedy to treat chronic arsenic poisoning. It reduced the amount of arsenic in the hair sample of patients and increased the urinary excretion of arsenic (Misbahuddin et al., 2006).

4.5.9 SUPPORTING GUT MICROFLORA

Spirulina improves gastrointestinal health by modulating gut microflora (Gupta et al., 2017). It enhanced the growth of probiotics such as *Lactobacillus acidophilus*, *Streptococcus thermophilus* and *Lactobacillus casei*, as well as inhibited pathogens such as *Proteus vulgaris*, *E. coli*, *Bacillus subtilis*, *Pseudomonas aeruginosa* and *Staphylococcus aureus* (Bhowmik et al., 2009). Likewise, Bensehaila et al. (2013) reported a positive effect of spirulina on the growth of probiotic *Bifidobacterium breve* and an inhibitory effect on pathogenic *Salmonella* and *E. coli*. It indicates the prebiotic ability of spirulina to assist the colonization of beneficial bacteria in the gut environment. A clinical study found that the oral supplementation of spirulina to healthy rats favorably modulated the microbial diversity and concentration in the colon that increased leptin content, an appetite-controlling hormone (Hu et al., 2019). Neyrinck et al. (2017) stated that the introduction of spirulina in the daily diet of mice changed the constitution of gut microbiota, particularly *Lactobacillus* and *Roseburia* genera, and enhanced the innate immunity by producing antimicrobial peptides in the gut.

4.5.10   OBESITY TREATMENT

As spirulina consumption enhances the leptin content, it could serve as a therapeutic agent for managing obesity (Hu et al., 2019). A study involving in vitro and in vivo obesity models found spirulina could inhibit adipogenesis observed in obesity by reducing triglyceride (TG) accumulation and protein expression of adipogenic regulators (Seo et al., 2018). A decreased blood lipid and cholesterol, and increased activities of antioxidant enzymes were observed on feeding spirulina and cinnamon extract mixture to fatty diet-induced overweight rats (Ibrahim and El-Rahman, 2018). Supplementation of spirulina (500 mg) twice a day to obese individuals around 12 weeks resulted in a significant reduction in body mass index, body weight and appetite level (Zeinalian et al., 2017). Supportive results have been obtained by Szulinska et al. (2017) in which the introduction of spirulina (2 g) in the daily diet of obese patients decreased body mass index, LDL cholesterol, and waist circumference significantly.

4.5.11   ANEMIA CURE

Iron and folic acid content in spirulina is generally much higher than the RDA (recommended dietary allowance) level; therefore, it would be the best source to treat anemia, especially for pregnant and lactating women (Visnegarwala and Mahesh, 2017). Spirulina feeding (300 mg/kg) for 30 days to rats increased hemoglobin, red, and white blood cells, packed cell volume, mean corpuscular volume mean corpuscular hemoglobin concentration, reticulocytes, and lymphocytes (Simsek et al., 2009). Similar findings have been reported in a human clinical study supplemented two 500 mg spirulina tablets for 12 weeks to anemic individuals (Jagathy and Divya, 2014). A study involving pregnant women from Dakar, Senegal found that spirulina supplementation significantly increased the hemoglobin level with an average gain of about 0.36 g/dl during the period between the 28th week of pregnancy and delivery. Spirulina fed group had higher hemoglobin levels than the group fed with iron and folic acid tablets. Moreover, the infant born from spirulina fed women had significantly higher hemoglobin levels and mean weight gain from birth to 42nd day postnatal (Niang et al., 2017).

4.5.12   ENHANCING BLOOD VITAMIN A AND VISION

Spirulina has 10 times higher β-carotene level than carrot, and it serves to be a preferred source for enhancing eye vision (Seema and Sonia, 2016). A clinical study found that the feeding of spirulina for a month increased serum retinol level in preschool children. The serum retinol level was higher in children fed with spirulina than those fed with vitamin A supplements (Annapurna et al., 1991). An inhibition of reduction in soluble protein, glutathione, and water content of lens was observed on spirulina supplementation to female rats having naphthalene induced cataract. The underlying mechanism of spirulina behind preventing or delaying cataract is because of its antioxidant activity (Haque and Gilani, 2005). Yu et al. (2012) found a significantly elevated level of zeaxanthin in human subjects supplemented with spirulina. Spirulina intake was found to prevent photo-induced visual dysfunctions by suppressing the decrement of rhodopsin, photoreceptor cell death, photoreceptor layer thinning and reactive oxygen species (ROS) (Okamoto et al., 2019). Apart from improving the vitamin A level of pregnant and lactating women, spirulina consumption enhanced the bioavailability of β-carotene to the fetus (Naik, 2001). A study found that the daily supplementation of spirulina (1 g) to 5000 preschool children for 150 days reduced Bitot's spot, caused due of vitamin A deficiency from 80% to 10% (Seshadri, 1993).

4.5.13   OTHER HEALTH BENEFITS

The radiation-induced impairment of thyroid and reproductive hormones was improved on consuming spirulina (300 mg/kg) for 15 days (Ebrahim, 2020). A rat-based study found that supplementing a combination of spirulina and amantadine helped to overcome Parkinson's disease by significantly increasing (78.3%) dopamine level and enhancing locomotory functions (Chattopadhyaya et al., 2015). Hwang et al. (2011) proposed the benefit of spirulina to prevent memory loss as evident from enhanced catalase activity and reduced accumulation of Aβ protein and lipid peroxidation in mice model. Spirulina and its isolated PC exhibited anti-allergenic properties and had shown improvement from asthma and allergic rhinitis (Labhe et al., 2001; Mao et al., 2005). Immulina extracted from spirulina shown to have anti-inflammatory action by inhibiting histamine, interleukin-4, leucitrienes, and tumor necrosis factor-α (Appel et al., 2018). Spirulina exhibited anti-pyretic effect by reducing the rectal temperature of the rat model (Somchit et al., 2014). The oral administration of spirulina to rat having rheumatoid arthritis resulted in reduced swelling, pain, redness, and other inflammation effects (Kumar et al., 2009). Spirulina was found to enhance the fertilizing ability and prevent testicular damage by free radicals in animal models (Djalil et al., 2019). An excellent antithrombotic and fibrinolytic activity of sodium spirulan isolated from spirulina has its application in arresting blood clot (Yamamoto et al., 2003).

4.6   APPLICATIONS OF SPIRULINA IN FOOD INDUSTRIES

4.6.1   EXTRUDED AND EXPANDED PRODUCTS

Food extrusion was the technique to process snack foods and ready to eat foods which are one of the key players in the global food products market. These extruded and expanded products mostly made from a low protein, carbohydrate-rich cereals. The nutrient profile of these products can be enhanced by incorporation or combination with nutrient-rich products. Many researchers explored the incorporation of spirulina in extruded and expanded products on their physical, nutritional, and functional properties. Santos et al. (2016) investigated the proximate composition, sensory, and shelf life of spirulina incorporated powdered food to combat nutrition deficiency in the aged people. The authors reported that the developed product was made into shakes. The proximate analysis proved that it tends to improve the nutritional composition when compared to shake without spirulina. It increased proteins, vitamins, and minerals which served to be an interesting alternative to a nutrient-deficient diet of the elderly population. The spirulina shakes were accepted by the aged people in sensory aspects and proven to have a shelf life of 19 months. Since spirulina has better digestibility than other proteins, it could be most suited for the elderly population.

An extruded product incorporated with 5, 10, and 15% of spirulina showed that 5% formulation was better when compared to others based on sensory analysis. Increased percentage of spirulina biomass may tend to influence the acceptability based on the color and flavor of the product (Vijayarani et al., 2012). The spirulina incorporation in the extruded product does not significantly affect the properties like expansion rate, density, grain index, water absorption and water solubility index (Morsy et al., 2014).

Joshi (2014) studied the characteristics of the expanded product obtained from maize and spirulina mixture. The results of their study indicated that an increase in spirulina proportion than maize greatly affected the expansion characteristics of the product. This is because of increased protein and decreased carbohydrate in the mixture. Also, 'a' value (indicating greenness) for color increased because of the dark green pigment present in spirulina biomass (Lucas et al., 2018). The color of the product, being one of the sensory attributes plays a major role in consumer acceptance. But nowadays, consumers do not entail on synthetic colors. Therefore, the addition of spirulina in food products promotes the natural color of the product (Özyurt et al., 2015).

4.6.2 BAKERY PRODUCTS

In recent days, the requirement of food products that serve entire nutrition to the consumers was preferred. Bakery products being the most consumed product among world population, researchers had explored to incorporate spirulina in bakery products like bread, biscuits, cookies, cakes, cupcakes, doughnuts, croissants, etc., on functional, nutritional, and quality characteristics of the product, because the above-mentioned characteristics play a predominant role in marketability of the products. The following paragraph discusses the various studies conducted on bakery products with spirulina addition.

The biscuits supplemented with spirulina biomass and PC extract showed better sensory properties in terms of color, flavor, and aroma, whereas the firmness of the biscuits increased with increase in spirulina biomass. This is attributed to its high protein content influencing the structure formation of biscuits. These biscuits also showed good oxidative stability during its storage period, which could be referred to the antioxidant property of spirulina (El-Baky et al., 2015). Similarly, nutritional, and sensory properties of bread made with the incorporation of spirulina indicated a significant increment when compared to control bread. But the physical properties of bread such as specific volume and density were affected negatively (Hafsa et al., 2014). In biscuits and cookies, spread ratio was affected (Bolanho et al., 2014; Onacik-Gür et al., 2018). Another research, in which the nutrient profile of cassava cake enriched with spirulina biomass showed an increment in protein, lipids, and mineral content. The cake was acceptable with a sensory score of 6.92 on a 7-point hedonic scale, making it a good consumable product with better marketability (Navacchi et al., 2012). Results of study about the development of croissant supplemented with spirulina indicated that microbial biomass increases the water holding capacity, thereby reducing the firmness of the product (Massoud et al., 2016). Although spirulina could supplement all the essential nutrients and bioactive compounds, the functional property of the spirulina supplemented products were highly influenced because of its high protein content. In products like bread, cakes, and doughnut, the structural integrity of the dough would be needed for final product characteristics. The biomass supplemented in higher levels affected the structural properties. Therefore, optimization of the ingredient composition would be necessary to avail scale up for industrial production of the product.

4.6.3 DAIRY PRODUCTS

Fermented dairy products, such as yogurt, cheese, cultured milk, and kefir are rich in probiotics. These products were consumed regularly to have improved health benefits. The high market value of these products made dairy industries to manufacture and distribute them widely. The viable counts of beneficial microbes should be maintained during its storage period that will be delivered during consumption. But the viability of the probiotic microorganism would be lost during the storage period because of the unfavorable environmental conditions. The growth of probiotics was stimulated by supplementation of prebiotics like inulin, fiber, glucan, glucose, etc. Apart from this, microalgal biomass could also be supplemented to improve the acidification and viability of probiotic microorganism (Beheshtipour et al., 2013).

One of the studies exhibited that yogurt supplemented with spirulina had lower buffering capacity, better nutritional properties (protein, fat, ash, and fiber) and textural characteristics (firmness, cohesiveness, and elasticity) whereas the sensory parameters got affected at increased spirulina addition (Barkallah et al., 2017). The spirulina incorporated fermented milk decreased the pH than control and it improves the viable count (6 log CFU/g) of probiotic bacteria (*Lactobacillus bulgaricus*) (Akalin et al., 2009). Similarly, spirulina powder substituted in yogurt indicated the increase in the viable count of lactic acid bacteria during the 30 days storage at 4°C (Guldas and Irkin, 2010). The major drawback of spirulina in dairy products was the acceptance of products based on sensory parameters. Mostly, sensory properties influenced greatly only on a higher proportion of spirulina. Therefore, minimal quantities would be adequate to enhance the nutritional and biofunctional properties of the product.

4.6.4   MISCELLANEOUS

Spirulina can be supplemented in most of the food products because of its biofunctional and nutraceutical property. For example, vegetable gelled dessert made with pea protein isolate and starch with the addition of spirulina, showed the increased polyunsaturated fatty acid profiles (DHA, GLA, EPA) of the product (Gouveia et al., 2008).

4.7   CONCLUSION

Consumer's nutrition requirements can be meet by the microalgae spirulina due to their rapid biomass production, potential nutrients, and health benefits. It has proven health benefits like improved growth, cardioprotective, antioxidant, antimicrobial, antidiabetic, anticancerous properties. Thus, Spirulina incorporated products occupy a significant market. Several studies can be exploited to preserve the sensory properties of spirulina incorporated products. Developing nutraceutical and functional foods from biofunctional rich spirulina encourage prospective opportunities for sustainable growth.

KEYWORDS

- anemia
- antidiabetic
- antimicrobial
- bakery products
- cardioprotective
- extruded products
- fortification
- functional foods
- immunomodulatory
- nutraceutical
- protein-rich
- spirulina

REFERENCES

Abed, E., Ihab, A., Suliman, E., & Mahmoud, A., (2016). Impact of spirulina on nutritional status, hematological profile, and anemia status in malnourished children in the Gaza strip: Randomized clinical trial. *Maternal and Pediatric Nutrition, 2*(2). https://doi.org/10.4172/2472-1182.1000110.

Acién-Fernández, F. G., Fernández, S. J. M., Sánchez, P. J. A., Molina, G. E., & Chisti, Y., (2001). Airlift-driven external-loop tubular photobioreactors for outdoor production of microalgae: Assessment of design and performance. *Chemical Engineering Science, 56*(8), 2721–2732. https://doi.org/10.1016/S0009-2509(00)00521-2.

Akalin, A., Ünal, G., & Dalay, M., (2009). Influence of *Spirulina platensis* biomass on microbiological viability in traditional and probiotic yogurts during refrigerated storage. *Italian Journal of Food Science, 21*(3), 357–364.

Akao, Y., Ebihara, T., Masuda, H., Saeki, Y., Akazawa, T., Hazeki, K., Hazeki, O., et al., (2009). Enhancement of antitumor natural killer cell activation by orally administered spirulina extract in mice. *Cancer Science, 100*(8), 1494–1501. https://doi.org/10.1111/j.1349-7006.2009.01188.x.

Alam, A., Siddiqui, M. A., Quamri, A., Fatima, S., Roqaiya, M., & Ahmad, Z., (2016). Efficacy of spirulina (Tahlab) in patients of type 2 diabetes mellitus (*Ziabetus shakri*): A randomized controlled trial. *Journal of Diabetes and Metabolism, 7*(10). https://doi.org/10.4172/2155-6156.1000710.

Al-Attar, A. M., (2010). Antilithiatic influence of spirulina on ethylene glycol induced nephrolithiasis in male rats. *American Journal of Biochemistry and Biotechnology, 6*(1), 25–31. https://doi.org/10.3844/ajbbsp.2010.25.31.

Al-Dhabi, N. A., & Arasu, M. V., (2016). Quantification of phytochemicals from commercial spirulina products and their antioxidant activities. *Evidence-Based Complementary and Alternative Medicine, 2016*, 1–13. https://doi.org/10.1155/2016/7631864.

Annapurna, V., Shah, N., Bhaskaram, P., Bamji, M. S., & Reddy, V., (1991). Bioavailability of spirulina carotenes in preschool children. *Journal of Clinical Biochemistry and Nutrition, 10*(2), 145–151. https://doi.org/10.3164/jcbn.10.145.

Appel, K., Munoz, E., Navarrete, C., Cruz-Teno, C., Biller, A., & Thiemann, E., (2018). Immunomodulatory and inhibitory effect of immulina®, and immunloges® in the Ig-E mediated activation of RBL-2H3 cells. A new role in allergic inflammatory responses. *Plants, 7*(1), 1–14. https://doi.org/10.3390/plants7010013.

Asghari, A., Fazilati, M., Latifi, A. M., Salavati, H., & Choopani, A., (2016). A review on antioxidant properties of spirulina. *Journal of Applied Biotechnology Reports, 3*(1), 345–351.

Avdagić, N., Ćosović, E., Nakaš-Ičindič, E., Mornjaković, Z., Začiragić, A., & Hadžović-Džuvo, A., (2008). *Spirulina platensis* protects against renal injury in rats with gentamicin-induced acute tubular necrosis. *Bosnian Journal of Basic Medical Sciences, 8*(4), 331–336. https://doi.org/10.17305/bjbms.2008.2892.

Azabji-Kenfack, M., Dikosso, S. E., Loni, E. G., Onana, E. A., Sobngwi, E., Gbaguidi, E., Kana, A. L. N., et al., (2011). Potential of *Spirulina platensis* as a nutritional supplement in malnourished HIV-infected adults in Sub-Saharan Africa: A randomized, single-blind study. *Nutrition and Metabolic Insights, 4*, 29–37. https://doi.org/10.4137/nmi.s5862.

Balzarini, J., (2007). Carbohydrate-binding agents: A potential future cornerstone for the chemotherapy of enveloped viruses. *Antiviral Chemistry and Chemotherapy, 18*(1), 1–11. https://doi.org/10.1177/095632020701800101.

Barkallah, M., Dammak, M., Louati, I., Hentati, F., Hadrich, B., Mechichi, T., Ayadi, M. A., et al., (2017). Effect of *Spirulina platensis* fortification on physicochemical, textural, antioxidant and sensory properties of yogurt during fermentation and storage. *LWT-Food Science and Technology, 84*, 323–330. https://doi.org/10.1016/j.lwt.2017.05.071.

Beheshtipour, H., Mortazavian, A. M., Mohammadi, R., Sohrabvandi, S., & Khosravi-Darani, K., (2013). Supplementation of *Spirulina platensis* and *Chlorella vulgaris* algae into probiotic fermented milks. *Comprehensive Reviews in Food Science and Food Safety, 12*(2), 144–154. https://doi.org/10.1111/1541-4337.12004.

Bensehaila, S., Benhadja, L., Doumandji, A., Bey, F., & Benzaiche, A., (2013). Effect of the concentration of *Spirulina platensis* on the growth of *Bifidobacterium breve*. *Journal of Life Sciences, 7*(12), 1225–1233.

Betoret, E., Betoret, N., Vidal, D., & Fito, P., (2011). Functional foods development: Trends and technologies. *Trends in Food Science and Technology, 22*(9), 498–508. https://doi.org/10.1016/j.tifs.2011.05.004.

Bhowmik, D., Dubey, J., & Mehra, S., (2009). Probiotic efficiency of *Spirulina platensis*-stimulating growth of lactic acid bacteria. *World Journal of Dairy and Food Sciences, 4*(2), 160–163.

Bishop, W. M., & Zubeck, H. M., (2012). Evaluation of microalgae for use as nutraceuticals and nutritional supplements. *Journal of Nutrition and Food Sciences, 2*(5). https://doi.org/10.4172/2155-9600.1000147.

Bolanho, B. C., Egea, M. B., Jácome, A. L. M., Campos, I., De Carvalho, J. C. M., & Danesi, E. D. G., (2014). Antioxidant and nutritional potential of cookies enriched with *Spirulina platensis* and sources of fiber. *Journal of Food and Nutrition Research, 53*(2), 171–179.

Braga, V. S., Mastrantonio, D. J. S., Costa, J. A. V., & Morais, M. G., (2018). Cultivation strategy to stimulate high carbohydrate content in spirulina biomass. *Bioresource Technology, 269*(8), 221–226. https://doi.org/10.1016/j.biortech.2018.08.105.

Campanella, L., Crescentini, G., & Avino, P., (1999). Chemical composition and nutritional evaluation of some natural and commercial food products based on spirulina. *Analysis, 27*(6), 533–540. https://doi.org/10.1051/analusis:1999130.

Capelli, B., & Cysewski, G. R., (2010). Potential health benefits of spirulina microalgae. *Nutrafoods, 9*(2), 19–26. https://doi.org/10.1007/bf03223332.

Chaiklahan, R., Chirasuwan, N., Loha, V., & Bunnag, B., (2008). Lipid and fatty acids extraction from the cyanobacterium spirulina. *ScienceAsia, 34*(3), 299–305. https://doi.org/10.2306/scienceasia1513-1874.2008.34.299.

Chakraborty, B., Varsale, A., Singh, V., Mali, S., Parihar, P., & Mane, R., (2019). Phytochemical analysis, antioxidant, and antifungal activity of different solvent extracts of *Spirulina platensis* collected from Rankala Lake, Kolhapur, Maharashtra. *Journal of Algal Biomass Utilization, 10*(1), 36–42.

Chattopadhyaya, I., Gupta, S., Mohammed, A., Mushtaq, N., Chauhan, S., & Ghosh, S., (2015). Neuroprotective effect of *Spirulina fusiform* and amantadine in the 6-OHDA induced Parkinsonism in rats. *BMC Complementary and Alternative Medicine, 15*(1), 1–11. https://doi.org/10.1186/s12906-015-0815-0.

Cho, J. A., Baek, S. Y., Cheong, S. H., & Kim, M. R., (2020). Spirulina enhances bone modeling in growing male rats by regulating growth-related hormones. *Nutrients, 12*(4), 1–19. https://doi.org/10.3390/nu12041187.

Christenson, L., & Sims, R., (2011). Production and harvesting of microalgae for wastewater treatment, biofuels, and bioproducts. *Biotechnology Advances, 29*(6), 686–702. https://doi.org/10.1016/j.biotechadv.2011.05.015.

Chu, W. L., Lim, Y. W., Radhakrishnan, A. K., & Lim, P., (2010). Protective effect of aqueous extract from *Spirulina platensis* against cell death induced by free radicals apoptotic assay. *BMC Complementary and Alternative Medicine, 10*(1), 53. https://doi.org/10.1186/1472.

Converti, A., Lodi, A., Del, B. A., & Solisio, C., (2006). Cultivation of *Spirulina platensis* in a combined airlift-tubular reactor system. *Biochemical Engineering Journal, 32*(1), 13–18. https://doi.org/10.1016/j.bej.2006.08.013.

Czerwonka, A., Kaławaj, K., Sławińska-Brych, A., Lemieszek, M. K., Bartnik, M., Wojtanowski, K. K., Zdzisińska, B., & Rzeski, W., (2018). Anticancer effect of the water extract of a commercial spirulina (*Arthrospira platensis*) product on the human lung cancer A549 cell line. *Biomedicine and Pharmacotherapy, 106*, 292–302. https://doi.org/10.1016/j.biopha.2018.06.116.

Dasgupta, T., Banerjee, S., Yadav, P. K., & Rao, A. R., (2001). Chemo modulation of carcinogen metabolizing enzymes, antioxidant profiles and skin and forestomach papilloma genesis by *Spirulina platensis*. *Molecular and Cellular Biochemistry, 226*(1), 17–26. https://doi.org/10.1023/A:1012721332221.

Djalil, O. I. A., Ferdinand, N., Adoum, I. Y., Herve, T., Djalal, A. K., Narcisse, V. B., & Herman, N. V., (2019). Effects of hydro-ethanolic extract of spirulina (*Spirulina platensis*) on semen characteristics and oxidative stress indicators in male rabbit (*Oryctolagus cuniculus*). *Journal of Veterinary Medicine and Animal Sciences, 2*(1), 1–6.

Ebrahim, R. M., (2020). Prophylactic effect of *Spirulina platensis* on radiation-induced thyroid disorders and alteration of reproductive hormones in female albino rats. *International Journal of Radiation Research, 18*(1), 83–90. https://doi.org/10.18869/acadpub.ijrr.18.1.83.

El-Baky, H. H. A., El-Baroty, G. S., & Ibrahem, E. A., (2015). Functional characters evaluation of biscuits sublimated with pure phycocyanin isolated from spirulina and spirulina biomass. *Nutricion Hospitalaria, 32*(1), 231–241. https://doi.org/10.3305/nh.2015.32.1.8804.

El-Baz, F. K., Aly, H. F., El-Sayed, A. B., & Mohamed, A. A., (2013). Role of *Spirulina platensis* in the control of glycemia in DM2 rats. *International Journal of Scientific and Engineering Research, 4*(12), 1731–1740.

El-Sayed, S. M., Hikal, M. S., El-Khair, B. E. A., El-Ghobashy, R. E., & El-Assar, A. M., (2018). Hypoglycemic and hypolipidemic effects of *Spirulina platensis*, phycocyanin, phycocyanopeptide and phycocyanobilin on male diabetic rats. *Arab Universities Journal of Agricultural Sciences, 26*(2A), 1121–1134.

Estrada, J. E. P., Bescós, P. B., & Del, F. A. M. V., (2001). Antioxidant activity of different fractions of *Spirulina platensis* protean extract. *Farmaco, 56*(5–7), 497–500. https://doi.org/10.1016/S0014-827X(01)01084-9.

Fayyad, R. J., Ali, A. N. M., Dwaish, A. S., & Al-Abboodi, A. K. A., (2019). Anticancer activity of *Spirulina platensis* methanolic extracts against l20b and mcf7 human cancer cell lines. *Plant Archives, 19*, 1419–1426.

Fayzunnessa, N., Morshedul, M. A., Uddin, A., Parvin, A., & Saifur, R., (2011). *In vivo* study on the efficacy of hypoglycemic activity of *Spirulina plantesis* in long Evan rats. *International Journal of Biomolecules and Biomedicine, 1*, 27–33.

Fehmida, V. I., Manoj, G. K., Balu, H., Mahesh, R., Hawaldar, S., Veena, R., Palled, V., & Kedar, S. C., (2018). Successful use of spirulina in combatting childhood undernutrition: A community intervention study as part of a mission project. *International Journal of Recent Scientific Research, 9*(12), 30007–30016. https://doi.org/10.24327/IJRSR.

Fujisawa, T., Narikawa, R., Okamoto, S., Ehira, S., Yoshimura, H., Suzuki, I., Masuda, T., et al., (2010). Genomic structure of an economically important cyanobacterium, *Arthrospira*

(*Spirulina*) *platensis* NIES-39. *DNA Research, 17*(2), 85–103. https://doi.org/10.1093/dnares/dsq004.

Gouveia, L., Batista, A. P., Raymundo, A., & Bandarra, N., (2008). *Spirulina maxima* and *Diacronema vlkianum* microalgae in vegetable gelled desserts. *Nutrition and Food Science, 38*(5), 492–501. https://doi.org/10.1108/00346650810907010.

Guldas, M., & Irkin, R., (2010). Influence of *Spirulina platensis* powder on the microflora of yoghurt and acidophilus milk. *Mljekarstvo, 60*(4), 237–243.

Gupta, S., Gupta, C., Garg, A. P., & Prakash, D., (2017). Prebiotic efficiency of blue-green algae on probiotics microorganisms. *Journal of Microbiology and Experimentation, 4*(4), 1–4. https://doi.org/10.15406/jmen.2017.04.00120.

Hafsa, Y. A., Amel, D., Samia, S., & Sidahmed, S., (2014). Evaluation of nutritional and sensory properties of bread enriches with spirulina. *Annals of Food Science and Technology, 15*(2), 270–275.

Han, L. K., Li, D. X., Xiang, L., Gong, X. J., Kondo, Y., Suzuki, I., & Okuda, H., (2006). Isolation of pancreatic lipase activity-inhibitory component of *Spirulina platensis* and it reduce postprandial triacylglycerolemia. *Journal of the Pharmaceutical Society of Japan, 126*(1), 43–49. https://doi.org/10.1248/yakushi.126.43.

Haque, S. E., & Gilani, K. M. A., (2005). Effect of ambroxol, spirulina and vitamin-E in naphthalene induced cataract in female rats. *Indian Journal of Physiology and Pharmacology, 49*(1), 57–64.

Hayashi, T., Hayashi, K., Maeda, M., & Kojima, I., (1996). Calcium spirulan, an inhibitor of enveloped virus replication, from a blue-green alga *Spirulina platensis*. *Journal of Natural Products, 59*(1), 83–87. https://doi.org/10.1021/np960017o.

Hernández-Corona, A., Nieves, I., Meckes, M., Chamorro, G., & Barron, B. L., (2002). Antiviral activity of *Spirulina maxima* against herpes simplex virus type 2. *Antiviral Research, 56*(3), 279–285. https://doi.org/10.1016/S0166-3542(02)00132-8.

Hernández-Lepe, M. A., Wall-Medrano, A., López-Díaz, J. A., Juárez-Oropeza, M. A., Hernández-Torres, R. P., & Ramos-Jiménez, A., (2019). Hypolipidemic effect of *Arthrospira* (*Spirulina*) *maxima* supplementation and a systematic physical exercise program in overweight and obese men: A double-blind, randomized, and crossover-controlled trial. *Marine Drugs, 17*(270), 1–13.

Hetta, M., Mahmoud, R., El-senousy, W., Ibrahim, M., El-taweel, G., & Ali, G., (2014). Antiviral and antimicrobial activities of spirulina. *World Journal of Pharmacy and Pharmaceutical Sciences, 3*(6), 31–39.

Hirahashi, T., Matsumoto, M., Hazeki, K., Saeki, Y., Ui, M., & Seya, T., (2002). Activation of the human innate immune system by spirulina: Augmentation of interferon production and NK cytotoxicity by oral administration of hot water extract of *Spirulina platensis*. *International Immunopharmacology, 2*(4), 423–434. https://doi.org/10.1016/S1567-5769(01)00166-7.

Hoseini, S. M., Khosravi-Darani, K., & Mozafari, M. R., (2013). Nutritional and medical applications of spirulina microalgae. *Mini-Reviews in Medicinal Chemistry, 13*(8), 1231–1237. https://doi.org/10.2174/1389557511313080009.

Hu, J., Li, Y., Pakpour, S., Wang, S., Pan, Z., Liu, J., Wei, Q., She, J., Cang, H., & Zhang, R. X., (2019). Dose effects of orally administered spirulina suspension on colonic microbiota in healthy mice. *Frontiers in Cellular and Infection Microbiology, 9*, 243. https://doi.org/10.3389/fcimb.2019.00243.

Huang, H., Liao, D., Pu, R., & Cui, Y., (2018). Quantifying the effects of spirulina supplementation on plasma lipid and glucose concentrations, body weight, and blood pressure. *Diabetes, Metabolic Syndrome and Obesity: Targets and Therapy, 11*, 729–742. https://doi.org/10.2147/DMSO.S185672.

Hwang, J. H., Lee, I. T., Jeng, K. C., Wang, M. F., Hou, R. C. W., Wu, S. M., & Chan, Y. C., (2011). Spirulina prevents memory dysfunction, reduces oxidative stress damage, and augments antioxidant activity in senescence-accelerated mice. *Journal of Nutritional Science and Vitaminology, 57*(2), 186–191. https://doi.org/10.3177/jnsv.57.186.

Ibrahim, G., & El-Rahman, A., (2018). Evaluation of the efficacy of combined mixture of *Spirulina platensis* and cinnamon extracts in overweight rats fed on a fatty diet. *Life Science Journal, 15*(7), 37–46. https://doi.org/10.7537/marslsj150718.06.

Jagathy, K., & Divya, D., (2014). Cultivation of spirulina (*Arthrospira platensis*) and its use as a food supplement to increase blood hemoglobin level. *International Journal of Advanced Research in Biological Sciences, 1*(5), 39–45.

Joshi, S. M. R., Bera, M. B., & Panesar, P. S., (2014). Extrusion cooking of maize/spirulina mixture: Factors affecting expanded product characteristics and sensory quality. *Journal of Food Processing and Preservation, 38*(2), 655–664. https://doi.org/10.1111/jfpp.12015.

Kalafati, M., Jamurtas, A. Z., Nikolaidis, M. G., Paschalis, V., Theodorou, A. A., Sakellariou, G. K., Koutedakis, Y., & Kouretas, D., (2010). Ergogenic and antioxidant effects of spirulina supplementation in humans. *Medicine and Science in Sports and Exercise, 42*(1), 142–151. https://doi.org/10.1249/MSS.0b013e3181ac7a45.

Khan, M., Shobha, J. C., Mohan, I. K., Naidu, M. U. R., Sundaram, C., Singh, S., Kuppusamy, P., & Kutala, V. K., (2005). Protective effect of spirulina against doxorubicin-induced cardio-toxicity. *Phytotherapy Research, 19*(12), 1030–1037. https://doi.org/10.1002/ptr.1783.

Khan, Z., Bhadouria, P., & Bisen, P., (2005). Nutritional and therapeutic potential of spirulina. *Current Pharmaceutical Biotechnology, 6*(5), 373–379. https://doi.org/10.2174/138920105774370607.

Kumar, N., Singh, S., Patro, N., & Patro, I., (2009). Evaluation of protective efficacy of *Spirulina platensis* against collagen-induced arthritis in rats. *Inflammopharmacology, 17*(3), 181–190. https://doi.org/10.1007/s10787-009-0004-1.

Labhe, R. U., Mani, U. V., Iyer, U. M., Mishra, M., Jani, K., & Bhattacharya, A., (2001). The effect of spirulina in the treatment of bronchial asthma. *Journal of Nutraceuticals, Functional and Medical Foods, 3*(4), 53–60. https://doi.org/10.1300/J133v03n04_06.

Layam, A., & Reddy, C. L. K., (2007). Antidiabetic property of spirulina. *Diabetologia Croatica, 35*(2), 29–33.

Løbner, M., Walsted, A., Larsen, R., Bendtzen, K., & Nielsen, C. H., (2008). Enhancement of human adaptive immune responses by administration of a high-molecular-weight polysaccharide extract from the cyanobacterium *Arthrospira platensis*. *Journal of Medicinal Food, 11*(2), 313–322. https://doi.org/10.1089/jmf.2007.564.

Lucas, B. F., De Morais, M. G., Santos, T. D., & Costa, J. A. V., (2018). Spirulina for snack enrich-ment: Nutritional, physical, and sensory evaluations. *LWT-Food Science and Technology, 90*, 270–276. https://doi.org/10.1016/j.lwt.2017.12.032.

Lupatini, A. L., Colla, L. M., Canan, C., & Colla, E., (2017). Potential application of microalgae *Spirulina platensis* as a protein source. *Journal of the Science of Food and Agriculture, 97*(3), 724–732. https://doi.org/10.1097/AOG.0b013e3181989578.

Ma, Z., & Gao, K., (2009). Photoregulation of morphological structure and its physiological relevance in the cyanobacterium *Arthrospira* (*Spirulina*) *platensis*. *Planta, 230*(2), 329–337. https://doi.org/10.1007/s00425-009-0947-x.

Mao, T. K., De Water, J. V., & Gershwin, M. E., (2005). Effects of a spirulina-based dietary supplement on cytokine production from allergic rhinitis patients. *Journal of Medicinal Food, 8*(1), 27–30. https://doi.org/10.1089/jmf.2005.8.27.

Massoud, R., Khosravi-Darani, K., Nakhsaz, F., & Varga, L., (2016). Evaluation of physico-chemical, microbiological, and sensory properties of croissants fortified with *Arthrospira platensis* (*Spirulina*). *Czech Journal of Food Sciences, 34*(4), 350–355. https://doi.org/10.17221/289/2015-CJFS.

Mathew, B., Sankaranarayanan, R., Nair, P. P., Varghese, C., Somanathan, T., Amma, B. P., Amma, N. S., & Nair, M. K., (1995). Evaluation of chemoprevention of oral cancer with *Spirulina fusiformis*. *Nutrition and Cancer, 24*(2), 197–202. https://doi.org/10.1080/01635589509514407.

Matondo, F. K., Takaisi, K., Nkuadiolandu, A. B., Lukusa, A. K., & Aloni, M. N., (2016). Spirulina supplements improved the nutritional status of undernourished children quickly and significantly: Experience from Kisantu, the Democratic Republic of the Congo. *International Journal of Pediatrics, 2016*, 1–5. https://doi.org/10.1155/2016/1296414.

Memije-Lazaro, I. N., Blas-Valdivia, V., Franco-Colín, M., & Cano-Europa, E., (2018). *Arthrospira maxima* (*Spirulina*) and C-phycocyanin prevent the progression of chronic kidney disease and its cardiovascular complications. *Journal of Functional Foods, 43*, 37–43. https://doi.org/10.1016/j.jff.2018.01.013.

Misbahuddin, M., Islam, A. M., Khandker, S., Al-Mahmud, I., Islam, N., & Anjumanara, (2006). Efficacy of spirulina extract plus zinc in patients of chronic arsenic poisoning: A randomized placebo-controlled study. *Clinical Toxicology, 44*(2), 135–141. https://doi.org/10.1080/15563650500514400.

Mishima, T., Murata, J., Toyoshima, M., Fujii, H., Nakajima, M., Hayashi, T., Kato, T., & Saiki, I., (1998). Inhibition of tumor invasion and metastasis by calcium spirulan (Ca-SP), a novel sulfated polysaccharide derived from a blue-green alga, *Spirulina platensis. Clinical and Experimental Metastasis, 16*(6), 541–550. https://doi.org/ 10.1023/A:1006594318633.

Morsy, O., Sharoba, A. M., Bahlol, H., & Abdel-mawla, E., (2014). Production and evaluation of extruded food products by using spirulina algae. *Annals Agric. Sci. Moshtohor, 52*(4), 329–342.

Mühling, M., Belay, A., & Whitton, B. A., (2005). Variation in fatty acid composition of *Arthrospira* (*Spirulina*) strains. *Journal of Applied Phycology, 17*(2), 137–146. https://doi.org/10.1007/s10811-005-7213-9.

Nagaoka, S., Shimizu, K., Kaneko, H., Shibayama, F., Morikawa, K., Kanamaru, Y., Otsuka, A., et al., (2005). A novel protein C-phycocyanin plays a crucial role in the hypocholesterolemic action of *Spirulina platensis* concentrate in rats. *The Journal of Nutrition, 135*(10), 2425–2430. https://doi.org/10.1093/jn/135.10.2425.

Naik, M. A., (2001). *Effect of Spirulina Supplementation on Vitamin A Status During Pregnancy and Lactation.* PhD Dissertation, S.N.D.T. Women's University, Mumbai.

Nasirian, F., Dadkhah, M., Moradi-Kor, N., & Obeidavi, Z., (2018). Effects of *Spirulina platensis* microalgae on antioxidant and anti-inflammatory factors in diabetic rats. *Diabetes, Metabolic Syndrome and Obesity: Targets and Therapy, 11*, 375–380. https://doi.org/10.2147/DMSO.S172104.

Navacchi, M. F. P., De Carvalho, J. C. M., Takeuchi, K. P., & Danesi, E. D. G., (2012). Development of cassava cake enriched with its own bran and *Spirulina platensis. Acta Scientiarum-Technology, 34*(4), 465–472. https://doi.org/10.4025/actascitechnol.v34i4.10687.

Nemoto-Kawamura, C., Hirahashi, T., Nagai, T., Yamada, H., Katoh, T., & Hayashi, O., (2004). Phycocyanin enhances secretory IgA antibody response and suppresses allergic IgE antibody response in mice immunized with antigen-entrapped biodegradable microparticles. *Journal of Nutritional Science and Vitaminology, 50*, 129–136.

Neyrinck, A. M., Taminiau, B., Walgrave, H., Daube, G., Cani, P. D., Bindels, L. B., & Delzenne, N. M., (2017). Spirulina protects against hepatic inflammation in aging: An effect related to the modulation of the gut microbiota. *Nutrients, 9*(6), 1–13. https://doi.org/10.3390/nu9060633.

Ngo-Matip, M. E., Pieme, C. A., Azabji-Kenfack, M., Moukette, B. M., Korosky, E., Stefanini, P., Ngogang, J. Y., & Mbofung, C. M., (2015). Impact of daily supplementation of *Spirulina platensis* on the immune system of native HIV-1 patients in Cameroon: A 12-months single-blind, randomized, multicenter trial. *Nutrition Journal, 14*, 1–7. https://doi.org/10.1186/s12937-015-0058-4.

Niang, K., Ndiaye, P., Faye, A., Tine, J. A. D., Diongue, F. B., Camara, M. D., Leye, M. M., & Tal-Dia, A., (2017). Spirulina supplementation in pregnant women in the Dakar region (Senegal). *Open Journal of Obstetrics and Gynecology, 7*(1), 147–154. https://doi.org/10.4236/ojog.2017.71016.

Norata, G. D., Raselli, S., Grigore, L., Garlaschelli, K., Vianello, D., Bertocco, S., Zambon, A., & Catapano, A. L., (2009). Small dense LDL and VLDL predict common carotid artery IMT and elicit an inflammatory response in peripheral blood mononuclear and endothelial cells. *Atherosclerosis, 206*(2), 556–562. https://doi.org/10.1016/j.atherosclerosis.2009.03.017.

Nuhu, A. A., (2013). Spirulina (*Arthrospira*): An important source of nutritional and medicinal compounds. *Journal of Marine Biology, 2013*, 1–8. https://doi.org/10.1155/2013/325636.

Ohmori, M., & Ehira, S., (2014). Spirulina: An example of cyanobacteria as nutraceuticals. In: Sharma, N. K., Rai, A. K., & Stal, L. J., (eds.), *Cyanobacteria: An Economic Perspective* (pp. 103–118). John Wiley and Sons, UK. https://doi.org/10.1002/9781118402238.ch7.

Okamoto, T., Kawashima, H., Osada, H., Toda, E., Homma, K., Nagai, N., Imai, Y., et al., (2019). Dietary spirulina supplementation protects visual function from photostress by suppressing retinal neurodegeneration in mice. *Translational Vision Science and Technology, 8*(6). https://doi.org/10.1167/tvst.8.6.20.

Okechukwu, P. N., Ekeuku, S. O., Sharma, M., Nee, C. P., Chan, H. K., Mohamed, N., & Froemming, G. R. A., (2019). *In vivo* and *in vitro* antidiabetic and antioxidant activity of spirulina. *Pharmacognosy Magazine, 15*(62), 17–29. https://doi.org/10.4103/pm.pm.

Onacik-Gür, S., Zbikowska, A., & Majewska, B., (2018). Effect of spirulina (*Spirulina platensis*) addition on textural and quality properties of cookies. *Italian Journal of Food Science, 30*(1), 1–12. https://doi.org/10.14674/IJFS-702.

Ou, Y., Lin, L., Yang, X., Pan, Q., & Cheng, X., (2013). Antidiabetic potential of phycocyanin: Effects on KKAy mice. *Pharmaceutical Biology, 51*(5), 539–544. https://doi.org/10.3109/13880209.2012.747545.

Özyurt, G., Uslu, L., Yuvka, I., Gökdoğan, S., Atci, G., Ak, B., & Işik, O., (2015). Evaluation of the cooking quality characteristics of pasta enriched with *Spirulina platensis*. *Journal of Food Quality, 38*(4), 268–272. https://doi.org/10.1111/jfq.12142.

Pizzino, G., Irrera, N., Cucinotta, M., Pallio, G., Mannino, F., Arcoraci, V., Squadrito, F., Altavilla, D., & Bitto, A., (2017). Oxidative stress: Harms and benefits for human health. *Oxidative Medicine and Cellular Longevity, 2017*, 1–13. https://doi.org/10.1155/2017/8416763.

Pugh, N., Ross, S. A., Elsohly, H. N., Elsohly, M. A., & Pasco, D. S., (2001). Isolation of three high molecular weight polysaccharide preparations with potent immunostimulatory activity from *Spirulina platensis, Aphanizomenon flos-aquae* and *Chlorella pyrenoidosa*. *Planta Medica, 67*(8), 737–742.

Rojas-Franco, P., Franco-Colín, M., Camargo, M. E. M., Carmona, M. M. E., Ortíz-Butrón, M. D. R. E., Blas-Valdivia, V., & Cano-Europa, E., (2018). Phycobiliproteins and phycocyanin of *Arthrospira maxima* (*Spirulina*) reduce apoptosis promoters and glomerular dysfunction in mercury-related acute kidney injury. *Toxicology Research and Application, 2*, 1–10. https://doi.org/10.1177/2397847318805070.

Safi, C., Charton, M., Pignolet, O., Silvestre, F., Vaca-Garcia, C., & Pontalier, P. Y., (2013). Influence of microalgae cell wall characteristics on protein extractability and determination of nitrogen-to-protein conversion factors. *Journal of Applied Phycology, 25*(2), 523–529. https://doi.org/10.1007/s10811-012-9886-1.

Samuels, R., Mani, U. V., Iyer, U. M., & Nayak, U. S., (2002). Hypocholesterolemic effect of spirulina in patients with hyperlipidemic nephrotic syndrome. *Journal of Medicinal Food, 5*(2), 91–96. https://doi.org/10.1089/109662002760178177.

Santos, T. D., De Freitas, B. C. B., Moreira, J. B., Zanfonato, K., & Costa, J. A. V., (2016). Development of powdered food with the addition of spirulina for food supplementation of the elderly population. *Innovative Food Science and Emerging Technologies, 37*(1), 216–220. https://doi.org/10.1016/j.ifset.2016.07.016.

Seema, Sonia & Mahipal, (2016). Spirulina as dietary supplement for health: A pilot study. *The Pharma Innovation Journal, 5*(4), 7–9.

Seo, Y. J., Kim, K. J., Choi, J., Koh, E. J., & Lee, B. Y., (2018). *Spirulina maxima* extract reduces obesity through suppression of adipogenesis and activation of browning in 3T3-L1 cells and high-fat diet-induced obese mice. *Nutrients, 10*(6), 1–15. https://doi.org/10.3390/nu10060712.

Seshadri, C. V., (1993). Large scale nutritional supplementation with spirulina alga. In: *Monograph Series on Engineering of Photosynthetic Systems* (Vol. 36, p. 79). Shri AMM Murugappa Chettiar Research Center (MCRC), Chennai.

Sharoba, A. M., (2014). Nutritional value of spirulina and its use in the preparation of some complimentary baby food formulas. *Journal of Agroalimentary Processes and Technologies, 20*(4), 330–350. https://doi.org/10.21608/jfds.2014.53033.

Shekharam, K. M., Venkataraman, L. V., & Salimath, P. V., (1987). Carbohydrate composition and characterization of two unusual sugars from the blue-green alga *Spirulina platensis*. *Phytochemistry, 26*(8), 2267–2269. https://doi.org/10.1016/S0031-9422(00)84698-1.

Shimamatsu, H., (2004). Mass production of spirulina, an edible microalga. *Hydrobiologia, 512*, 39–44. https://doi.org/10.1023/B:HYDR.0000020364.23796.04.

Simsek, N., Karadeniz, A., Kalkan, Y., Keles, O. N., & Unal, B., (2009). *Spirulina platensis* feeding inhibited the anemia and leucopenia-induced lead and cadmium in rats. *Journal of Hazardous Materials, 164*(2, 3), 1304–1309. https://doi.org/10.1016/j.jhazmat.2008.09.041.

Somchit, M. N., Mohamed, N. A., Ahmad, Z., Zakaria, Z. A., Shamsuddin, L., Fauzee, M. S. O., & Kadir, A. A., (2014). Anti-inflammatory and antipyretic properties of *Spirulina platensis* and *Spirulina lonar*: A comparative study. *Pakistan Journal of Pharmaceutical Sciences, 27*(5), 1277–1280.

Soni, R. A., Sudhakar, K., & Rana, R. S., (2017). Spirulina – From growth to nutritional product: A review. *Trends in Food Science and Technology, 69*(11), 157–171. https://doi.org/10.1016/j.tifs.2017.09.010.

Szulinska, M., Gibas-Dorna, M., Miller-Kasprzak, E., Suliburska, J., Miczke, A., Walczak-Gałezewska, M., Stelmach-Mardas, M., et al., (2017). *Spirulina maxima* improves insulin sensitivity, lipid profile, and total antioxidant status in obese patients with well-treated hypertension: A randomized double-blind placebo-controlled study. *European Review for Medical and Pharmacological Sciences, 21*(10), 2473–2481.

Torres-Duran, P. V., Ferreira-Hermosillo, A., & Juarez-Oropeza, M. A., (2007). Antihyper-lipemic and antihypertensive effects of *Spirulina maxima* in an open sample of Mexican population: A preliminary report. *Lipids in Health and Disease, 6*(1), 1–8. https://doi.org/10.1186/1476-511X-6-33.

Udayan, A., Arumugam, M., & Pandey, A., (2017). Nutraceuticals from algae and cyanobacteria. In: Rastogi, R. P., Madamwar, D., & Pandey, A., (eds.), *Algal Green Chemistry: Recent Progress in Biotechnology* (pp. 65–89). Elsevier, Netherlands. https://doi.org/10.1016/B978-0-444-63784-0.00004-7.

Ugwu, C. U., Aoyagi, H., & Uchiyama, H., (2008). Photobioreactors for mass cultivation of algae. *Bioresource Technology, 99*(10), 4021–4028. https://doi.org/10.1016/j.biortech.2007.01.046.

Usharani, G., Saranraj, P., & Kanchana, D., (2012). Spirulina cultivation: A review. *International Journal of Pharmaceutical and Biological Archives, 3*(6), 1327–1341.

Vijayarani, D., Ponnalaghu, S., & Rajathivya, J., (2012). Development of value-added extruded product using spirulina. *International Journal of Health Sciences and Research, 2*(4), 42–47.

Visnegarwala, F. I., & Mahesh, R., (2017). Spirulina: A panacea for iron-deficiency anemia of pregnancy (a hypothesis-based review). *Journal of Alternative Medical Research, 3*(1), 1–3. https://doi.org/10.1016/j.hoc.2015.11.002.3.

Voltarelli, F. A., Araújo, M. B., De Moura, L. P., Garcia, A., Silva, C. M. S., Junior, R. C. V., Melo, F. C. L., & De Mello, M. A. R., (2011). Nutrition recovery with spirulina diet improves body growth and muscle protein of protein-restricted rats. *International Journal of Nutrition and Metabolism, 3*(3), 22–30.

Vonshak, A., & Tomaselli, L., (2000). *Arthrospira (Spirulina)*: Systematics and ecophysiology. In: Whitton, B. A., & Potts, M., (eds.), *The Ecology of Cyanobacteria* (pp. 505–522). Springer, Dordrecht. https://doi.org/10.1007/0-306-46855-7_18.

Yamamoto, C., Nakamura, A., Shimada, S., Kaji, T., Lee, J. B., & Hayashi, T., (2003). Differential effects of sodium spirulan on the secretion of fibrinolytic proteins from vascular endothelial cells: Enhancement of plasminogen activator activity. *Journal of Health Science, 49*(5), 405–409. https://doi.org/10.1248/jhs.49.405.

Yu, B., Wang, J., Suter, P. M., Russell, R. M., Grusak, M. A., Wang, Y., Wang, Z., et al., (2012). Spirulina is an effective dietary source of zeaxanthin to humans. *British Journal of Nutrition, 108*(4), 611–619. https://doi.org/10.1017/S0007114511005885.

Zeinalian, R., Farhangi, M. A., Shariat, A., & Saghafi-Asl, M., (2017). The effects of *Spirulina platensis* on anthropometric indices, appetite, lipid profile and serum vascular endothelial growth factor (VEGF) in obese individuals: A randomized double-blinded placebo-controlled trial. *BMC Complementary and Alternative Medicine, 17*(1), 225. https://doi.org/10.1186/s12906-017-1670-y.

CHAPTER 5

ALGAL METABOLITES IN NUTRACEUTICAL INDUSTRIES: CURRENT TRENDS AND FUTURE PERSPECTIVES

HARAM SARFRAZ and IFFAT ZAREEN AHMAD

Natural Products Laboratory, Department of Bioengineering, Integral University, Dasauli, Lucknow, Uttar Pradesh–226026, India

ABSTRACT

Nutraceutical is any food item with augmented health benefits besides the usual health benefits of regular foods. Due to the current worldwide interest in finding novel and economical health supplements, there is an increment in the consumption of these nutraceuticals with amplified benefits. Algae endowed with favorable health benefits are amongst the novel entries in this sector. Further, based on the abundance of high-value metabolites algae has gained an important position in the health and nutraceutical markets. Apart from being a protein source, algae have various metabolites such as essential fatty acids (EFAs), essential amino acids (EAAs), vitamins, minerals, carbohydrates, bioactive peptides, pigments, etc. All of these compounds hold valuable therapeutic activities (anticancer, antioxidant, anti-inflammatory, and antimicrobial) and have commercial applications in nutraceutical industries. Keeping in view the great potential of algal metabolites along with its importance in the development of nutraceuticals as a way to improve and maintain health is discussed comprehensively.

5.1   INTRODUCTION

Nutraceutical is any food item with enhanced medical advantages which beats the standard medical advantages of common nourishments (Borowitzka, 2013) or it is described as "items that are gotten from characteristic sources which can likewise be sustained, whose ingestion is probably going to advantage human wellbeing" (Burja et al., 2008). The term nutraceutical was given by Stephen DeFelice, organizer of the Foundation for Innovation in Medicine in Cranford, New Jersey. This word nutraceutical is a hybrid of two terms "nutrition" and "pharmaceutical," which alludes to stuff associated with sustenance, blessed with biological aids-related with assurance or potentially anticipation against persistent ailments. Also, nutraceuticals are frequently characterized likely with practical nourishments in the media which include isolated nutrients, nutritious biomass, dietary supplements, innately altered food, herbal products, also, braced nourishments with basic nutrients, amino acids, cancer prevention agents, minerals, EFAs, and extraordinary worth included bioactive molecules. Studies likewise show that a generally acknowledged term for nutraceuticals as "food supplements." However, it is a generally held view that there is, by all accounts, a limit between utilitarian nourishments and nutraceuticals as these compound(s) might be focused to offer the asserted wellbeing welfares. In this way, the significance of looking for nutraceuticals that finance to enhanced human wellbeing has raised around the world.

Lately, because of an upsurge in the utilization of enhancements and nutraceuticals exchanged drug stores, a prosperous market has been created. That is the reason among the novel passages in the segment of nourishment supplements. Algae have shown up as nutraceutical components, talented with positive impacts on wellbeing. Between promising wellsprings of chemopreventive eating regimens of premium and utilitarian nourishments, microalgae are increasing general consideration, given their lavishness in high-esteem items, including nutrients, basic amino acids, carotenoids, proteins, omega-rich oils, and in general, anti-inflammatory, and antioxidant compounds. Furthermore, the usage of algal biomass has interested consideration since they develop quickly regardless of the land's reasonableness for cultivating. By and large, the development of algae can be completed liberated from freshwater flexibly and does not fight with biodiverse scenes or arable land.

Most of the microalgae with health aids are marine (Lim et al., 2012). Algae are subsequently viewed as a perfect hotspot for the maintainable physiologically dynamic compound creation (Abdelaziz et al., 2013; Hudek et al., 2014). As each product of nutraceutical esteem is relied upon to have distinct medical advantages, because of the event of either antioxidant, other bioactive metabolites, or signaling-pathway modulators. Thus, algae epitomize promising opportunities in this direction because they create valued bioactive fixings, for example, carotenoids with effectively natural medical advantages (Galasso et al., 2017), as well as additional antioxidant compounds (Smerilli et al., 2017). This factor implies an additional in addition to esteem as synergism between assorted bioactive compounds families unequivocally upgrades their significant consequences for wellbeing (Lordan and Ross, 2011; Hamed et al., 2015).

Various algal bioactive metabolites include antioxidant pigments chlorophyll, β-carotene, xanthophylls, and phycobilin pigments; polyunsaturated fatty acids (PUFAs); polyphenols and α-tocopherols; and ascorbate, glutathione, and ergothioneine. Additionally, many specific metabolites have also been mined from algae such as natural food colorants, supplement added substances for use in a few food complexes (Priyadarshani and Rath, 2012; Murray et al., 2013). Those fractionated bioactive molecules, in the normal circumstances, also have extraordinary essential unsaturated fats [just in case of docosahexaenoic acid (DHA)] from *Crypthecodinium cohnii*; eicosapentaenoic acid (EPA) from *Phaeodactylum tricornutum* or extraordinary antioxidant roles [in the case of β-carotene (*Dunaliella salina*); astaxanthin (*Haematococcus pluvialis*); fucoxanthin (*Phaeodactylum tricornutum*); C-phycocyanin (*Spirulina platensis*); B-phycoerythrin (*Porphyridium cruentum*) or lutein (*Chlorella protothecoides*)].

Hence, we can say that one of the attractive misuses of algae is aligned with their cancer prevention agent highlights, which has allowed the microalgae advancement into wellbeing just as nourishment markets. In such a manner, two distending models are astaxanthin (*Haematococcus pluvialis*) and β-carotene (*Dunaliella salina*). Furthermore, the carotenoid astaxanthin is a multipurpose extraordinary worth included bioactive molecule, as it shows multiple functionalities and activities. Previous studies have mentioned the algal nutraceuticals' health benefits, which include control of heftiness and cholesterol, diminished circulatory strain, and kept up ideal heart condition. Algal nutraceuticals were likewise discovered to be successful in neurological turn of events, improving resistance, upgrading the strength of different organs including teeth, bones, digestive tract, and so on. Various investigations have additionally archived certain anticancer properties and antiviral (Venugopal, 2008). This part gives a short outline of the algal nutraceuticals alongside the hugeness of various algal metabolites concerning their nutraceutical potential.

5.2   ALGAE IN NUTRACEUTICAL INDUSTRY

As of now, there is an upsurged worldwide enthusiasm for the hint of nutritious food on wellbeing manageability. On this part, nutraceuticals have picked up head significance dependent on their possible nourishing, remedial, and rumored security esteem. In such a manner, algae are among the potential living beings that are useful to humankind in innumerable manners and speak to the best encouraging asset for imaginative items and applications (Pulz and Gross, 2004). Attributable to their incidental substance properties, algae can be used as a source of natural food tints as well as a nutritional supplement (Apt and Bahrens, 1999; Borowitzka, 1999; Soletto et al., 2005). As measures that can be considered incorporate healthfully significant biomolecules, for example, carotenoids, and other pigments, EFAs, nutrients, fundamental amino acids, minerals, sugars as dietary strands and bioactivities, proteins, and antimicrobial compounds (Figure 5.1). These high valuable compounds hold antiviral, antimicrobial, antioxidant, anticancer, anti-inflammatory, anti-fungal activity, and an immune modulator.

Marine-based mixes are known to include a broad scope of sub-atomic focuses with an observable selectivity, which upgrades drug interest (Guedes et al., 2011). Algae (full scale and miniature in nature both) have been rehearsed by unique populaces around the globe, exclusively in China, the Republic of Korea and Japan. Algae are a usual reservoir that plays a vital healthful part in the biosphere by loading both earthly and amphibian evolved ways of life. The utilization of algae for upgrading physiological wellbeing is expanding in response to requests of the market, which are genuinely reliant (Frost and Sullivan, 2011). The use of algae in food or food supplements, aquaculture (indirect human nutrition), high value-added biomolecules, renewable energy, and bioremediation areas have picked up energy. Algae under marketable cultivation for the production of the precise bioactive molecule are morphologically varied. As a base 15 algal species, types are now developed productively, while others with bizarre potential are just at access-level. Novel algal species are either beneath attractive development advancement or are forthcoming innovation transmission from research focuses and the scholarly community to industry for business examination. Algal biotechnology, similar to most extraordinary ventures, is market-driven and matter to reasonable cost-serious, and conservative creation. Up until now, algae are ordinarily utilized as whole biomass as prosperity supplements.

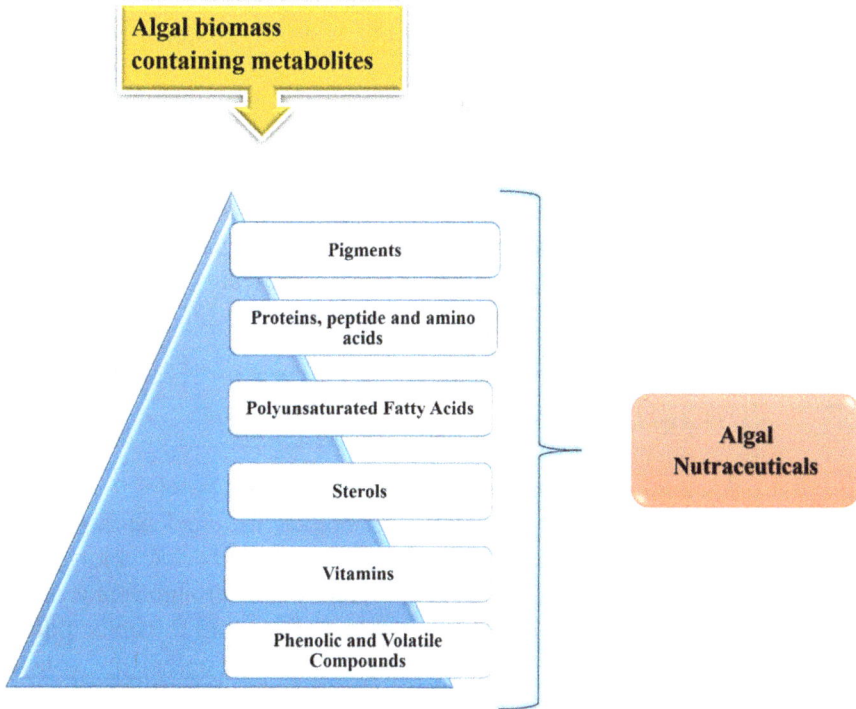

FIGURE 5.1   Major algal metabolites used in nutraceutical industries.

Though certain algae such as *Arthrospira* and *Nostoc* have been mistreated for centuries, the attractive capability of algae is known uniquely in the current past. Just an uncommon 100 among 30,000 species that accepted to exist are investigated for chemical content (Chaumont, 1993; Radmer and Parker, 1994), out of which quite a few are cultivated in manufacturing grade (Guerrin, 2003). Currently, the use of algal biomass in the wellbeing food souk, as well as its attractive potential, is extensively perceived. The expanding utilization of biomass of algae for nutraceutical designs should offer a striking income stream for algal ventures. Best biotechnologically pertinent algal species include Chlorophyceae: *Haematococcus pluvialis*, *Dunaliella salina*, *Chlorella vulgaris*, and cyanobacteria: *Spirulina maxima* plus *Spirulina platensis* (Gouveia et al., 2008; Guedes et al., 2011). Algal species consumed as whole biomass include *Chlorella vulgaris*, *Spirulina platensis*, and *Aphanizomenon flos-aquae*. These species are mostly indorsed for consumption in the entire-biomass form because of their absence of toxicity and expressed wholesome advantages, for example, essential nutrients and minerals, great content of protein and antioxidant pigment, as well as dietary fiber.

Nearly 75% of algae species are microalgae, contributing around 40% of the oxygen in the atmosphere. Despite the deceptive simplicity of their cells, as a minimum of 40,000 microalgae, phytoplankton species have been recognized (Singh, 2011). The key to this achievement is in the metabolism, viz., in the wholeness of substances existing in the cytoplasm. This is the reason for their fundamental importance in nutraceuticals.

5.2.1   THE DOMINANCE OF CYANOBACTERIA IN NUTRACEUTICALS

As discussed earlier, cyanobacteria are one of the major species involved in the production of nutraceuticals; hence the market for algal nutraceuticals is controlled by two cyanobacteria, commonly known as the chloropicrin *Chlorella* and *Spirulina*. The work of the chlorophyll in autotrophic algae is the starting base model of the progression line leading to contemporary plants. The evolutionary pathways emphasize improved organisms, considering them a chief source of medicinal drugs and food. It is now time to inverse this approach. Once more, the chief signal of originality comes from nutraceuticals. Algae are profoundly used as underdone materials in nutrition supplements. The entitlement is to find both equilibria in the food and an exact activity, in conformity with the appeal of the functional food.

5.2.1.1   SPIRULINA

Spirulina, a cyanobacterium present as free-floating filaments, is principally known worldwide for its impending nutritional value. It is among one of the unusual edible bacteria, owing to its low concentration of purine, which permits it to pose a negligible risk of uric acid accumulation in the body (Pittman et al., 2011). The food industry categorizes it as a single-celled protein, denoting a comestible microbe with an extraordinary food value (Habib et al., 2008). The nutritional worth of *Spirulina* was previously known to the *Aztecs*, who yielded the alga from Texcoco Lake, close to Mexico City (Nyenje and Ndip, 2013). This cyanobacterium is rich in vitamins, carotene, EFAs, minerals, and antioxidants, all of which have simplified its commercial production as a human nutrition supplement over the passage of the earlier decade (Juarez-Oropeza et al., 2009; Capelli and Cysewski, 2010). Its intake has been shown to have positive cardiovascular effects, reducing cholesterol, and lowering blood pressure (Juarez-Oropeza et al., 2009). In 1986, it was also used for the treatment of radiation sickness in individuals that were affected by the Chernobyl nuclear mishap, which determines its anti-carcinogenic properties (Hirahashi et al., 2002; Mišurcová et al., 2012; Mosulishvili et al., 2012; Ku et al., 2013).

Currently, it is used in diet supplements, as powder or tablets, alone or combined with plant extracts or other algae, to use humans or animals. Essentially, *Spirulina* reflects an honorable source of essential amino acids (EAAs) and vitamins. Correspondingly it has a well-balanced composition of excellent protein content, making it even more necessary as a food supplement (Campanella et al., 2002; Habib et al., 2008). It also has amazing oil content, in quantity (7%) and quality [linoleic acid (LNA), α-linolenic acid (ALA), stearidonic acid, DHA, EPA, and arachidonic acid (AA)]. *Spirulina* and *Klamath* are promoted with indicative adjectives, such as the food of the future or superfood. Numerous activities are described and, to some extent, confirmed by various types of testing and clinical trials. In the major position, the anti-obesity effect and the nutritional value are considerable marketing requests, despite the ailing taste.

5.2.1.2 CHLORELLA

Chlorella is a tiny unicellular seaweed relating to the green algae (*Chlorophyta*). The apparent green pigmentation is owing to the occurrence of the two chlorophylls a and b, the similar ones in land-dwelling plants. It can be effortlessly cultivated in modest conditions, generating massive quantities of biomass in petite time. It requires only carbon dioxide, water, light, and a trivial quantity of minerals. However, the flavor of this seaweed, along with the other ones, is not pleasurable in comparison with usual foods. In this case, also, there is an extended list of assigned health properties and activities, including the stimulation of the immune system and detoxifying action. Sideways, evidence of improved digestion, wellbeing, and normalization of sugar metabolism has been stated.

5.3 ALGAL METABOLITES AND THEIR POTENTIAL HEALTH BENEFITS

Algae have an abundant, diverse class of metabolites of interest for diet supplements/nutraceuticals claims. Here, we examine and condense chosen huge bioactive metabolites with potential chemopreventive impacts present in algae. Metabolic compounds that are synthesized by algae, which are under deliberation for commercial extraction or presently being commercialized, include carotenoids, astaxanthin, β-carotene, chlorophylls, phycobilins, long-chain polysaccharides, proteins, sterols, vitamins, phenolic compounds. Based on the algal sources and specific activity, these metabolic compounds may serve as anti-inflammatory agents, antioxidants, antimicrobial agents, antitumoral agents, and antiviral agents (Apt and Bahrens, 1999; Borowitzka, 1999) (Table 5.1).

5.3.1 PIGMENTS

Pigments are fundamental molecules that use electromagnetic radiation's visible spectrum to absorb light. An average human eye can see the wavelength reflected by these molecules and hence, manifest the adjacent color. Pigments are utilized for various products like aquaculture, food additives/colorants, nutraceutical, and pharmaceutical products (Begum et al., 2013). Whereas, due to the increment in demands for the pigments prepared naturally because of environmental and safety hazards related to artificially prepared pigments (Kobylewski and Jacobson, 2010; Begum et al., 2013; Parmar and Singh, 2018). The food industry takes special care of using naturally made pigments as food colorants instead of artificial ones to avoid the harmful effects of synthetic pigments (Parmar and Singh, 2018). The natural pigment production sources comprise flowers, fruits, insects, vegetables, and photosynthetic micro-organisms, similar to cyanobacteria (Farré et al., 2010). Photosynthetic microorganisms (algae) that produce a variety of pigments specific to their corresponding species and colors. For example, red, and blue algae own phycobiliprotein (PBP), green algae comprise chlorophyll, and red, yellow, and orange algae synthesize carotenoids. Out of different sources, algae are a serious wellspring of pigments attributable to their capability to create regular colors in more prominent focus as coordinated to others (Mulders, 2014).

TABLE 5.1 Certain Algal Metabolites Along with Their Health Benefits and Sources

Algal Metabolites	Species	Health Benefits	References
Pigments (β-carotene, lycopene, astaxanthin, zeaxanthin, lutein, fucoxanthin, chlorophyll)	*Dunaliella salina, Nannochloropsis gaditana, Chlorella zofingiensis, Haematococcus pluvialis, Dunaliella salina, Scenedesmus almeriensis*	Prevention of cancer, protection from macular degeneration, cognitive impairment and protein degradation, antioxidant, and anti-inflammatory properties, protection against Parkinson's disease, reduced vision, rheumatoid arthritis, dyslipidemia, and diabetes	Macias et al. (2007); Guedes et al. (2011a, b); Stengel et al. (2011); Borowitzka (2013)
Protein, peptides, and amino acids (essential amino acids, mycosporine-like amino acids, glycoprotein, lectins, *diatoms*, DMSP)	*Haematococcus pluvialis, Chlorella* sp., *Chlamydomonas reinhardtii, Spirulina* sp., *Scenedesmus* sp., *Dunaliella* sp., *Navicula incerta*	Anticancer, hepatoprotective, immunomodulating, antioxidant, antimicrobial, and anti-inflammatory properties, DNA protection, recombinant proteins production, angiotensin I inhibitory and antihypertensive activities, induction of apoptosis	Sunda et al. (2002); Mayfield et al. (2007); Sheih et al. (2009); Agyei and Danquah (2011); Gong et al. (2011); Himaya et al. (2012); Dewapriya and Kim (2014); Mustopa et al. (2016); Ejike and Collins (2017)
Polyunsaturated fatty acids (ALA, DHA, EPA, GLA, ARA)	*Coelastrella* sp. F50, *Dunaliella salina, Isochrysis galbana, Crypthecodinium cohnii, Pavlova salina, Monodus subterraneus, Arthrospira platensis* PCC9108, *Porphyridium cruentum* 1380 la, *Parietochloris incisa, Thalassiosira pseudonana*	Antiviral, anti-inflammatory, antioxidant, antitumor, antihyperlipidemic, and anticoagulant activities, biogenetic precursor of leukotrienes and prostaglandin, important for visual sharpness ingredient in baby food formulations, role in neurodegeneration and neural development	Pratoomyot et al. (2005); Abe et al. (2007); Gang-Guk et al. (2008); Hu et al. (2013); Janssen and Kiliaan (2014)

TABLE 5.1 *(Continued)*

Algal Metabolites	Species	Health Benefits	References
Sterols (brassicasterol, stigmasterol, sitosterol, isofucosterol, fucosterol)	*Dunaliella salina, Dunaliella tertiolecta, Giraudyopsis* sp., *Chrysowaernella* sp.	Cholesterol-lowering activities, anti-inflammatory, anticancer, neurotransmission, and development, protection against autoimmune encephalomyelitis, amyotrophic lateral sclerosis	Billard et al. (1990); Francavilla et al. (2012); Ostad et al. (2012); Hwang et al. (2014)
Vitamins [vitamin A, B1 (thiamin), B2 (riboflavin), B3 (Niacin), B6 (pyridoxine), B12 (cobalamin), C (ascorbic acid), E (tocopherols), pantothenic acid, folic acid]	*Dunaliella tertiolecta, Chlorella* sp., *Spirulina* sp., *Tetraselmis suecica, Macrocystis pyrifera, Porphyridium cruentum*	Antioxidant activities, role in body function, immunity, digestive system, inhibits prostatic, hepatic, bladder cancers	Fabregas and Herrero (1990); Durmaz et al. (2007); Bryan et al. (2010), Kumudha et al. (2015); Islam et al. (2017)
Phenolic and volatile compounds (flavonoids, isoflavonoids, phenolic acids, stilbenes, phenolic polymers, lignans)	*Caespitella pascheri, Euglena cantabrica, Phaeodactylum tricornutum*	Anticancer, antidiabetic action, lessen the risks of cardiovascular and neurodegenerative diseases, protection from biotic and abiotic stress	Li et al. (2007); Kim et al. (2009); Stengel et al. (2011); Rico et al. (2013); Lopez et al. (2015)

Stengel and his co-workers identified a trivial of the chief categories of algal pigments, particularly microalgae, which are tetrapyrroles as porphyrins (chlorophyll c), chlorophylls a and b (chlorins), carotenoids (polyisoprenoids and carotenes and xanthophylls), and open-chain tetrapyrroles (phycobilin pigments). Amongst them, the incomparable focused on pigment bunches are phycobilins as well as carotenoids, which are so far extensively utilized by ventures (Stengel et al., 2011). Specific species of algae assemble high levels of extremely very viable anti-oxidative scavenger developments, perhaps, carotenoids (lycopene, astaxanthin, zeaxanthin, β-carotene, and lutein). Algal species synthesizes all these bio-molecules and have been proven to rectify the damage caused by ultraviolet (UV)-oxidation to skin and eyes.

Substantially, isoprenoid lipophilic pigments such as carotenoids originate in non-photosynthetic organisms, higher plants, and algae (Takaichi, 2011). They curatively affect animals and humans as they own robust properties (antioxidant); hence, shielding the organisms from oxidative stresses caused by free-radicals (Sathasivam et al., 2017). More than 620 carotenoids have been perceived and assigned, yet just an uncommon was financially utilized (Guerrin et al., 2003). In the human body, carotenoids go about as provitamin-A and are generally present in the scope of 0.1–0.2% of overall microalgae (dry matter) (Christaki et al., 2011). These compounds are characterized by functional groups present. Among them, carotenes barely have hydrocarbon groups and carotenoids with hydroxyl, oxo, and epoxy groups are identified as xanthophylls. Carotenoids can be distributed into two groups: (i) oxygen-containing xanthophyll derivatives and (ii) hydrocarbon-based carotenoids.

In a commercial sense, for natural coloring materials, algal carotenoids have been used for a long time. Carotenoids are reported to possess physiological significance in metabolism functionality in humans. Therefore, among more than 400 carotenoids that are identified, only astaxanthin and β-carotene are commercialized technically (Spolaore et al., 2006). Specifically, β-carotene is transformed into provitamin-A (retinol) *in vivo*, which plays a vital part in preparations of multivitamins (Spolaore et al., 2006; Guedes et al., 2011a). Inside the body, one β-carotene molecule may be transformed into two molecules of vitamins A. This provides all the health perks of vitamin A, like boosted immune response, better eyesight, and the defense against certain cancers by scavenging destructive free radicals (Paiva and Russell, 1999).

The β-carotene is a crucial carotenoid pigment as it contains provitamin-A, an add-on in multivitamin tablets, and a supplement. It is also used as a food color in margarine, butter, and cheese (Spolaore et al., 2006). *D.*

salina (total dry matter) comprises up to 10–14% of β-carotene (Stewart et al., 2016). The latest studies have shown that enhancing the life medium conditions brings about higher productivity of β-carotene, as well as the production of 9-cis isomers (Borowitzka, 2013). Positive results have been shown on plasma lipids by the 9-cis isomer (β-carotene); thus, it hinders the growth of atherosclerosis in humans (Maria et al., 2015). An investigation performed to check the impacts of β-carotene on the cardiovascular soundness of 640 people affirmed that atherosclerosis illnesses could be secured with an appropriate expansion of different cell reinforcements and β-carotene in diet, expected that different unnecessary factors, for example, dyslipidemia, hypertension, diabetes, and smoking are monitored (Riccioni et al., 2009).

Another highly sought carotenoid is astaxanthin, which is another vital carotenoid pigment, and most numerous companies extricate it from *Haematococcus pluvialis* (Lorenz and Cysewski, 2000). It is an ordinary pigment that has antioxidant properties related to another carotenoid, for example, lutein, zeaxanthin, lycopene, as well as β-carotene or Vitamin E, C (Martins et al., 2010). This carotenoid has very sturdy properties (anti-inflammatory and antioxidant). Consequently, astaxanthin acts as protection pigment against macular degeneration, rheumatoid arthritis, cancer, reduced vision, Parkinson's disease, etc., (Nakajima et al., 2008; Hudek et al., 2014). β-carotene and Astaxanthin have captured better attention owing to their medical-benefiting attributes and their non-synthetic or 'natural' characterization when obtained from algae. Characteristic astaxanthin is the lone structure financially exchanged the human nutraceutical market. While manufactured astaxanthin could be modest than the normally extricated ones, in any case, their combination into groceries that make specific food and wellbeing privileges are restricted.

Although algal astaxanthin is classically synthetically produced and correspondingly commercially used. Furthermore, when algal biomass obtained in trivial amounts arising out of the genera *Scenedesmus*, *Chlorella*, or else *Spirulina* were mixed with cattle feed, improved the animals' immune system significantly (Guerin et al., 2003; Pulz and Gross, 2004; Plaza et al., 2009). Currently, numerous studies have been done to validate that astaxanthin addition in the human diet decrease oxidative stress, inflammation, and additional improvement in the patients' immune system having cardiovascular problems is seen (Guerin et al., 2003; Park et al., 2010; Kim et al., 2011). It was also reported that *Haematococcus* astaxanthin intake helps in reducing and preventing oxidative damage by overpowering the peroxidation ailments (Lavie et al., 2009). Distinct advantageous functions of PUFAs include antimicrobial, free radical foraging, anti-inflammation, anticancer, as well as antiviral activities. Reduced levels of asthma, chronic disruptive pulmonary illness, atherosclerosis, rheumatoid arthritis also has been seen. Next to these, further, it was observed that these lighten the signs of cystic fibrosis as well as Crohn's ailment (Stengel et al., 2011). Similarly, AA (ω-6) and DHA (ω-3) are two prominent components of the eye, nervous system, and brain tissues. Henceforth, they perform an extremely vital role in child growth and neuronal development.

DHA has been raised as an IQ-related unsaturated fat, owing to its participation with brain functioning and development during life, especially concerning cognitive and visual development (Auestad et al., 2003). A small number of researches demonstrate the improved behavior of children when they intake augmented PUFAs diet (Janssen and Kiliaan, 2014). Also, some studies have revealed that DHA and EPA have a contribution to the deterrence of hypertension, stroke, Alzheimer's, dementia, coronary heart disease, as well as hopelessness (Kris-Etherton et al., 2002; Das, 2008). Current researches tinted the anti-inflammatory properties of DHA and EPA as well (Wall et al., 2010). The nutritional supplementation of DHA and EPA (2 g day^{-1}) overpowers stimulation of interleukins (IL6, IL8) and toll-like receptor 4 in trophoblast and adipose cells quarantined from the placenta of lipids in heavy smokers (Park et al., 2010). Reports suggest that with 12–18 mg daily dose of natural astaxanthin (nASX) resulted in an upsurge in adiponectin and serum high-density lipoprotein in 61 non-obese subjects (Yoshida et al., 2010).

Moreover, algal carotenoids like β-carotene, zeaxanthin, and lutein were found to have a role against premenopausal breast cancer; however, α-carotene and cryptoxanthin exhibited success against cervical cancer. It has been correspondingly verified that lycopene pigment helps treat plus averting stomach and prostate tumor (Stengel et al., 2011). Numerous *Chlorella* strains are allegedly decent sources of β-carotene, α-carotene, lutein, antheraxanthin, astaxanthin, violaxanthin, as well as zeaxanthin. It was suggested that carotenoid extracted from *C. vulgaris* and *C. ellipsoidea* presented properties like anticancer activity by encouraging cell apoptosis in colon cancer (Cha et al., 2008). Furthermore, defense from cognitive impairment and macular degeneration has been observed in transgenic mice as soon as they were nursed β-carotene and lutein-rich extracts of *Chlorella* (Guedes et al., 2011a).

A pigment produced from *Cylindrotheca closterium*, which, when dissolved in ether, generated a fluorescent solution blue in color, has been recognized as one of the xanthophylls, i.e., fucoxanthin. It is among the greatest naturally copious carotenoids, which has been described to be active at encouraging apoptosis in colon cancer and mammalian leukemic cells. Besides, fucoxanthin is an oxygenated carotenoid showing multifunctional properties as well, such as it has antioxidant, antiobesity, antidiabetic, and anti-inflammatory properties (Maeda et al., 2005). In a particularly unique case, an *in vitro* study stated that in response to fucoxanthin, Graffi myeloid tumor cell propagation was inhibited in a dose-dependent manner (Minkova et al., 2011).

In addition to their antioxidant role, algae possess other biological metabolites related to photosynthesis with described health aids, such as chlorophyll, which is reflected as an active nutrients source for humans owing to its foretold antioxidant properties and other health aids (Simonich et al., 2007, 2008; Bai et al., 2011). The intake of chlorophyll upsurge bile secretion and promotes the liver (Bishop and Zubeck, 2012). It also has anticarcinogenic, antimutagenic, and antigenotoxic properties (García et al., 2017).

Chlorophyll is a renowned phytonutrient, as well as a detoxifying agent. It enhances carbohydrates, lipids, and protein metabolism in the human body and possesses affirmative effects in human reproduction (Seyidoglu et al., 2014). It comprises of chlorophyllin (CHL), which is easily engrossed by the human body. A study stated that CHL addition to diet inhibits cancer progression as it aims at various carcinogens pathways (Nagini et al., 2015). During a study by Das et al. (2014), CHL introduction in mice helped sluggish lung cancer progression. Further, it was established that because the CHL can cross the blood-brain-barrier, it will possibly be effective in human use (Das et al., 2014). Dietary intake of chlorophyll in animals was stated to avert multi-organ carcinogenesis in rainbow trout and rats (Simonich et al., 2007, 2008).

Generally, *Spirulina platensis* is an outstanding source for vitamin E and B, co-enzyme, and antioxidants (Plavsic et al., 2004). It is moreover a water-soluble phycocyanin (PC) source, with observed antioxidant properties (Cohen et al., 1993; Romay et al., 2003; Benedetti et al., 2004). Water-soluble pigment phycoerythrin (PE) is synthesized by the *Porphyridium cruentum*, which has renowned antioxidant properties (Spolaore et al., 2006; Minkova et al., 2011). These PBPs are hydrophilic protein multiplexes that arrest energy from light and therefore support in the process of photosynthesis of red microalgae as well as cyanobacteria (Sonani et al., 2016). The key producers of PC are *Spirulina* sp., *A. flos-aquae*, and *Porphydrium* sp. (Gouveia et al., 2007). A luminous blue-colored PBP, PC, mined out of *Spirulina* sp., is used as an ordinary dye. It is moreover used in confectionery, chewing gum, popsicles, dairy products, soft drinks, and wasabi (Spolaore et al., 2006). Nevertheless, in the nutraceutical industry, they are commercialized as anti-oxidative, anti-inflammatory, neuroprotective, antiviral, as well as hepatoprotective agents (Sekar and Chandramohan, 2008; Bleakley and Hayes, 2017; García et al., 2017).

5.3.2   PROTEINS, PEPTIDES, AND AMINO ACIDS

Several algal genera, for example, *Scenedesmus*, *Chlorella*, *Spirulina*, *Dunaliella*, *Chlamydomonas*, *Euglena*, *Oscillatoria*, as well as *Micractinium*, comprise of proteins in reasonable quantities (above 50% of their DW (distilled water)) (Becker, 2007). Thus, algae serve as a fascinating protein-luscious feedstock for nutraceutical industries. Indeed, even the left-over algal biomass after extraction of oil can be reasonably reused for bioactive proteins (Dewapriya and Kim, 2014). As stated by WHO/United Nations University/FAO, *Spirulina*, and *Chlorella* were suggested for animal and human nutritional consumption, owing to its extraordinary nutritional quality in essential amino acid content (Chronakis and Madsen, 2011). Algae is recognized as a substitute rich-protein source which may encounter the malnourished population necessities (Christaki et al., 2011) by feasting algae employing a nutritional addition through tablets, pills, paste or powder (Pulz and Gross, 2004). Though, in current years, proteins derived from algae have also been fused in other food products (Liang et al., 2004). Hence, because of added nutritional properties and high content of protein, *Spirulina* sp. are extensively consumed all over the world.

Various EAAs are present in high content in algae, such as isoleucine, leucine, valine, and lysine. All these amino acids represent about 35% of the general fundamental amino acids in muscle protein of human. In light of a few nourishing prerequisite investigations, 10 amino acids were chosen as basic amino acids for the creature and human sustenance, together with arginine, isoleucine, histidine, leucine, lysine, threonine, phenylalanine, methionine, tryptophan just as valine. Fascinatingly, all the above crucial amino acids may be traced out of algal protein resulting from *Dunaliella bardawil*, *Scenedesmus obliquus*, *Chlorella vulgaris*, *Arthrospira maxima*, as well as *Spirulina platensis*. One of the facts related to commercial interest is that, as logical investigation concerning the practical, healthful, and nutraceutical properties of algal proteins has been increased, the algal protein market has begun to progress throughout the most recent 5 years.

A study was conducted in which shrimps were enhanced by defatted *H. pluvialis*. They saw that the algae-supplemented shrimps remained a lot more beneficial contrasted with the benchmark group served with a viable diet (Ju et al., 2012). Certain cyanobacteria, such as *Porphyridium* sp. and *S. platensis*, were also stated to partake as anti-inflammatory, hepatoprotective, immune-modulating, antioxidant, and anticancer roles. In an investigation, it was discovered that when algae are presented to pressure conditions, for example, supplement starvation and UV radiation (Gage et al., 1997; Sunda et al., 2002; Jäpelt and Jakobsen, 2013), a metabolite of methionine, dimethylsulphoniopropionate (DMSP) (Gage et al., 1997) shows an upsurge in its focus. In microalgae, DMSP, as well as dimethylsulfide (its enzymatic cleavage product), have been established to exhibit antioxidant properties (Sunda et al., 2002).

Moreover, marine proteins have exclusive and pertinent properties on the cardiovascular system, for instance, Angiotensin I (AI) inhibitory and antihypertensive properties (Suetsuna and Chen, 2001; Sheih et al., 2009; Ejike and Collins, 2017). Algal protein bunch incorporates various bioactive peptides, 2–20 amino acids proteins that can go through the layer of cells, and thus offering them by a hormone-like function in the human body (Fan et al., 2014). A 7.5 μM *Chlorella vulgaris* trivial peptide displays a scavenging effect contrary to cellular damage instigated by hydroxyl radicals and has protective activity on deoxyribonucleic acid (DNA) because of its antioxidant activity (Sheih et al., 2009). Furthermore, this peptide has a role in gastrointestinal enzyme-confrontation, but no cytotoxic outcome was perceived on normal human fibroblasts (Wi38 cells) *in vitro* (Sheih et al., 2009). In a current study, it was found that apoptosis of lung cancer cell line is activated by a glycoprotein mined from the *Alexandrium minutum*, deprived of some toxic result on Wi38 (normal human cells) (Galasso et al., 2018). Low atomic proteins, lectins, have been found in macroalgae and microalgae, which are engaged with various natural cycles, for example, cell-cell correspondence, have microbe connections, and acceptance of apoptosis (Takebe et al., 2013; Mustopa et al., 2016).

Plus, marine peptides utilize a considerable amount of other advantageous impacts on wellbeing, for example, antihypertensive, antimutagenic, against tyrosinase, anticoagulant, and some more (Kim and Kim, 2013; Vo et al., 2013; Rizzello et al., 2016). Thus, curiosity in marine bioactive peptides increases (Fan et al., 2014; Caporgno and Matthys, 2018). Bioactive peptides can be described as short fragments of protein with definite amino acid sequences that influence the metabolism of cells and eventually on wellbeing

(Kitts and Weiler, 2003; Walsh and FitzGerald, 2004). These peptides have drug- or hormone-corresponding action that ultimately modifies functional role by binding to definite receptors on target cells, heading to the initiation of physiological retorts (Fitzgerald and Murray, 2006). Furthermost bioactive peptides are shaped and kept in a passive structure, which is set off when required. It is very much reported that, other than their rudimentary nourishing job, endless food proteins contain peptide successions capable of altering specific physiological capacities (FitzGerald et al., 2004).

They likewise go about as signaling particles and show imperative functions in pathogenesis and physiological roles. The diatom *Navicula incerta* has the tetramer EDKR that holds antioxidative events contrary to hydroxyl as well as superoxide radicals and 2,2-diphenyl-1-picrylhydrazyl (Kang et al., 2011), although additional peptides, for instance, NIPP-2 (VEVLPPAEL) and NIPP-1 (PGWNQWFL), quarantined out of the *Navicula incerta* (diatom) shows the inhibitory influence of hepatic fibrosis on Transforming Growth Factor-β1 that induce human hepatic stellate cells (LX-2) activation (Kang et al., 2013). A microalga *Diacronema lutheri* gained peptide MPGPLSPL, persuades myofibroblasts differentiation in human skin fibroblasts cells after fermentation with *Candida rugopelliculosa* (Ryu, 2011). The cyanobacterial peptides, from *Lyngbya* sp., denoted as Lyngbyastatins, were stated to have anti-chymotrypsin (Matthew et al., 2007) and anti-elastase (Matthew et al., 2007; Taori et al., 2007) activities. Another peptide, sequestered from *Lyngbya* sp., called Kempopeptins A and B, have anti-chymotrypsin and anti-elastase activities as well (Taori et al., 2008). Pompanopeptins A, a novel cyclic peptide, quarantined out of *Lyngbya confervoides*, has anti-trypsin roles (Matthew et al., 2008).

Another group of bioactive molecules that are synthesized by microalgae are polar, water-soluble compounds, i.e., mycosporine-like amino acids (MAAs) having a low molecular weight (Llewellyn and Airs, 2010) that protect cells from the damage induced by UV rays because of the presence of the double bond, providing them photostability and antioxidant properties (Lawrence et al., 2017).

The additional forthcoming area is the therapeutic recombinant protein development from algae that offers relatively easy and cost-effective methods that are appropriate for the biotechnology business coordinated to other open protein articulation frameworks and does exclude costly yeast and bacteria-based bioreactors. Also, algal-based bioreactors use recombinant protein to grant post-translational and post-transcriptional changes of biosynthesized proteins (Dewapriya and Kim, 2014). Additionally, different recombinant proteins developed from *C. reinhardtii*, for example, human glutamic acid decarboxylase 65 as well as HSV8-scFv and HSV8-lsc (a human IgA anti-herpes monoclonal antibody), a crucial autoantigen for identifying type 1 diabetes (Gong et al., 2011).

5.3.3   POLYUNSATURATED FATTY ACIDS (PUFA)

In the nutraceutical industry, algae play a crucial role as they are reflected as fair wellsprings of unsaturated fats having medical advantages. Fatty acids are considered the structural gears of glycolipids, phospholipids, and triacylglycerols (TAG) (Wall et al., 2010); besides, a huge pool of FA is required by all living organisms for health and survival. Algal unsaturated fats are mostly ironic long-chain polyunsaturated fatty acids (PUFAs) just as they have a more prominent level of unsaturation. PUFA are EFA explicitly required for fitting cells working as they comprise of additional 18 carbon molecules and beyond one-two fold bond in their arrangements, for example, AA (20:4 omega-6), γ-linolenic corrosive (GLA) (18:3 omega-6), DHA (22:6 omega-3) as well as EPA (20:5 omega-3) (Ratledge, 2010; Borowitzka, 2013).

Under usual conditions, both EPA and DHA are synthesized by the body out of ALA; however, this process exerts pressure on the body's schemes while limiting the ALA usage somewhere else in the body. Algae can work as a feedstock for 'veggie lover' wellsprings of the above expressed PUFAs (Martins et al., 2013). The quantity and type of PUFAs exist in algae; in any case, they differ concerning culture conditions and species type. Interestingly, humans and higher organisms cannot produce PUFA, and therefore they require nutritional supplements of PUFA for retaining decent wellbeing (Wall et al., 2010). They are an obligatory sound eating regimen part as they have a prominent function in neurodegeneration and neural turn of events (Janssen and Kiliaan, 2014).

Furthermore, PUFAs have a significant part in the cell and tissue metabolism, together with oxygen and electron transport, membrane fluidity regulation, and thermal adaptation (Funk, 2001). They also have a role in lowering lipid levels affecting hyperlipidemia and further reducing the risk of atherosclerosis and heart diseases. Therefore, ω-3 PUFA treatment has been a focal point of investigation for the preclusion of cardiovascular and adipose tissue of treated pregnant women (Haghiac et al., 2015).

Generally, ω-6 PUFA is GLA, which is a significant precursor used for prostaglandin synthesis. Intake of GLA has been associated with a reduction of levels of low-density lipoprotein (LDL) in formerly detected hypercho-lesteremic patients, mitigation of symptoms of pre-menstrual syndrome, plus in the enhancement of atopic eczema (Horrobin, 2000). Particularly, the cyanobacterium '*Spirulina*' is a rich wellspring of GLA and along these lines assumes a key function in the therapy and avoidance of various sicknesses like cardiovascular arrhythmia, pulse, prostate malignancy, atherosclerosis, liver disease, and glioma, just as the immune system illnesses, for example, rheumatoid joint pain, intense respiratory misery condition, and diabetes including premenstrual disorder and skin infections. It likewise helps in the best possible working of the cerebrum and boosts in susceptibility (Rakesh and Harikumar, 2005). Consequently, EPA, DHA, and GLA are especially compelling for social wellbeing (Richmond, 1986; Kerby et al., 1987).

At present, there is an expanding interest for EFAs which are from veggie lover sources. A considerable amount of algal strains have been chronicled as remarkable PUFAs sources. DHA-rich oil (equal to 39% of total fatty acids) production by *Crypthecodinium cohnii* is one such example, which is widely used in baby formulation (Borowitzka, 2013). Others are *Pavlova salina* and *Isochrysis galbana*, which also produces substantial DHA quanti-ties (Guedes et al., 2011). DHA and EPA produced by *Schizochytrium* strains are presently used as an adult nutritive addition in healthy diets, nutraceuti-cals, maricultural products, as well as animal feeds. The marine green algae *Chlorella minutissima* produces EPA in high amounts, which can further be increased by increasing salinity and reducing temperature (Seto et al., 1984). Other useful microalgae that produce EPA include *Monodus subterraneus* (about 24% EPA of its overall content of fatty acid) and *Nannochloropsis* sp. (up to 40% of the overall content of fatty acid) (Cohen, 1999; Borowitzka, 2013; Sharma and Schenk, 2015).

Additionally, *Dunaliella salina* PUFAs for instance, linolenic acid, oleic acid, and palmitic acid, could subsidize equal to 85% of its overall matter of fatty acid (Herrero et al., 2006) while the finest source of γ-linolenic acid (GLA) (Tanticharoen et al., 1994), palmitoleic lauric acid, and DHA (Tokuşoglu and Unal, 2003) is *Spirulina platensis* (also known as *Arthro-spira platensis*). Green algae *Parietochloris incisa* has elevated amounts of AA and overall lipid contents (Solovchenko et al., 2008). The maximum production of comparatively unusual existence of GLA has attained out of *Nostoc* strains as well as *Spirulina* strains. Hence, WHO, and FAO reports suggest that a fair eating routine proportion of PUFAs: immersed unsaturated fat is one overhead 0.4, and particularly microalgae comprise the endorsing scope of these proportions, specifically, 1.65 to 3.71 (Milovanovic et al., 2012).

5.3.4 STEROLS

Sterols, mainly phytosterols (C28 and C29 sterols), are well-known for their key role in pharmaceutical and nutraceutical industries as they are precursors of certain bioactive molecules, for example, vitamins; furthermore, in humans, sterols have unveiled to reduce total and LDL cholesterol levels by inhibiting the absorption of cholesterol from the intestine. Moreover, rather than their cholesterol-lowering properties, sterols seem to have anti-inflammatory and antiatherogenic activity as well as antioxidant plus anticancer events (Francavilla et al., 2010). Scientists highlight the significance of phytosterols, which may merely be taken up solely by eating regimens. Thus, given their cholesterol-lowering activities, phytosterols are supplemented to various marketable foodstuffs such as yogurts, milk, and margarine. Also, sterols partake a starring function in nervous system irregularities, for example, amyotrophic lateral sclerosis, autoimmune encephalomyelitis and so on (Francavilla et al., 2012). Likewise, phytosterols rich animal feed may interrupt or prevent the commencement of Alzheimer's disease in some creature models (Francavilla et al., 2012).

Algae consist of an elevated sterol concentration (Fábregas et al., 1997; Ahmed et al., 2015) and exhibit a huge variety of phytosterols, for instance, brassicasterol, stigmasterol, as well as sitosterol (Fábregas et al., 1997; Volkman et al., 1999). Few algal species (for example, *Tetraselmis* sp., *Pavlova* sp., and *Nannochloropsis* sp.) comprise a combination of 10 or even additional phytosterols (Volkman et al., 1999). The composition of sterols varies based on algae strain plus modified by temperature, light intensity, or development stage (Fábregas et al., 1997). These properties make algae a promising source of phytosterols (mostly unknown), for medical advantages. Hence, valuable impacts of algal-derivative of phytosterols that were experimentally stated are anti-inflammatory, anticancer, anti-cholesteroligenic, or antioxidant properties (Khanavi et al., 2012; Hwang et al., 2014). The anti-metastatic role was also described for fucosterol (earthy colored algal phytosterol) (Khanavi et al., 2012; Kim et al., 2013). Isofucos-terol and fucosterol are likewise present in algal species [*Chrysomeris* sp., *Chrysoderma* sp., *Giraudyopsis* sp., as well as *Chrysowaernella* sp. (Billard et al., 1990)], though their bioactivity has not yet been clarified. Phytosterols can also turn into secondary messengers in a hormone-like style, prompting pertinent cellular changes, for example, neurotransmission, and development. In a study, a prominent quantity of phytosterols, such as sitosterol, brassicasterol, stigmasterol, and many more, was observed to show *in vivo* neuromodulatory action (Francavilla et al., 2012).

5.3.5 VITAMINS

Vitamins are forerunners of significant enzyme cofactors that are obligatory for crucial metabolic purposes and exhibit robust antioxidant action, hence are extremely important for health. As humans cannot endogenously synthesize maximum vitamins, they need to be consumed orally. Since algal products are rich in vitamins, a study was conducted (Fabregas and Herrero, 1990) for determining the content of vitamins in numerous algal species. Algae epitomize an uncharted basis of approximately all renowned vitamins [A, B1 (thiamin), B2 (riboflavin), B3 (Niacin), B6 (pyridoxine), B12 (cobalamin), C (ascorbic acid), E (tocopherols), pantothenic acid and folic acid], which are present at extraordinary levels as compared to other monotonous foods (Becker, 2004).

Certain algae comprise both lipid- and water-soluble vitamins in elevated amounts. Accordingly, consumption of these algae may encounter the obligation of a few vitamins in animals as well as humans (Fabregas and Herrero, 1990). Vitamin B6, B9, plus B12 fundamentally control the enzymes of mitochondria to uphold the one-carbon transferal cycle of amino acid metabolism in mitochondria (Depeint et al., 2006). The researchers observed that algae, particularly microalgae, contained vitamin B1, provitamin A, folic acid, and vitamin E in elevated concentration compared to other conventional nutrient sources. Also, vitamin B12, B2, E, as well as provitamin A was synthesized by *Dunaliella tertiolecta*. Moreover, it was observed that an outstanding vitamin B1, vitamin B3, vitamin B5 (thenic acid), vitamin B6, and vitamin C source is *Tetraselmis suecica* (Fabregas and Herrero, 1990) while *Chlorella* sp. contains vitamin B7 (biotin) in great amounts. An experiment was conducted, which concluded that *Chlorella* strains are also an ironic source of vitamin B12 (around 9–18%) (Islam et al., 2017). Further, it was reported that even though *Spirulina* species are proficient in producing vitamin B12, better bioavailability was seen in *Chlorella* species (Watanabe et al., 2002).

Additionally, red microalga *Porphyridium cruentum* was found to have vitamin E (tocopherol) content in a significant amount, particularly, α-tocopherol and γ-tocopherol contents (Durmaz et al., 2007). *Macrocystis pyrifera* (Kelp), have α-tocopherol at higher concentration (Ortiz et al., 2009; Skrovánková, 2011). Furthermore, the β-carotene levels in *Gracilaria chilensis* and *Codium fragile* were more than that of carrots (Ortiz et al., 2009). Interestingly, algal vitamin content also differs with external forces, such as nutritional status of the culture, light (in terms of intensity and spectrum) (Smerilli et al., 2017), genotype, and growth phase (Roeck-Holtzhauer et al., 1991; Brown and Miller, 1992; Brown et al., 1999).

5.3.5.1   PRO-VITAMIN A

Algae can produce precursors as it cannot produce vitamin A, i.e., α- and β-carotene. Although records are previously existing on carotenoids in algae, there is still scantiness of apocarotenoids (Ahrazem et al., 2016). The biosynthesis of apocarotenoid begins with the activity of carotenoid cleavage dioxygenases, also comprises intriguing biotechnological compounds, for instance, retinol (Ahrazem et al., 2016). This compound inhibits the development and growth of different kinds of tumors, for example, breast, skin, gastrointestinal, lung, prostatic, oral cavity, hepatic, and bladder cancers (Niles, 2000; Altucci and Gronemeyer, 2001; Arrieta et al., 2010; Bryan et al., 2010; Siddikuzzaman et al., 2010).

5.3.5.2   VITAMIN B12

This vitamin is water-soluble, existing in particular macroalgae and microalgae (Croft et al., 2005). Among people ensuing strict vegan or vegetarian diets, deficiency of this vitamin is common (Watanabe et al., 2014). Albeit cobalamin from few algae does not appear to be bioavailable (Dagnelie et al., 1991), but in some microalgae such as *Pleurochrysis carterae* (coccolithophid) (Watanabe et al., 2002) or *Chlorella* sp. (Kumudha et al., 2015), vitamin B12 seems to be bioavailable, and used as nutrition supplements. Furthermore, vitamin B12 may act on histone methylation and DNA repair mechanism, and increased levels of vitamin B12 and folate may reduce breast cancer risk (Gruber, 2016).

5.3.5.3   VITAMIN C

It is also a water-soluble vitamin with antioxidant features required by humans for numerous compounds (Padayatty et al., 2003). Also, it has been described as a regulator (Knowles et al., 2003), a chief microenvironmental Hypoxia-Inducible Factor 1 driver of tumor angiogenesis and carcinogenesis. Further, it acts on the extracellular matrix, impaction on collagen deposition, also biosynthesis (Sharma et al., 2008) and has beneficial health effects, comprising prevention of atherosclerosis and cancer. Vitamin C also prevents some severe infections (e.g., tuberculosis) by acting as an immunomodulatory agent (Boyera et al., 1998; Nunes-Alves et al., 2014).

5.3.5.4 VITAMIN D

This vitamin exists in several forms, i.e., D1 to D5, but in humans' chief forms are D2 and D3, which are generally associated with calcium metabolism and absorption, a vital cycle for homeostasis and bone health. Several reports show the vitamin D health benefits in preventing cancer and anti-neurodegenerative effects (Lappe et al., 2007; Annweiler et al., 2012; Feldman et al., 2014). Even though below par documentation, it is eminent that algae consist of vitamins D2 and/or D3 (Atsuko et al., 1991; Rao and Raghuramulu, 1996), together with provitamin D3. In addition to their renowned calcium-related functions, this vitamin has chemoprevention activities through immune-modulatory and antiproliferative effects on cancer cells *in vitro* as well as *in vivo* growth (Giammanco et al., 2015). Vitamin D also has activities related to cancer chemoprevention, by hindering cell proliferation through Wnt/β-catenin-signaling pathways (Larriba et al., 2013), delaying cell cycle progression (Colston and Hansen, 2002), curbing the insulin growth factor expression (Teegarden and Donkin, 2009), and raising autophagy or apoptosis (Giammanco et al., 2015).

5.3.5.5 VITAMIN E

Vitamin E (tocopherols and tocotrienols) act as a liposoluble antioxidant by shielding membrane lipids from oxidative destruction; subsequently, they can prevent the lipid peroxidation propagation acting as chain-breaking molecules. They also obstruct the reactive oxygen species (ROS) production and inhibit the oxidation of LDL, a procedure identified to show part in atherosclerosis growth, further enhances vascular health and endothelial function and lessens vascular damage (Corina et al., 2019). The pathway of phosphoinositide 3-kinase, which has a part in the action of vitamin E, inhibits the growth of prostate cancer cells as well (Ni et al., 2005). A study shows this vitamin can reduce the pancreatic cancer risk in mice (200 mg/kg) (Husain et al., 2017), thus also have a chemoprotective role (Kline et al., 2007).

5.3.6 PHENOLIC AND VOLATILE COMPOUNDS

These compounds are secondary metabolites that show protective activities in algae against external factors (biotic and abiotic stresses), for example,

UV radiation, herbivorous foraging, and metal corruption (Smerilli et al., 2017). Numerous certainly occurring phytochemicals are made up of phenolic compounds. Phenolic compounds are alienated into flavonoids, isoflavonoids, phenolic acids, stilbenes, phenolic polymers, and lignans (Manach et al., 2004). Generally, these compounds in humans hail from fruits, vegetables, tea, coffee, or wine. Moreover, one-third of the overall intake comprises phenolic acids, and lingering two-thirds are flavonoids (Manach et al., 2004). The core bioactivity allied to algal phenolic compounds is antioxidant activity (Kumar et al., 2008). Besides, these compounds also exhibit a wide variety of biological actions, for instance, anticancer, anti-inflammatory, anti-allergic, anti-aging, antimicrobial, and anti-diabetes roles (Li et al., 2007). It has additionally been guaranteed that these compounds may reduce the dangers of neurodegenerative and cardiovascular ailments (Kim et al., 2009).

Macroalgae have extensive data on polyphenols (Freile-Pelgrin and Robledo, 2014). Flavonoids and phenolic acids have also been exposed in varied microalgae (*Spirogyra* sp., *Ankistrodesmus* sp., *Caespitella pascheri*, and *Euglena cantabrica*) and cyanophytes, emphasizing their extraordinary unpredictability among strains (Jerez-Martel et al., 2017). Phenolic acids were observed in a great amount in two dissimilar algal species: diatom *Phaeodactylum tricornutum* and the green alga *Dunaliella tertiolecta* (Rico et al., 2013; Lopez et al., 2015). Furthermore, it has been detected that numerous algal species possess the overall content of phenolics alike or greater than quite a few fruits and vegetables (Manach et al., 2004; Li et al., 2007). Preceding studies correspondingly displayed that phenolic-rich extract of *Spirulina maxima* guards' liver from carbon tetrachloride generated lipid peroxidation (El-Baky et al., 2009). *Spirulina* holds antimicrobial action because of the occurrence of phenolic compounds, for example, tetradecane, and heptadecane (Ozdemir et al., 2004). Besides, α- and β-ionone, phytol, β-cyclocitral, as well as neophytadiene are impulsive compounds having antimicrobial functions that are shared in *D. salina* (Herrero et al., 2006).

The profile and content of phenolic compounds in marine algae fluctuate with the species. Perhaps, in marine brown algae, phlorotannins (phlorethols, fucols, fuhalols, fucophlorethols, and sulphited and halogenated phlorotannins) are the chief phenolic compounds. Certain authors have described the impending protection of phlorotannins in contradiction of oxidative degradation and its probable role in the treatment and prevention of free radical associated diseases (Kumar et al., 2008). Particular algal phenolic substances spotted in red macroalgae (for example, *Porphyra* genus) shown anti-inflammatory activity, such as hesperidin, rutin, morin, catechol, caffeic acid, epigallocatechin, and catechin in gallate. Records also enlightened the anti-inflammatory use and potential mechanism of phenolic substances in nitric oxide synthase concerned diseases (Kumar et al., 2008). Another intriguing example is shown by the marennin (Gastineau et al., 2015), which owns fascinating bioactivities: antibacterial, antioxidant, antiviral, and preventing actions on the growth of tumor cell (lung and kidney carcinoma), melanoma (Carbonnelle et al., 1999). This class of phenolic compound synthesized by the diatom *Haslea ostrearia* (Lebeau et al., 2002) needs further studies to explore their products in additional algae.

5.4   CURRENT SCENARIO

Due to the leading-edge, comprehensively accessible electronic and correspondence skills, customers are knowledgeable about logical indications of progress than previously. They are turning out to be slowly additional wellbeing cognizant. In current years, issues related to health have been increasing, plus there is a thriving interest in the consumption of natural nutritional supplements that improve health, which is further increasing the demand for 'nutraceuticals' or 'superfoods.' Precisely, the last span has seen an increment in the numerous scientific researches, linking the usage of nutraceutical with the prevention of chronic illness progression and expansion. This has driven novel examination possibilities for evaluating assorted hotspots for solid, useful nourishments creation (Seyidoglu et al., 2014). Similarly, healthcare costs are rising, and both the younger generations and the aging population are picking alternative anticipatory measures to avoid the costs related to conventional treatments (Frost and Sullivan, 2011).

Among the algal-determined items, the dried *Spirulina* sp. creation is the greatest with just about 12,000 tons for each year, trailed by *Chlorella* sp. (5000 tons per year), followed by *Dunaliella salina* (3000 tons for carotene), *A. flos-aquae* (1500 tons for diet), *Haematococcus pluvialis* (700 tons for astaxanthin), *C. cohnii* (500 tons of DHA) just as *Shizochytrium* (20 tons of DHA) for each year, exclusively (García et al., 2017). The overall nutraceuticals industry must have devoured the significant volume of *H. pluvialis* normal astaxanthin in the year 2017, which is practically 54.8%, and it is conceivable to spread 190 metric tons by 2024. The worldwide carotenoids souk is extended to be evaluated at the United States Dollar (USD) 1.24 billion each 2016 besides is required to arrive at USD 1.53 billion by 2021 (Kobylewski and Jacobson, 2010).

Further, the compound annual growth rate (CAGR) of algae-based items is likely going to cross 5.2%, plus the souk worth will hold at USD 44.6 Billion by 2023, states the newly distributed Credence Research market report on algal items (https://www.credenceresearch.com/report/algae-products-market). Because of the increasing demand for microalgae, mainly *Spirulina*, for natural colorants and in cosmetics applications, the CAGR of the international *Spirulina* souk was projected to be nearly 10% with a predictable estimation of USD 2000 million constantly 2026. In contrast, market trends of *Chlorella* ingredients are predicted to achieve a CAGR of 25.4%, attaining USD 700 million till 2022. The international market for common astaxanthin sources in beauty care products, aquafeed, beverages, and food, and nutraceuticals additionally displays the capability of exploiting algae to offer the necessities of the separate interest of the market (García et al., 2017).

5.5 FUTURE PROSPECTS

Nutraceuticals are presently getting universal appreciation owing to their considerable therapeutic and nutritional value. Thus, in the current chapter, information has been delivered about the importance of algal bioactive metabolites as pharmaceutical products or functional food ingredients. Considering all the facts, in the last span, algae have gained importance for the nutraceuticals production at an industrial scale. Furthermore, another advantage of using algae is their potential to attain great cell density at optimum conditions and rapid growth rate. The recognition of algal biomass and its metabolites has given rise to the development of numerous innovative food supplements or nutraceuticals in the recent past. Notwithstanding, from the biotechnology perspective, even though the algal biotechnology area is rising quicker than any time in recent memory, the guides from up-and-coming algae are inconsistent. They are completely subject to innovative turns of events and modern interests as the upside of algae is the event of plenty of bioactive metabolites in the cell, made out of the climate and high species-subordinate grouping of essential and optional metabolites of concern. Additionally, different optional metabolites from algae have been appeared to hold possible restorative criticalness. These outcomes show further encouraging future algal biotechnology advancement.

One of the perspectives to be improved is the screening of algal biodiversity, finding the useful characters of algae that control their capability to biochemically change formerly in this manner produce bioactive particles. The primary test is the improved recuperation activity while doing practical creation. The complete perception of algae-based bioprocesses requests the improved presentation of algae through metabolic designing, hereditary designing, measure designing, improvement of strain for better return and quicker development involving the inclusive selection of current algal species for fresh metabolites alongside separation of unique, overwhelming algal species having the possibility for original metabolites creation. Additional research is also needed for comprehensive screening of bioactive metabolites synthesized by algae to handle environmental stress. A number of them are comparatively well known while others are less recognized or evidence on their bioactivity is limited.

When qualitative as well as quantitative data, will be obtainable on the valuable functions of bioactive pools in algae, investigation on their interactions along with their molecular pathways has to be progressive. Autonomously, their roles and function in the cells and their cellular localization have to be illustrated.

To be a promising and competent functional drinks/food ingredient or nutraceutical, an alga needs to deliver bioactivity to the food or supplement, introducing an increased antioxidant action and/or capability to trigger biological cycles of repair, protection, as well as immunological reactions. A device to enhance supplements is utilizing a diverse blend of a range of compounds from various algae. Another aspect that needs to be additionally amended is concerned with the flavor of the dietary supplements and functional foods/drinks mined from algae along with maintenance of bioactivity capacity with time in the course of conservation.

5.6   CONCLUSION

As a prominent basis of nutraceuticals, algae are the rapid rising section of today's nutrition plus health sectors because it has a rich source of disease-overpowering and health-endorsing metabolites. Besides, algae augment the nutritional worth of human as well as animal diets as it also has multiple groups of amino acids, antioxidants, fatty acids, carotenoids, and other significant metabolites. In recent years, the growth of algae in mass culture and its commercial production has improved. Several diseases in humans and animals are cured with the help of the medical specialty of algae as it has effective metabolites, which can act as hepatoprotective, anti-carcinogenic, antihypertensive, and antioxidative agents. In developing nations, food/ nutraceutical products derived from algae have proven to slow down the rate of malnourishment.

Moreover, studies in recent decades clearly show that the products derived from algae are being utilized exponentially. Nevertheless, the vital obstacle regarding the success of nutraceuticals derived from algae is the absence of inducements for the production of algal-based foods connected with unfamiliarity about its benefits on health. If somehow, we can settle these obstacles, it will profit the health plus wellbeing of humans by incorporating algae in the nutraceutical and food industry. Furthermore, it will challenge issues associated with climate variation, also outplay the nutrition need for expanding inhabitants of the world. To conclude, it can be supposed that the evolution in the investigation of algal nutraceuticals is right towards the quantum leap in the biotechnology industry. In the coming years, algae will further mark a substantial alteration in fields of pharmaceutical, cosmetics, energy, as well as nutraceutical industry because of its versatility and massive potential. Therefore, this chapter accentuates the significance of algal metabolites for future developments in the nutraceutical industries.

KEYWORDS

- anti-carcinogenic
- anti-inflammatory
- antioxidant
- astaxanthin
- carotenoids
- cyanobacteria
- docosahexaenoic acid
- metabolites
- microalgae
- nutraceutical
- polyunsaturated fatty acids
- sterols
- vitamins

REFERENCES

Abdelaziz, A. E., Leite, G. B., & Hallenbeck, P. C., (2013). Addressing the challenges for sustainable production of algal biofuels: I. Algal strains and nutrient supply. *Environmental Technology, 34*(13–16), 1783–1805.

Abe, K., Hattori, H., & Hirano, M., (2007). Accumulation and antioxidant activity of secondary carotenoids in the aerial microalga *Coelastrella striolata* var. *multistriata*. *Food Chemistry, 100*, 656–661.

Agyei, D., & Danquah, M. K., (2011). Industrial-scale manufacturing of pharmaceutical-grade bioactive peptides. *Biotechnology Advances, 29*(3), 272–277.

Ahmed, F., Zhou, W., & Schenk, P. M., (2015). *Pavlova lutheri* is a high-level producer of phytosterols. *Algal Research, 10*, 210–217.

Ahrazem, O., Gómez-Gómez, L., Rodrigo, M. J., Avalos, J., & Limón, M. C., (2016). Carotenoid cleavage oxygenases from microbes and photosynthetic organisms: Features and functions. *International Journal of Molecular Sciences, 17*(11), 1781.

Altucci, L., & Gronemeyer, H., (2001). The promise of retinoids to fight against cancer. *Nature Reviews Cancer, 1*(3), 181–193.

Annweiler, C., Rolland, Y., Schott, A. M., Blain, H., Vellas, B., Herrmann, F. R., & Beauchet, O., (2012). Higher vitamin D dietary intake is associated with lower risk of Alzheimer's disease: A 7-year follow-up. *Journals of Gerontology Series A: Biomedical Sciences and Medical Sciences, 67*(11), 1205–1211.

Apt, K. E., & Behrens, P. W., (1999). Commercial developments in microalgal biotechnology. *Journal of Phycology, 35*(2), 215–226.

Arrieta, O., González-De, L. R. C. H., Aréchaga-Ocampo, E., Villanueva-Rodríguez, G., Cerón-Lizárraga, T. L., Martínez-Barrera, L., Vázquez-Manríquez, M. E., et al., (2010). Randomized phase II trial of all-trans-retinoic acid with chemotherapy based on paclitaxel and cisplatin as first-line treatment in patients with advanced non-small-cell lung cancer. *J. Clin. Oncol., 28*, 3463–3471.

Atsuko, T., Toshio, O., Makoto, T., & Tadashi, K., (1991). Possible origin of extremely high contents of vitamin D3 in some kinds of fish liver. *Comparative Biochemistry and Physiology Part A: Physiology, 100*(2), 483–487.

Auestad, N., Scott, D. T., Janowsky, J. S., Jacobsen, C., Carroll, R. E., Montalto, M. B., Halter, R., et al., (2003). Visual, cognitive, and language assessments at 39 months: A follow-up study of children fed formulas containing long-chain polyunsaturated fatty acids to 1 year of age. *Pediatrics, 112* (3), 177–183.

Bai, M. D., Cheng, C. H., Wan, H. M., & Lin, Y. H., (2011). Microalgal pigments potential as by products in lipid production. *Journal of the Taiwan Institute of Chemical Engineers, 42*(5), 783–786.

Becker, E. W., (2004). 18 microalgae in human and animal nutrition. In: Richmond, A., & Becker, W., (eds.), *Handbook of Microalgal Culture: Biotechnology and Applied Phycology* (pp. 312–351). Blackwell Science, Oxford.

Becker, E. W., (2007). Micro-algae as a source of protein. *Biotechnology Advances, 25*(2), 207–210.

Begum, H., Yusoff, F. M., Banerjee, S., Khatoon, H., & Shariff, M., (2016). Availability and utilization of pigments from microalgae. *Critical Reviews in Food Science and Nutrition, 56*(13), 2209–2222.

Benedetti, S., Benvenuti, F., Pagliarani, S., Francogli, S., Scoglio, S., & Canestrari, F., (2004). Antioxidant properties of a novel phycocyanin extract from the blue-green alga *Aphanizomenon flos-aquae*. *Life Sciences, 75*(19), 2353–2362.

Billard, G., Dauguet, J. C., Maume, D., & Bert, M., (1990). Sterols and chemotaxonomy of marine chrysophyceae. *Botanica Marina, 33*(3), 225–228.

Bishop, W. M., & Zubeck, H. M., (2012). Evaluation of microalgae for use as nutraceuticals and nutritional supplements. *J. Nutr. Food Sci., 2*(5), 1–6.

Bleakley, S., & Hayes, M., (2017). Algal proteins: Extraction, application, and challenges concerning production. *Foods, 6*(5), 33.

Borowitzka, M. A., (1999). Commercial production of microalgae: Ponds, tanks, tubes, and fermenters. *Progress in Industrial Microbiology, 35*, 313–321.

Borowitzka, M. A., (2013). High-value products from microalgae - their development and commercialization. *Journal of Applied Phycology, 25*(3), 743–756.

Boyera, N., Galey, I., & Bernard, B. A., (1998). Effect of vitamin C and its derivatives on collagen synthesis and cross-linking by normal human fibroblasts. *International Journal of Cosmetic Science, 20*(3), 151–158.

Brown, M. R., & Miller, K. A., (1992). The ascorbic acid content of eleven species of microalgae used in mariculture. *Journal of Applied Phycology, 4*(3), 205–215.

Brown, M. R., Mular, M., Miller, I., Farmer, C., & Trenerry, C., (1999). The vitamin content of microalgae used in aquaculture. *Journal of Applied Phycology, 11*(3), 247–255.

Bryan, M., Pulte, E. D., Toomey, K. C., Pliner, L., Pavlick, A. C., Saunders, T., & Wieder, R., (2010). A pilot phase II trial of all-trans retinoic acid (Vesanoid) and paclitaxel (Taxol) in patients with recurrent or metastatic breast cancer. *Investigational New Drugs, 29*(6), 1482–1487.

Burja, A. M., & Radianingtyas, H., (2008). Nutraceuticals and functional foods from marine microbes: An introduction to a diverse group of natural products isolated from marine macroalgae, microalgae, bacteria, fungi, and cyanobacteria. In: Barrow, C. J., & Shahidi, F., (eds.), *Marine Nutraceuticals and Functional Foods* (pp. 367–403). CRC Press, Boca Raton.

Campanella, L., Russo, M. V., & Avino, P., (2002). Free and total amino acid composition in blue-green algae. *Ann. Chim., 92*, 343–352.

Capelli, B., & Cysewski, G. R., (2010). Potential health benefits of *Spirulina* microalgae. *Nutrafoods, 9*(2), 19–26.

Caporgno, M. P., & Matthys, A., (2018). Trends in microalgae incorporation into innovative food products with potential health benefits. *Frontiers in Nutrition, 5*, 58.

Carbonnelle, D., Pondaven, P., Morançais, M., Massé, G., Bosch, S., Jacquot, C., Briand, G., Robert, J., & Roussakis, C., (1999). Antitumor and antiproliferative effects of an aqueous extract from the marine diatom *Haslea ostearia* (Simonsen) against solid tumors: Lung carcinoma (NSCLC-N6), kidney carcinoma (E39) and melanoma (M96) cell lines. *Anticancer Research, 19*, 621–624.

Cha, K. H., Koo, S. Y., & Lee, D. U., (2008). Antiproliferative effects of carotenoids extracted from *Chlorella ellipsoidea* and *Chlorella vulgaris* on human colon cancer cells. *Journal of Agricultural and Food Chemistry, 56*(22), 10521–10526.

Chaumont, D., (1993). Biotechnology of algal biomass production: A review of systems for outdoor mass culture. *Journal of Applied Phycology, 5*(6), 593–604.

Christaki, E., Florou-Paneri, P., & Bonos, E., (2011). Microalgae: A novel ingredient in nutrition. *International Journal of Food Sciences and Nutrition, 62*(8), 794–799.

Chronakis, I. S., & Madsen, M., (2011). Algal proteins. *Handbook of Food Proteins* (pp. 353–394). Woodhead Publishing.

Cohen, Z., (1999). Production of polyunsaturated fatty acids by the microalga. In: Cohen, Z., (ed.), *Production of Chemicals by Microalgae* (pp. 1–24). Taylor and Francis, London.

Cohen, Z., Reungjitchachawali, M., Siangdung, W., & Tanticharoen, M., (1993). Production and partial purification of γ-linolenic acid and some pigments from *Spirulina platensis*. *Journal of Applied Phycology, 5*, 109–115.

Colston, K. W., & Hansen, C. M., (2002). Mechanisms implicated in the growth regulatory effects of vitamin D in breast cancer. *Endocrine-Related Cancer, 9*(1), 45–59.

Corina, A., Rangel-Zúñiga, O. A., Jiménez-Lucena, R., Alcalá-Díaz, J. F., Quintana-Navarro, G., Yubero-Serrano, E. M., López-Moreno, J., et al., (2019). Low intake of vitamin E accelerates cellular aging in patients with established cardiovascular disease: The CORDIOPREV study. *The Journals of Gerontology: Series A, 74*(6), 770–777.

Croft, M. T., Lawrence, A. D., Raux-Deery, E., Warren, M. J., & Smith, A. G., (2005). Algae acquire vitamin B 12 through a symbiotic relationship with bacteria. *Nature, 438*(7064), 90–93.

Dagnelie, P. C., Van, S. W. A., & Van, D. B. H., (1991). Vitamin B-12 from algae appears not to be bioavailable. *The American Journal of Clinical Nutrition, 53*(3), 695–697.

Das, J., Samadder, A., Mondal, J., Abraham, S. K., & Khuda-Bukhsh, A. R., (2016). Nano-encapsulated chlorophyllin significantly delays progression of lung cancer both in *in vitro* and *in vivo* models through activation of mitochondrial signaling cascades and drug-DNA interaction. *Environmental Toxicology and Pharmacology, 46*, 147–157.

Das, U. N., (2008). Folic acid and polyunsaturated fatty acids improve cognitive function and prevent depression, dementia, and Alzheimer's disease-but how and why? *Prostaglandins, Leukotrienes and Essential Fatty Acids, 78*, 11–19.

De Roeck-Holtzhauer, Y., Quere, I., & Claire, C., (1991). Vitamin analysis of five planktonic microalgae and one macroalga. *Journal of Applied Phycology, 3*(3), 259–264.

Depeint, F., Robert, B. W., Shangari, N., Mehta, R., & O'Brien, P. J., (2006). Mitochondrial function and toxicity: Role of the B vitamin family on mitochondrial energy metabolism. *Chemico-Biological Interactions, 163*(1, 2), 94–112.

Dewapriya, P., & Kim, S. K., (2014). Marine microorganisms: An emerging avenue in modern nutraceuticals and functional foods. *Food Research International, 56*, 115–125.

Durmaz, Y., Monteiro, M., Bandarra, N., Gökpinar, Ş., & Işik, O., (2007). The effect of low temperature on fatty acid composition and tocopherols of the red microalga, *Porphyridium cruentum*. *Journal of Applied Phycology, 19*(3), 223–227.

Ejike, C. E. C. C., Collins, S. A., Balasuriya, N., Swanson, A. K., Mason, B., & Udenigwe, C. C., (2017). Prospects of microalgae proteins in producing peptide-based functional foods for promoting cardiovascular health. *Trends in Food Science and Technology, 59*, 30–36.

El-Baky, H. H. A., El Baz, F. K., & El-Baroty, G. S., (2009). Production of phenolic compounds from *Spirulina maxima* microalgae and its protective effects *in vitro* toward hepatotoxicity model. *African Journal of Pharmacy and Pharmacology, 3*(4), 133–139.

Fabregas, J., & Herrero, C., (1990). Vitamin content of four marine microalgae. Potential use as source of vitamins in nutrition. *Journal of Industrial Microbiology, 5*(4), 259–263.

Fábregas, J., Arán, J., Morales, E. D., Lamela, T., & Otero, A., (1997). Modification of sterol concentration in marine microalgae. *Phytochemistry, 46*, 1189–1191.

Fan, X., Bai, L., Zhu, L., Yang, L., & Zhang, X., (2014). Marine algae-derived bioactive peptides for human nutrition and health. *Journal of Agricultural and Food Chemistry, 62*(38), 9211–9222.

Farré, G., Sanahuja, G., Naqvi, S., Bai, C., Capell, T., Zhu, C., & Christou, P., (2010). Travel advice on the road to carotenoids in plants. *Plant Science, 179*(1–2), 28–48.

Feldman, D., Krishnan, A. V., Swami, Giovannucci, E., & Feldman, B. J., (2014). The role of vitamin D in reducing cancer risk and progression. *Nature Reviews Cancer, 14*(5), 342–357.

FitzGerald, R. J., Murray, B. A., & Walsh, D. J., (2004). Hypotensive peptides from milk proteins. *The Journal of Nutrition, 134*, 980S–988S.

FitzGerald, R., & Murray, B. A., (2006). Bioactive peptides and lactic fermentations. *International Journal of Dairy Technology, 59*, 118–125.

Francavilla, M., Colaianna, M., Zotti, M., Morgese, M., Trotta, P., Tucci, P., Schiavone, S., Cuomo, V., & Trabace, L., (2012). Extraction, characterization and *in vivo* neuromodulatory activity of phytosterols from microalga *Dunaliella tertiolecta*. *Current Medicinal Chemistry, 19*(18), 3058–3067.

Francavilla, M., Trotta, P., & Luque, R., (2010). Phytosterols from *Dunaliella tertiolecta* and *Dunaliella salina*: A potentially novel industrial application. *Bioresource Technology, 101*, 4144–4150.

Freile-Pelgrin, Y., & Robledo, D., (2014). Bioactive phenolic compounds from algae. In: Hernandez-Ledesma, B., & Herrero, M., (eds.), *Bioactive Compounds from Marine Foods: Plant and Animal Sources* (pp. 113–130). Wiley Blackwell, Chichester, UK.

Frost & Sullivan, (2011). *Global Nutraceutical Industry: Investing in Healthy Living*. http://www.frost.com/prod/servlet/cio/236145272 (accessed on 2 July 2021).

Funk, C. D., (2001). Prostaglandins and leukotrienes: Advances in eicosanoids biology. *Science, 294*, 1871–1875.

Gage, D. A., Rhodes, D., Nolte, K. D., Hicks, W. A., Leustek, T., Cooper, A. J., & Hanson, A. D., (1997). A new route for synthesis of dimethylsulphoniopropionate in marine algae. *Nature, 387*(6636), 891–894.

Galasso, C., Corinaldesi, C., & Sansone, C., (2017). Carotenoids from marine organisms: Biological functions and industrial applications. *Antioxidants, 6*, 96.

Galasso, C., Nuzzo, G., Brunet, C., Ianora, A., Sardo, A., Fontana, A., & Sansone, C., (2018). The marine dinoflagellate *Alexandrium minutum* activates a mitophagic pathway in human lung cancer cells. *Marine Drugs, 16*(12), 502.

Gang-Guk, C., Bae, M. S., Ahn, C. Y., & Oh, H. M., (2008). Enhanced biomass and γ-linolenic acid production of mutant strain *Arthrospira platensis*. *Journal of Microbiology and Biotechnology, 18*(3), 539–544.

García, J. L., De Vicente, M., & Galán, B., (2017). Microalgae, old sustainable food, and fashion nutraceuticals. *Microbial Biotechnology, 10*(5), 1017–1024.

Gastineau, R., Turcotte, F., Pouvreau, J. B., Morançais, M., Fleurence, J., Windarto, E., Prasetiya, F. S., et al., (2014). Marennine, promising blue pigments from a widespread *Haslea* diatom species complex. *Marine Drugs, 12*(6), 3161–3189.

Giammanco, M., Di Majo, D., La Guardia, M., Aiello, S., Crescimannno, M., Flandina, C., Tumminello, F. M., & Leto, G., (2015). Vitamin D in cancer chemoprevention. *Pharmaceutical Biology, 53*(10), 1399–1434.

Gong, Y., Hu, H., Gao, Y., & Gao, H., (2011). Microalgae as platforms for production of recombinant proteins and valuable compounds: Progress and prospects. *Journal of Industrial Microbiology and Biotechnology, 38*(12), 1879–1890.

Gruber, B. M., (2016). B-group vitamins: Chemoprevention? *Advances in Clinical and Experimental Medicine, 25*(3), 561–568.

Guedes, A. C., Amaro, H. M., & Malcata, F. X., (2011a). Microalgae as sources of carotenoids. *Marine Drugs, 9*(4), 625–644.

Guedes, A. C., Amaro, H. M., & Malcata, F. X., (2011b). Microalgae as sources of high added-value compounds: A brief review of recent work. *Biotechnology Progress, 27*(3), 597–613.

Guerin, M., Huntley, M. E., & Olaizola, M., (2003). *Haematococcus* astaxanthin: Applications for human health and nutrition. *Trends in Biotechnology, 21*(5), 210–216.

Habib, M. A. B., Parvin, M., Huntington, T. C., & Hasan, M. R., (2008). *A Review on Culture, Production and Use of Spirulina as Food for Humans and Feeds for Domestic Animals.* FAO Fisheries and Aquaculture Circular (FAO).

Haghiac, M., Yang, X. H., Presley, L., Smith, S., Dettelback, S., Minium, J., Belury, M. A., Catalano, P. M., & Hauguel-De, M. S., (2015). Dietary omega-3 fatty acid supplementation reduces inflammation in obese pregnant women: A randomized double-blind controlled clinical trial. *PLoS One, 10*(9), e0137309.

Hamed, I., Özogul, F., Özogul, Y., & Regenstein, J. M., (2015). Marine bioactive compounds and their health benefits: A review. *Comprehensive Reviews in Food Science and Food Safety, 14*(4), 446–465.

Herrero, M., Ibanez, E., Cifuentes, A., Reglero, G., & Santoyo, S., (2006). *Dunaliella salina* microalga pressurized liquid extracts as potential antimicrobials. *Journal of Food Protection, 69*(10), 2471–2477.

Himaya, S. W. A., Ngo, D. H., Ryu, B., & Kim, S. K., (2012). An active peptide purified from gastrointestinal enzyme hydrolysate of Pacific cod skin gelatin attenuates angiotensin-1 converting enzyme (ACE) activity and cellular oxidative stress. *Food Chemistry, 132*(4), 1872–1882.

Hirahashi, T., Matsumoto, M., Hazeki, K., Saeki, Y., Ui, M., & Seya, T., (2002). Activation of the human innate immune system by *Spirulina*: Augmentation of interferon production and NK cytotoxicity by oral administration of hot water extract of *Spirulina platensis*. *International Immunopharmacology, 2*(4), 423–434.

Horrobin, D. F., (2000). Essential fatty acid metabolism and its modification in atopic eczema. *The American Journal of Clinical Nutrition, 71*, 367S–372S.

Hu, C. W., Chuang, L. T., Yu, P. C., & Chen, C. N. N., (2013). Pigment production by a new thermotolerant microalga *Coelastrella* sp. F50. *Food Chemistry, 138*, 2071–2078.

Hudek, K., Davis, L. C., Ibbini, J., & Erickson, L., (2014). Commercial products from algae. In: Bajpai, R., Prokop, A., & Zappi, M., (eds.), *Algal Biorefineries* (pp. 275–295). Springer, Dordrecht.

Husain, K., Centeno, B. A., Coppola, D., Trevino, J., Sebti, S. M., & Malafa, M. P., (2017). Tocotrienol, a natural form of vitamin E, inhibits pancreatic cancer stem-like cells and prevents pancreatic cancer metastasis. *Oncotarget, 8*(19), 31554–31567.

Hwang, E., Park, S. Y., Sun, Z. W., Shin, H. S., Lee, D. G., & Yi, T. H., (2014). The protective effects of fucosterol against skin damage in UV β-irradiated human dermal fibroblasts. *Marine Biotechnology, 16*, 361–370.

Islam, M. N., Alsenani, F., & Schenk, P. M., (2017). Microalgae as a sustainable source of nutraceuticals. *Microbial Functional Foods and Nutraceuticals*, 1–19.

Janssen, C. I., & Kiliaan, A. J., (2014). Long-chain polyunsaturated fatty acids (LCPUFA) from genesis to senescence: The influence of LCPUFA on neural development, aging, and neurodegeneration. *Progress in Lipid Research, 53*, 1–17.

Jäpelt, R. B., & Jakobsen, J., (2013). Vitamin D in plants: A review of occurrence, analysis, and biosynthesis. *Plant Science, 4*, 136.

Jerez-Martel, I., García-Poza, S., Rodríguez-Martel, G., Rico, M., Afonso-Olivares, C., & Gómez-Pinchetti, J. L., (2017). Phenolic profile and antioxidant activity of crude extracts from microalgae and cyanobacteria strains. *Journal of Food Quality, 2017,* 2924508. https://doi.org/10.1155/2017/2924508.

Ju, Z. Y., Deng, D. F., & Dominy, W., (2012). A defatted microalgae (*Haematococcus pluvialis*) meal as a protein ingredient to partially replace fishmeal in diets of Pacific white shrimp (*Litopenaeus vannamei*, Boone, 1931). *Aquaculture, 354,* 50–55.

Juarez-Oropeza, M. A., Mascher, D., Torres-Durán, P. V., Farias, J. M., & Paredes-Carbajal, M. C., (2009). Effects of *Spirulina* on vascular reactivity. *Journal of Medicinal Food, 12*(1), 15–20.

Kang, K. H., Qian, Z. J., Ryu, B., & Kim, S. K., (2011). Characterization of growth and protein contents from microalgae *Navicula incerta* with the investigation of antioxidant activity of enzymatic hydrolysates. *Food Science and Biotechnology, 20,* 183–191.

Kang, K. H., Qian, Z. J., Ryu, B., Karadeniz, F., Kim, D., & Kim, S. K., (2013). Hepatic fibrosis inhibitory effect of peptides isolated from *Navicula incerta* on TGF-β1 induced activation of LX-2 human hepatic stellate cells. *Preventive Nutrition and Food Science, 18,* 124–132.

Kapoor, R., & Nair, H., (2005). Gamma linolenic acid: Sources and functions. *Bailey's Industrial Oil and Fat Products,* 1–45.

Kerby, N. W., Niven, G. W., Rowell, P., & Stewart, W. D., (1987). Photoproduction of amino acids by mutant strains of N2-fixing cyanobacteria. *Applied Microbiology and Biotechnology, 25*(6), 547–552.

Khanavi, M., Gheidarloo, R., Sadati, N., Ardekani, M. R. S., Nabavi, S. M. B., Tavajohi, S., & Ostad, S. N., (2012). Cytotoxicity of fucosterol containing fraction of marine algae against breast and colon carcinoma cell line. *Pharmacognosy Magazine, 8*(29), 60.

Kim, G. N., Shin, J. G., & Jang, H. D., (2009). Antioxidant and antidiabetic activity of dangyuja (*Citrus grandis* Osbeck) extract treated with *Aspergillus saitoi. Food Chemistry, 117*(1), 35–41.

Kim, J. A., & Kim, S. K., (2013). Bioactive peptides from marine sources as potential anti-inflammatory therapeutics. *Current Protein and Peptide Science, 14*(3), 177–182.

Kim, J. H., Chang, M. J., Choi, H. D., Youn, Y. K., Kim, J. T., Oh, J. M., & Shin, W. G., (2011). Protective effects of *Haematococcus astaxanthin* on oxidative stress in healthy smokers. *Journal of Medicinal Food, 14*(11), 1469–1475.

Kim, M. S., Oh, G. H., Kim, M. J., & Hwang, J. K., (2013). Fucosterol inhibits matrix metalloproteinase expression and promotes type-1 procollagen production in UVB-induced HaCaT cells. *Photochemistry and Photobiology, 89*(4), 911–918.

Kitts, D. D., & Weiler, K. A., (2003). Bioactive proteins and peptides from food sources applications of bioprocesses used in isolation and recovery. *Current Pharmaceutical Design, 9,* 1309–1323.

Kline, K., Lawson, K. A., Yu, W., & Sanders, B. G., (2007). *Vitamin E and Cancer* (pp. 435–461). Elsevier, Amsterdam, The Netherlands.

Knowles, H. J., Raval, R. R., Harris, A. L., & Ratcliffe, P. J., (2003). Effect of ascorbate on the activity of hypoxia-inducible factor in cancer cells. *Cancer Research, 63*(8), 1764–1768.

Kobylewski, S., & Jacobson, M. F., (2010). *Food Dyes: A Rainbow of Risks.* Center for Science in the Public Interest.

Kris-Etherton, P. M., Harris, W. S., & Appel, L. J., (2002). Fish consumption, fish oil, omega-3 fatty acids, and cardiovascular disease. *Circulation, 106,* 2747–2757.

Ku, C. S., Pham, T. X., Park, Y., Kim, B., Shin, M. S., Kang, I., & Lee, J., (2013). Edible blue-green algae reduce the production of pro-inflammatory cytokines by inhibiting NF-κB pathway in macrophages and splenocytes. *Biochimica et Biophysica Acta (BBA)-General Subjects, 1830*(4), 2981–2988.

Kumar, C. S., Ganesan, P., Suresh, P. V., & Bhaskar, N., (2008). Seaweeds as a source of nutritionally beneficial compounds: A review. *Journal of Food Science and Technology, 45*, 113.

Kumudha, A., Selvakumar, S., Dilshad, P., Vaidyanathan, G., Thakur, M. S., & Sarada, R., (2015). Methyl cobalamin A form of vitamin B12 identified and characterized in *Chlorella vulgaris. Food Chemistry, 170*, 316–320.

Lappe, J. M., Travers-Gustafson, D., Davies, K. M., Recker, R. R., & Heaney, R. P., (2007). Vitamin D and calcium supplementation reduces cancer risk: Results of a randomized trial. *The American Journal of Clinical Nutrition, 85*(6), 1586–1591.

Larriba, M. J., González-Sancho, J. M., Barbáchano, A., Niell, N., Ferrer-Mayorga, G., & Muñoz, A., (2013). Vitamin D is a multilevel repressor of wnt/b-catenin signaling in cancer cells. *Cancers (Basel), 5*(4), 1242–1260.

Lavie, C. J., Milani, R. V., Mehra, M. R., & Ventura, H. O., (2009). Omega-3 polyunsaturated fatty acids and cardiovascular diseases. *Journal of the American College of Cardiology, 54*(7), 585–594.

Lawrence, K. P., Long, P. F., & Young, A. R., (2018). Mycosporine-like amino acids for skin photoprotection. *Current Medicinal Chemistry, 25*(40), 5512–5527.

Lebeau, T., Gaudin, P., Moan, R., & Robert, J. M., (2002). A new photobioreactor for continuous marennin production with a marine diatom: Influence of the light intensity and the immobilized-cell matrix (alginate beads or agar layer). *Applied Microbiology and Biotechnology, 59*(2, 3), 153–159.

Li, H. B., Cheng, K. W., Wong, C. C., Fan, K. W., Chen, F., & Jiang, Y., (2007). Evaluation of antioxidant capacity and total phenolic content of different fractions of selected microalgae. *Food Chemistry, 102*(3), 771–776.

Liang, S., Liu, X., Chen, F., & Chen, Z., (2004). Current microalgal health food R and D activities in China. *Asian Pacific Phycology in the 21st Century: Prospects and Challenges* (pp. 45–48), Springer, Dordrecht.

Lim, D. K., Garg, S., Timmins, M., Thomas-Hall, S. R., Schuhmann, H., Li, Y., & Schenk, P. M., (2012). Isolation and evaluation of oil-producing microalgae from subtropical coastal and brackish waters. *PLoS One, 7*(7), e40751.

Llewellyn, C. A., & Airs, R. L., (2010). Distribution and abundance of MAAs in 33 species of microalgae across 13 classes. *Marine Drugs, 8*, 1273–1291.

Lopez, A., Rico, M., Santana-Casiano, J. M., Gonzalez, A. G., & Gonzalez-Davila, M., (2015). Phenolic profile of *Dunaliella tertiolecta* growing under high levels of copper and iron. *Environmental Science and Pollution Research, 22*(19), 14820–14828.

Lordan, S., Ross, R. P., & Stanton, C., (2011). Marine bioactives as functional food ingredients: Potential to reduce the incidence of chronic diseases. *Marine Drugs, 9*, 1056–1100.

Lorenz, R. T., & Cysewski, G. R., (2000). Commercial potential for *Haematococcus* microalgae as a natural source of astaxanthin. *Trends in Biotechnology, 18*(4), 160–167.

Macias-Sanchez, M. D., Mantell, C., Rodriguez, M., Ossa, E. M., Lubian, L. M., & Montero, O., (2007). Supercritical fluid extraction of carotenoids and chlorophyll a from *Synechococcus* sp. *The Journal of Supercritical Fluids, 39*(3), 323–329.

Maeda, H., Hosokawa, M., Sashima, T., Funayama, K., & Miyashita, K., (2005). Fucoxanthin from edible seaweed, *Undaria pinnatifida*, shows antiobesity effect through UCP1 expression in white adipose tissues. *Biochemical and Biophysical Research Communications, 332,* 392–397.

Manach, C., Scalbert, A., Morand, C., Remesy, C., & Jimenez, L., (2004). Polyphenols: Food sources and bioavailability. *American Journal of Clinical Nutrition, 79*(5), 727–747.

Maria, A. G., Graziano, R., & Nicolantonio, D. O., (2015). Carotenoids: Potential allies of cardiovascular health? *Food and Nutrition Research, 59*(1), 26762.

Martins, A., Caetano, N. S., & Mata, T. M., (2010). Microalgae for biodiesel production and other applications: A review. *Renewable and Sustainable Energy Reviews, 14*(1), 217–232.

Martins, D. A., Custódio, L., Barreira, L., Pereira, H., Ben-Hamadou, R., Varela, J., & Abu-Salah, K. M., (2013). Alternative sources of n-3 long-chain polyunsaturated fatty acids in marine microalgae. *Marine Drugs, 11,* 2259–2281.

Matthew, S., Ross, C., Paul, V. J., & Luesch, H., (2008). Pompanopeptins A and B, new cyclic peptides from the marine cyanobacterium *Lyngbya confervoides. Tetrahedron, 64,* 4081–4089.

Matthew, S., Ross, C., Rocca, J. R., Paul, V. J., & Luesch, H., (2007). Lyngbyastatin 4, a dolastatin 13 analogue with elastase and chymotrypsin inhibitory activity from the marine cyanobacterium *Lyngbya confervoides. Journal of Natural Products, 70,* 124–127.

Mayfield, S. P., Manuell, A. L., Chen, S., Wu, J., Tran, M., Siefker, D., Muto, M., & Marine-Navarro, J., (2007). *Chlamydomonas reinhardtii* chloroplasts as protein factories. *Current Opinion in Biotechnology, 18*(2), 126–133.

Milovanovic, I., Misan, A., Saric, B., Kos, J., Mandić, A., & Simeunović, J., (2012). Evaluation of protein and lipid content and determination of fatty acid profile in selected species of cyanobacteria. *Proceedings of 6th Central European Congress on Food.*

Minkova, K. M., Toshkova, R. A., Gardeva, E. G., Tchorbadjieva, M. I., Ivanova, N. J., Yossifoya, L. S., & Gigova, L. G., (2011). Antitumor activity of B-phycoerythrin from *Porphyridium cruentum. Journal of Pharmacy Research, 4,* 1480–1482.

Mišurcová, L., Škrovánková, S., Samek, D., Ambrožová, J., & Machů, L., (2012). Health benefits of algal polysaccharides in human nutrition. *Advances in Food and Nutrition Research, 66,* 75–145.

Mosulishvili, L. M., Kirkesali, E. I., Beiokobylsky, A. I., Khizanishvili, A. I., Frontasyeva, M. V., Pavlov, S. S., & Gundorina, S. F., (2012). Experimental substantiation of the possibility of developing selenium and iodine containing pharmaceuticals based on blue-green algae *Spirulina platensis. Journal of Pharmaceutical and Biomedical Analysis, 30*(1), 87–97.

Mulders, K. J., Lamers, P. P., Martens, D. E., & Wijffels, R. H., (2014). Phototrophic pigment production with microalgae: Biological constraints and opportunities. *Journal of Phycology, 50*(2), 229–242.

Murray, P. M., Moane, S., Collins, C., Beletskaya, T., Thomas, O. P., Duarte, A. W., Nobre, F. S., et al., (2013). Sustainable production of biologically active molecules of marine based origin. *New Biotechnology, 30*(6), 839–850.

Mustopa, A. Z., Isworo, R., Nurilmala, M., & Susilaningsih, D., (2016). Molecular identification of microalgae *BTM 11* and its lectin isolation, characterization, and inhibition activity. *Ann. Bogor., 20,* 37.

Nagini, S., Palitti, F., & Natarajan, A. T., (2015). Chemopreventive potential of chlorophyllin: A review of the mechanisms of action and molecular targets. *Nutrition and Cancer, 67*(2), 203–211.

Nakajima, Y., Inokuchi, Y., Shimazawa, M., Otsubo, K., Ishibashi, T., & Hara, H., (2008). Astaxanthin, a dietary carotenoid, protects retinal cells against oxidative stress *in-vitro* and in mice *in-vivo*. *Journal of Pharmacy and Pharmacology, 60*(10), 1365–1374.

Ni, J., Wen, X., Yao, J., Chang, H. C., Yin, Y., Zhang, M., Xie, S., et al., (2005). Tocopherol-associated protein suppresses prostate cancer cell growth by inhibition of the phosphoinositide 3-kinase pathway. *Cancer Research, 65*(21), 9807–9816.

Niles, R. M., (2000). Vitamin A and cancer. *Nutrition, 16*, 573–576.

Nunes-Alves, C., Booty, M. G., Carpenter, S. M., Jayaraman, P., Rothchild, A. C., & Behar, S. M., (2014). In search of a new paradigm for protective immunity to TB. *Nature Reviews Microbiology, 12*(4), 289–299.

Nyenje, M. E., & Ndip, N., (2014). The challenges of foodborne pathogens and antimicrobial chemotherapy: A global perspective. *African Journal of Microbiology Research, 7*(14), 1158–1172.

Ortiz, J., Uquiche, E., Robert, P., Romero, N., Quitral, V., & Llanten, C., (2009). Functional and nutritional value of the Chilean seaweeds *Codium fragile, Gracilaria chilensis* and *Macrocystis pyrifera*. *European Journal of Lipid Science and Technology, 111*(4), 320–327.

Ozdemir, G., Karabay, N. U., Dalay, M. C., & Pazarbasi, B., (2004). Antibacterial activity of volatile component and various extracts of *Spirulina platensis*. *Phytotherapy Research, 18*(9), 754–757.

Padayatty, S. J., Katz, A., Wang, Y., Eck, P., Kwon, O., Lee, J. H., Chen, S., et al., (2003). Vitamin C as an antioxidant: Evaluation of its role in disease prevention. *Journal of the American College of Nutrition, 22*(1), 18–35.

Paiva, S. A. R., & Russell, R. M., (1999). Antioxidants and their clinical applications. β-carotene and other carotenoids as antioxidants. *Journal of the American College of Nutrition, 18*, 426–433.

Park, J. S., Chyun, J. H., Kim, Y. K., Line, L. L., & Chew, B. P., (2010). Astaxanthin decreased oxidative stress and inflammation and enhanced immune response in humans. *Nutrition and Metabolism, 7*(1), 18.

Parmar, R. S., & Singh, C., (2018). A comprehensive study of eco-friendly natural pigment and its applications. *Biochemistry and Biophysics Reports, 13*, 22–26.

Pittman, J. K., Dean, A. P., & Osundeko, O., (2011). The potential of sustainable algal biofuel production using wastewater resources. *Bioresource Technology, 102*(1), 17–25.

Plavšić, M., Terzic, S., Ahel, M., & Van, D. B. C. M. G., (2002). Folic acid in coastal waters of the Adriatic Sea. *Marine and Freshwater Research, 53*, 1245–1252.

Plaza, M., Herrero, M., Cifuentes, A., & Ibanez, E., (2009). Innovative natural functional ingredients from microalgae. *Journal of Agricultural and Food Chemistry, 57*(16), 7159–7170.

Pratoomyot, J., Srivilas, P., & Noiraksar, T., (2005). Fatty acids composition of 10 microalgal species. *Songklanakarin Journal of Science and Technology, 27*, 1179–1187.

Priyadarshani, I., & Rath, B. J., (2012). Commercial and industrial applications of micro algae: A review. *Journal of Algal Biomass Utilization, 3*, 89–100.

Pulz, O., & Gross, W., (2004). Valuable products from biotechnology of microalgae. *Applied Microbiology and Biotechnology, 65*(6), 635–648.

Radmer, R. J., & Parker, B. C., (1994). Commercial applications of algae: Opportunities and constraints. *Journal of Applied Phycology, 6*(2), 93–98.

Rao, D. S., & Raghuramulu, N., (1996). Food chain as origin of vitamin D in fish. *Comparative Biochemistry and Physiology Part A: Physiology, 114*(1), 15–19.

Ratledge, C., (2010). Single cell oils for the 21st century. In: Cohen, Z., & Ratledge, C., (eds.), *Single Cell Oils, Microbial and Algal Oils* (pp. 3–26). AOCS Press, Urbana.

Riccioni, G., D'Orazio, N., Palumbo, N., Bucciarelli, V., Ilio, E., Bazzano, L. A., & Bucciarelli, T., (2009). Relationship between plasma antioxidant concentrations and carotid intima-media thickness: The asymptomatic carotid atherosclerotic disease in Manfredonia study. *European Journal of Cardiovascular Prevention and Rehabilitation, 16*(3), 351–357.

Richmond, A., (1986). Outdoor mass cultures of microalgae. *CRC Handbook of Microalgal Mass Culture* (p. 528). CRC Press, Boca Raton, Florida.

Rico, M., López, A., Santana-Casiano, J. M., Gonzalez, A. G., & Gonzalez-Davila, M., (2012). Variability of the phenolic profile in the diatom *Phaeodactylum tricornutum* growing under copper and iron stress. *Limnology and Oceanography, 58*(1), 144–152.

Rizzello, C. G., Tagliazucchi, D., Babini, E., Rutella, G. S., Taneyo, S. D. L., & Gianotti, A., (2016). Bioactive peptides from vegetable food matrices: Research trends and novel biotechnologies for synthesis and recovery. *Journal of Functional Foods, 27*, 549–569.

Romay, C., Gonazelz, R., Ledon, N., Ramirez, D., & Rimbu, V., (2003). C-phycocyanin: A biliprotein with antioxidant, anti-inflammatory, and neuroprotective effects. *Current Protein and Peptide Science, 4*, 207–216.

Ryu, B. M., (2011). *A Peptide Derived from Microalga, Pavlova lutheri, Fermented by Candida rugopelliculosa Induces Myofibroblasts Differentiation in Human Dermal Fibroblasts.* PhD Thesis, Department of Chemistry, Pukyoung National University, Busan, South Korea.

Sathasivam, R., Radhakrishnan, R., Hashem, A., & Abd, A. E. F., (2019). Microalgae metabolites: A rich source for food and medicine. *Saudi Journal of Biological Sciences, 26*(4), 709–722.

Sekar, S., & Chandramohan, M., (2008). Phycobiliproteins as a commodity: Trends in applied research, patents, and commercialization. *Journal of Applied Phycology, 20*(2), 113–136.

Seto, A., Wang, H. L., & Hesseltine, C. W., (1984). Culture conditions affect eicosapentaenoic acid content of *Chlorella minutissima. Journal of the American Oil Chemists' Society, 61*(5), 892–894.

Seyidoglu, N., Inan, S., & Aydin, C., (2016). A prominent superfood: *Spirulina platensis. Superfood and Functional Food: The Development of Superfoods and Their Roles as Medicine*, 1–27.

Sharma, K., & Schenk, P. M., (2015). Rapid induction of omega-3 fatty acids (EPA) in *Nannochloropsis* sp. by UV-C radiation. *Biotechnology Bioengineering, 112*, 1243–1249.

Sharma, S. R., Poddar, R., Sen, P., & Andrews, J. T., (2008). Effect of vitamin C on collagen biosynthesis and degree of birefringence in polarization sensitive optical coherence tomography (PS-OCT). *African Journal of Biotechnology, 7*(12).

Sheih, I. C., Fang, T. J., & Wu, T. K., (2009). Isolation and characterization of a novel angiotensin I-converting enzyme (ACE) inhibitory peptide from the algae protein waste. *Food Chemistry, 115*(1), 279–284.

Sheih, I. C., Wu, T. K., & Fang, T. J., (2009). Antioxidant properties of a new antioxidative peptide from algae protein waste hydrolysate in different oxidation systems. *Bioresource Technology, 100*(13), 3419–3425.

Siddikuzzaman, Guruvayoorappan, C., & Berlin, G. V. M., (2011). All trans retinoic acid and cancer. *Immunopharmacology and Immunotoxicology, 33*(2), 241–249.

Simonich, M. T., Egner, P. A., Roebuck, B. D., Orner, G. A., Jubert, C., Pereira, C., Groopman, J. D., et al., (2007). Natural chlorophyll inhibits aflatoxin B1-induced multi-organ carcinogenesis in the rat. *Carcinogenesis, 28*, 1294–1302.

Simonich, M. T., McQuistan, T., Jubert, C., Pereira, C., Hendricks, J. D., Schimerlik, M., Zhu, B., et al., (2008). Low-dose dietary chlorophyll inhibits multi-organ carcinogenesis in the rainbow trout. *Food and Chemical Toxicology, 46*, 1014–1024.

Singh, N. K., & Dhar, D. W., (2011). Phylogenetic relatedness among *Spirulina* and related cyanobacterial genera. *World Journal of Microbiology and Biotechnology, 27*(4), 941–951.

Škrovankova, S., (2011). Seaweed vitamins as nutraceuticals. *Advances in Food and Nutrition Research* (Vol. 64, p. 357–369), Academic Press.

Smerilli, A., Orefice, I., Corato, F., Gavalás, O. A., Ruban, A. V., & Brunet, C., (2017). Photoprotective and antioxidant responses to light spectrum and intensity variations in the coastal diatom *Skeletonema marinoi*. *Environmental Microbiology, 19*(2), 611–627.

Soletto, D., Binaghi, L., Lodi, A., Carvalho, J. C. M., & Converti, A., (2005). Batch and fed batch cultivations of *Spirulina platensis* using ammonium sulphate and urea as nitrogen sources. *Aquaculture, 243*(1–4), 217–224.

Solovchenko, A. E., Khozin-Goldberg, I., Didi-Cohen, S., Cohen, Z., & Merzlyak, M. N., (2008). Effects of light intensity and nitrogen starvation on growth, total fatty acids, and arachidonic acid in the green microalga *Parietochloris incisa*. *Journal of Applied Phycology, 20*(3), 245–251.

Sonani, R. R., Rastogi, R. P., Patel, R., & Madamwar, D., (2016). Recent advances in production, purification, and applications of phycobiliproteins. *World Journal of Biological Chemistry, 7*(1), 100.

Sousa, I., Gouveia, L., Batista, A. P., Raymundo, A., & Bandarra, N. M., (2008). Microalgae in novel food products. *Food Chemistry Research Developments, 75*–112.

Spolaore, P., Joannis-Cassan, C., Duran, E., & Isambert, A., (2006). Commercial applications of microalgae. *Journal of Bioscience and Bioengineering, 101*(2), 87–96.

Stengel, D. B., Connan, S., & Popper, Z. A., (2011). Algal chemodiversity and bioactivity: Sources of natural variability and implications for commercial application. *Biotechnology Advances, 29*(5), 483–501.

Stewart, B. W., Bray, F., Forman, D., Ohgaki, H., Straif, K., Ullrich, A., & Wild, C. P., (2016). Cancer prevention as part of precision medicine: 'Plenty to be done.' *Carcinogenesis, 37*, 2–9.

Suetsuna, K., & Chen, J. R., (2001). Identification of antihypertensive peptides from peptic digest of two microalgae, *Chlorella vulgaris* and *Spirulina platensis*. *Marine Biotechnology, 3*(4), 305–309.

Sunda, W., Kieber, D. J., Kiene, R. P., & Huntsman, S., (2002). An antioxidant function for DMSP and DMS in marine algae. *Nature, 418*, 317–320.

Takaichi, S., (2011). Carotenoids in algae: Distributions, biosynthesis, and functions. *Marine Drugs, 9*(6), 1101–1118.

Takebe, Y., Saucedo, C. J., Lund, G., Uenishi, R., Hase, S., Tsuchiura, T., Kneteman, N., et al., (2013). Antiviral lectins from red and blue-green algae show potent *in vitro* and *in vivo* activity against hepatitis C virus. *PLoS One, 8*(5), pe64449.

Tanticharoen, M., Reungjitchachawali, M., Boonag, B., Vonktaveesuk, P., Vonshak, A., & Cohen, Z., (1994). Optimization of γ-linolenic acid (GLA) production in *Spirulina platensis*. *Journal of Applied Phycology, 6*(3), 295–300.

Taori, K., Matthew, S., Rocca, J. R., Paul, V. J., & Luesch, H., (2007). Lyngbyastatins 5-7, potent elastase inhibitors from Floridian marine cyanobacteria, *Lyngbya* spp. *Journal of Natural Products, 70,* 1593–1600.

Taori, K., Paul, V. J., & Luesch, H., (2008). Kempopeptins A and B, serine protease inhibitors with different selectivity profiles from a marine cyanobacterium, *Lyngbya* sp. *Journal of Natural Products, 71,* 1625–1629.

Teegarden, D., & Donkin, S. S., (2009). Vitamin D: Emerging new roles in insulin sensitivity. *Nutrition Research Reviews, 22*(1), 82–92.

Tokuşoglu, O., & Unal, M. K., (2003). Biomass nutrient profiles of three microalgae: *Spirulina platensis, Chlorella vulgaris,* and *Isochrisis galbana. Journal of Food Science, 68*(4), 1144–1148.

Venugopal, V., (2008). *Marine Products for Healthcare: Functional and Bioactive Nutraceutical Compounds from the Ocean.* Boca Raton, CRC Press.

Vo, T. S., Ryu, B., & Kim, S. K., (2013). Purification of novel anti-inflammatory peptides from enzymatic hydrolysate of the edible microalgal *Spirulina maxima. Journal of Functional Foods, 5*(3), 1336–1346.

Volkman, J. K., Barrett, S. M., & Blackburn, S. I., (1999). Eustigmatophyte microalgae are potential sources of C29 sterols, C22–C28 n-alcohols and C28–C32 n-alkyl diols in freshwater environments. *Organic Geochemistry, 30*(5), 307–318.

Wall, R., Ross, R. P., Fitzgerald, G. F., & Stanton, C., (2010). Fatty acids from fish: The anti-inflammatory potential of long-chain omega-3 fatty acids. *Nutrition Reviews, 68*(5), 280–289.

Walsh, D. J., & FitzGerald, R. J., (2004). Health related functional value of dairy proteins and peptides. In: Yada, R. Y., (ed.), *Proteins in Food Processing* (pp. 559–606), Woodhead Publishing Limited, Cambridge, UK.

Watanabe, F., Takenaka, S., Kittaka-Katsura, H., Ebara, S., & Miyamoto, E., (2002). Characterization and bioavailability of vitamin B12-compounds from edible algae. *Journal of Nutritional Science and Vitaminology, 48*(5), 325–331.

Watanabe, F., Yabuta, Y., Bito, T., & Teng, F., (2014). Vitamin B12-containing plant food sources for vegetarians. *Nutrients, 6*(5), 1861–1873.

Yoshida, H., Yanai, H., Ito, K., Tomono, Y., Koikeda, T., Tsukahara, H., & Tada, N., (2010). Administration of natural astaxanthin increases serum HDL-cholesterol and adiponectin in subjects with mild hyperlipidemia. *Atherosclerosis, 209.*

CHAPTER 6

ALGACULTURE AND ITS APPLICATIONS IN THE PHARMACEUTICAL AND NUTRACEUTICAL INDUSTRY

MEHWISH JAFFER and SHABNUM SHAHEEN

Department of Plant Sciences, Lahore College for Woman University, Lahore–54000, Pakistan

ABSTRACT

Algae are often used for their medicinal properties. These eukaryotic unicellular (or in some cases multicellular) organisms possess several biochemicals that can be used as medicines. Since algae are a vital component of aquatic ecosystems, they have acquired several adaptations that are highly prized in the nutraceutical and pharmaceutical industry. Algae are often harvested for commercial use via several techniques that are compiled under the name of algaculture. These techniques employ several different mechanisms, such as bioreactors or ponds to create algal biomass for harvesting. The harvest of such techniques is variable, and many researchers are trying to perfect the process to get more and more algal biomass. Algae find their applications not only in medicines but also in several other areas. As a food-based medicine, algae are a solid competitor, as it can be used to address several medical issues, some as extreme as cancer. All in all, as this chapter will highlight, it is suffice to say that algae are very useful in the pharmaceutical and nutraceutical industry.

6.1 INTRODUCTION

Algae are eukaryotic photosynthetic organisms that may be unicellular or simple multicellular. They are mostly aquatic, and although they differ from plants, which evolved from them, in certain aspects, the basic process of photosynthesis, which has been acquired by both from a distant ancestor - cyanobacteria (ancestral to algae, according to the endosymbiont theory) - is more or less identical.

They act as major producers in aquatic ecosystems and are hence responsible for providing for the whole ecosystem. More than half of the global photosynthetic activity that happens on this planet is accredited to algae and cyanobacteria (which are often collectively referred to as phytoplankton).

From time immemorable, algae have been used as food-with the belief that they impart medicinal benefits, i.e., as nutraceuticals. They have also been proven effective in their use as fertilizers, medicines, animal feed, bioindicators (in association with fungi-lichens), biofuels, and so much more. Microalgae have advanced to be pervasive all through the globe, and their diverse dispersions and phylogenies are reflected in amazingly differing inter-species metabolic abilities.

These varied biochemical machineries produce a huge variety of biochemicals with human interest, including nutraceuticals. Consider, for example, the carotenoids extracted from *Dunaliella* and *Haematococcus* (Borowitzka, 2013); valuable proteins and carbohydrates that are so profound in *Spirulina* (*Arthrospira*) and *Chlorella* (Khan et al., 2005; Gors et al., 2010); or the use of various microalgae for single cell oil (poly-unsaturated FAs) production as done by Ratledge (2004).

Many chemicals with exciting biotechnological applications are also accredited to algae. Examples include fluorescent biochemicals such as phycoerythrin (PE) and isoprenoids that find immense use in the food industry (Andersen, 2013). Additionally, they have also been recognized as appealing wellsprings of biofuel because various species can deliver an assortment of fuel items.

Different microalgae can deliver enormous amounts of triacylglycerols (TAG) (the most common form of lipids) while using CO_2. These in turn can be changed over to fatty acid methyl esters (FAMEs), the principal ingredients of algal biofuels (Hossain et al., 2008).

Various biomass components alongside lipids can be changed over into unrefined oil. For this, thermo-chemical procedures are employed (Barreiro et al., 2013). Their carbohydrates can also be used in brewing industries to brew alcohol (or in laboratories as well); a few species can even deliver biohydrogen (Radakovits et al., 2010).

6.1.1 ALGACULTURE

The plethora of applications that algae provide are tempting enough for any researcher to try and figure out a way of harvesting their innate potential. Algaculture or the practice of growing algae at a commercial level is gaining attention because of these prospects.

Algaculture employs a potent mixture of traditional land-based agriculture and aquaculture. Elements of both overlap quite considerably, but the activity itself remains to be accorded in its own right. Sunlight, CO_2, and water are the necessities for getting the wheels of the photosynthetic machines up and running but additional requirements, those associated with aquaculture, also need fulfillment, such as water quality assessment, water changes, etc. With the advent and widespread industrial use of bioreactors, microbial cultivation is easier and more practical than ever before.

As a result of their capacity to deliver products that cover numerous business sectors, including those of functional food (health-based diet), nutraceuticals, medicines, feed, and fodder, synthetics, and fuels, algae are excellent candidates for being among the most flexible crops (Rosenberg et al., 2008).

6.1.2 MICRO- AND MACRO-ALGAE

When looking over algae, one needs to recognize two separate groups: the microscopic micro-algae and the macroscopic macro-algae. While macro-algae are essentially what we commonly refer to as seaweeds, microalgae are infinitesimal life-forms, which make up the world's phytoplankton and contribute to quickly developing populaces of algae in water when provided with the vital supplements such as, but not limited to, nitrogen, and phosphorus.

Macro-algae show resemblance to plants, they are macroscopic and multicellular; their thallus contains hints of structures that have slight similarity with the body parts of modern vascular plants (or tracheophytes). Leaf-like blades, stem-like stipes, and root-like holdfasts (which anchor them on rocks or other substrates in the oceans) are the major parts of these algae. They display a wide range of colors, from brown to green, depending on their depth. Yet, from a myriad of different species, only a handful is edible for humans.

Micro-algae, on the other hand, are unicellular. *Arthrospira* (*Spirulina*), *Chlorella*, and cyanobacteria (which are technically not algae, but do have a superficial resemblance) are quite mainstream in this category. Mostly, freshwater tanks are employed for their cultivation. The thick layer of scum that is often spotted floating atop ponds and lakes, also known scientifically as algal bloom is a consequence of eutrophication (over-richness of limiting nutrients in a water body) which drive micro-algae to divide at a rapid pace, pushing their population curve steeply upwards at an exponential rate.

6.1.3   APPLICATIONS OF ALGACULTURE

The applications of algae are too much to be discussed with justice in this chapter, but the major highlights are as given in subsections.

6.1.3.1   BIOREMEDIATION

It is the use of microorganisms such as algae to degrade pollutants within a contaminated medium, in which they are grown. By and large, bioremediation is more affordable and more economical than other alternatives (EPA, 2011).

Bioremediation mostly involves redox reactions where either an electron acceptor (mostly oxygen) is added as an oxidant for a pollutant (i.e., hydrocarbons) which are in turn oxidized. Conversely, an electron-rich compound is added as a reducing agent to the pollutants (i.e., nitrate, perchlorate, oxidized metals, chlorinated solvents, explosives, and propellants) which are in turn reduced (EPA, 2013).

Extra supplements, vitamins, minerals, and pH buffers can be added, in both of the aforementioned methods, to streamline conditions for the microorganisms. Now and again, specific microbial cultures are also included to additionally improve biodegradation, this process is called bioaugmentation. Techniques like phytoremediation, bioventing, and bioleaching, alongside many more, are common examples of bioremediation.

Micro-algae and Macro-algae are often called to use for removing or detoxifying toxins including heavy metals, organic matter, and xenobiotics from wastewater, or CO_2 from exhaust vents. The algal biomass thus attained can be put to even further use to create synthetic chemicals, biodiesel, and assorted fuels as by-products (Munoz and Guieysse, 2006).

6.1.3.2   USE OF NON-ARABLE LAND

Other than the myriad of medicinal and nutraceutical products that algae provide us with, the sheer diversity of which remains to be accounted for, algaculture also has another trick up its sleeve. Normally crops require arable land, loaded with organic matter and inorganic salts of potassium, nitrogen, phosphorus, magnesium, iodine, and similar elements-these are quintessential for their growth.

The case with algae however is quite on the contrary. Not only do algae not require a consistent sunlight exposure, but they are also quite flexible in their nutrient requirements. They can thrive in nutrient-depleted soils and will do significantly better than any traditional crop would under similar conditions.

Not only are they indifferent to the influx of those organic and inorganic nutrients that traditional crops crave (although they do need small amounts of them), they are much more resilient too. This resilience has allowed them to not only tolerate but also thrive in the most inhospitable soil types (Trentacoste and Zenk, 2015).

6.1.3.3   CHEMICAL PRODUCTS

A wide variety of chemical compounds are produced by algae, their prices vary according to their worth. β carotene is extensively used as a food coloring agent and for its ability to boost the health of grain-fed cattle. Until the 1980s, it was synthesized synthetically, but a breakthrough changed everything. In the 1970s, several research specialists (Borowitzka and Brown, 1974; Ben-Amotz and Avron, 1980) understood that nutritional stress, increased salinity, and similar environmental stresses can cause Micro-algae such as *Dunaliella salina* to accumulate β carotene as a substantial percentage of its dry mass (14%). This revelation prompted the commercialization of this product through algae instead of synthetic means, as it was previously acquired.

Phycobiliproteins (PBPs) are yet another important algal product with a wide scope of use in diagnostic instruments. Biliproteins extracted from micro-algae, which are also simply referred to as PBPs, find immense use in cellular analysis as they are used as fluorescent tags for DNA tagging and screening (Pulz and Gross, 2004). Another noteworthy product is agar, which is a derivative of the agarose polysaccharide, is extracted from certain marine macro-algae, and is used as a solid culture medium in microbiology labs (Williams et al., 2000).

6.1.3.4   FOOD AND FEED PRODUCTS

Many algal species like kelps (macro-algae) are edible and often possess a fine mixture of dietary components. Many microalgae, for instance, are equipped with proteins, carbohydrates, non-saturated fatty acids, vitamins, and minerals. This makes them an ideal balanced diet.

Numerous microalgae have a high dietary benefit. They are loaded with proteins, vitamins, minerals, unsaturated fats, etc. These prospects prompted the commercialization of *Arthrospira* (*Spirulina*), a micro-algae grown in isolation, i.e., in a monoculture in the absence of other species; similarly, *Dunaliella salina* is also grown in monocultures. It is vastly successful, commercially, as a functional or health-based food on its own and is an ingredient to many such food items as well. Around the world, *Arthrospira* (*Spirulina*) has found much application as a nutritional supplement for humans and as a feed for animals, for this reason, it is grown extensively in many countries (FAO, 2010).

6.1.3.5   FOOD COLORING

Microalgae find utility as a wellspring of numerous natural food coloring options. As Borowitzka and Borowitzka (1987) demonstrated decades earlier, certain microalgae possess significant measures of different kinds of carotenes, other than the famous one: β-carotene. Different kinds of color shades show up in microalgae, depending on species and growth conditions. In unadulterated conditions, such products can easily cost as much as 1000 USD for one kilogram.

6.1.3.6   OSMOREGULATORS

Algae possess certain sugars that can influence the osmotic potential of the cell, relative to its surroundings. Glycerol is the most remarkable puzzle piece in this compound classification, which encompasses over other economically feasible items too. A significant percentage of the dry mass (as much as 50%) of a few micro-algal species, such as *Dunaliella salina*, can be changed to osmoregulators under fitting conditions. Micro-algae are one of the most effective sources of such osmoregulation, the only other alternatives being prokaryotes and animal fat. Research ought to and is probably going to find important osmoregulators that can be extracted from micro-algae and can be commercialized (Hochman and Zilberman, 2014).

6.1.3.7   BIODIESEL AND OTHER ALGAL BIOFUELS

Utilizing micro-algae to deliver biodiesel is anything but a novel thought and much research has been already designated to this pursuit (Gallaghar, 2011). In the United States of America, a task at the National Renewable Energy Laboratory has gathered about 3,000 strains of algal species for this very purpose (Sheehan et al., 1998).

6.1.3.8   MISCELLANEOUS USES

Microalgae possess numerous valuable biochemicals from which certain derivatives can be obtained; these derivatives will find utility in the times to come for some different applications, notwithstanding the ones referenced previously. They incorporate cosmetics and skin-care items, supplements for human food and animal feed, vitamins, composts, and various forms of biofuel. Microalgae might be less profitable than prokaryotes, and our capacity to control their cultures is comparatively slimmer.

Be that as it may, microalgae possess exceptional and complex items that are unique only to them. Accordingly, science conceivably can be moved to offer extensive research concerning the utilization and control of micro-algae. Today, notwithstanding the aforementioned items, *Arthrospira* (*Spirulina*) is utilized to deliver phycocyanin (PC) and biomass (Lee, 2001; Costa et al., 2003) while *Chlorella vulgaris* (a green algae of the division Chlorophyta) finds utility in the production of biomass (Lee, 2001).

6.1.4   CULTIVATION PRACTICES

6.1.4.1   RACEWAY PONDS

A raceway pond is a shallow fake lake or pond utilized in growing algae. The lake is made up of a grid of rectangular frameworks, with every compartment hosting one oval-shaped channel, similar to a car race-track. From atop, numerous ponds resemble a labyrinth. Every compartment contains a paddle-wheel to make the water stream consistently around the circuit. Numerous entrepreneurs of *Spirulina* use raceway lakes, to this day, as their essential technique for algal cultivation. Raceway lakes were utilized for the expulsion of Lead from wastewater utilizing live *Arthospira* sp. (*Spirulina*) (Siva et al., 2015).

6.1.4.2   PHOTOBIOREACTORS (PBRS)

These frameworks are various kinds of tanks or shut frameworks, in which algal cultivation is carried out (Richmond, 2004). In these frameworks, essential ingredients such as water, supplements, and CO_2 are provided in a regulated manner, while oxygen, a product of photosynthesis, is expelled.

6.1.4.3   HETEROTROPHIC BIOREACTORS

Micro-algae can also just as effectively be cultivated in regular "old school" bioreactors, in a heterotrophic manner, rather than the aforementioned photo-bioreactors to deliver high-worth items (Wen and Chen, 2003). Rather than the utilization of light which drives photosynthesis, this heterotrophic method puts to use carbon sources in the culture-medium: these carbon sources not only provide the carbon atoms but also the energy to drive algal cellular machinery (Ward and Singh, 2005).

6.2   ROLE OF ALGAE IN THE PHARMACEUTICAL INDUSTRY

The worldwide market of medicinal enterprises is on a gigantic ascent. All around, other ventures are dwarfed by pharmaceutical businesses and the market size is expanding annually. Algae were consistently useful to humanity; particularly the utilization of cyanobacteria (distantly related to algae, but technically prokaryotes) for antibiotics, and active medical chemicals have attracted an evenly expanding interest.

A myriad of items is being extracted from algae including, but not limited to: antimicrobial agents, antiviral agents, medicinal proteins, medicinal drugs, antifungal agents, etc. Generally, the bases of these activities (antioxidant, antiviral, anticancer, antibiotic, etc.), are subject to the attributes of the algal species being cultivated (Ana et al., 2016).

6.2.1 ANTIOXIDANT ALGAL PRODUCTS

The most remarkable water dissolvable antioxidizing agents that abound in algae are polyphenols, PBPs, and vitamins. Antioxidants assist in the hindrance of oncogenesis and metastasis by forcing a relapse of premalignant lesions (Prakash, 2016). Numerous algal species have assisted in the counteraction of oxidative harm via searching free radicals and dynamic oxygen which aids in the prevention of cancer.

Such chemicals are instrumental in combatting different illnesses including chronic disorders, heart-related issues, and inflammation. Polyphenols, which are quite abundant in marine algae (otherwise called phlorotannins) possess acceptable antioxidant properties.

Sulfate polysaccharides extracted from marine algae discharge radical rummaging exercises (Saleh, 2016). A few strategies for extraction are structured out by research scientists in this regard. Filamentous algae have incredible cancer prevention abilities. Macro-algae possess a myriad of biochemicals, which possess this ability and find much utility in industries (Blaga et al., 2015).

6.2.2 ANTI-CANCER ALGAL PRODUCTS

Generally, marine algae are associated with anticancer treatments as they have a huge scope of abilities. They possess great anti-microbial properties that manage to restrain numerous perilous infections. Oral oncogenesis can be dealt with by the utilization of algae which show cell reinforcement properties. For example, β-carotenes and assorted algal biochemical have been employed in a fight against carcinomas.

The aforementioned segment is under enthusiastic investigation by analysts and individuals from medical businesses. Algal extracts exhibit anticancer properties; cyanobacterium *S. platensis*, for instance, exhibits the most elevated cancer-prevention-agents which prompts effectiveness against cancer. Phyllocobiliproteins, such as those of *Sargassum* sp., are another model under investigation in this regard (Amouzgar et al., 2015).

6.2.3 ANTIVIRAL ALGAL PRODUCTS

At the point when constraints emerged on vaccination, it drove the approach to numerous artificial antiviral compounds for dealing with dynamic herpetic contaminations, yet this proved to be too ineffective (Kostal et al., 2013).

But then, researchers discovered the antiviral capacity in members of the class Phaeophyceae (also colloquially known as brown algae). They possess a huge variety of antiviral activities, which can repress infection. This revelation prompted antiviral chemotherapy (Jia, 2015). Algal polysaccharides are employed to fight against specific viral infections, examples include carrageen, galactan, alginate, nostaflan, fucan, etc.

6.2.4 ALGAL DRUGS

Algae find utility in making complex and directed malignant growth medications. Chloroplasts, the organelles responsible for carrying out photosynthesis, make it progressively accommodating. Researchers and analysts are tirelessly pursuing genetic engineering in micro-algae. These prove to be useful in destroying unsafe malignant growth cells, prompting tumor medications. This is a significant breakthrough in the improvement of malignant growth medicate treatment.

Micro-algae assume a major role in developing anti-malignant growth medicates; for instance, Cryptophycin, a biochemical extracted from algae, forms a solid segment in the process of development of anti-cancer medications. Algae possess an incredible capacity of collapsing proteins into complicated 3-D conformations. Human antibodies can be effectively created via algae (Visconti et al., 2015). Research facilities were keen on fusing algal proteins which help in antibody development; these were created in the chloroplasts.

Alkaloidal neurotoxins (toxins that target neurons), i.e., saxitoxin and polyketide are also produced by them. In regulated doses, these chemicals have anti-inflammatory action and are hostile to malignant growth (Vinale, 2015). Macro-algae, on the other hand, possess alkaloids offering an approach to anti-malignancy medications.

Various algal species possess varying levels and different natures of pharmaceutical attributes. This makes each of them special from others. They find utility in a wide variety of medical treatments; the sheer variety of these drugs is staggering as it is, examples of genera that host species that produce commonly used drugs include: *Enteromorpha, Acetabularia, Laminaria, Sargassum, Gelidium, Corallina, Grateloupia*, etc.

6.3   ALGAE AS NUTRACEUTICALS

The expression "nutraceutical" alludes to the type of nourishment that is bestowed with physiological advantages related to counteraction or potential assurance against constant infections. Nutraceuticals incorporate blends and unadulterated chemicals separated from plants or algae; examples include cereals, spices, condiments, and beverages.

Among prominent wellsprings of health-based nourishments and premium chemo-preventive foods, micro-algae are attracting more and more consideration from all over the world. In light of their extravagance in high-esteem items such as carotenoids, proteins, vitamins, basic amino acids, omega-rich lipids, and compounds that have proven themselves in being effective in mitigating and preventing cancer. Advantageous impacts of micro-algae on human well-being as well as health could, later on, help forestall or postpone the beginning of malignancy and cardiovascular maladies.

In the world of nutraceuticals, microalgae have surfaced as one of the most effective and potent contenders. They are enriched with advantageous impacts on our general well-being. Microalgae are an enormous aggregate of unicellular prokaryotic and eukaryotic living beings that are predominantly autotrophic and photosynthetic.

Likewise, microalgae find utility in high-esteem nourishment, well-being nourishment for humans, polysaccharides, additives in human food and animal feed, beauty care products, cancer prevention agents, anti-inflammatory drugs, arrangement of biofilms, etc. (Varfolomeev et al., 2011).

Cyanophyceae (blue-green growth), Chlorophyceae (green growth), Bacillariophyceae (counting diatoms), and Chrysophyceae (counting brilliant green growth) are among the most generally utilized micro-algae. For the past two decades, biotechnological, and nutraceutical utilization of micro-algae has concentrated explicitly on four significant algal species: (a) *Arthrospira* sp., (b) *Chlorella* sp., (c) *Dunaliella salina*, and (d) *Haematococcus pluvialis*.

Marine microalgae have attracted much attention from the realm of biotechnology owing to their high growth and production rate. This interest has been quickly developing in the most recent decade as a wellspring of economic biochemicals, proteins, and metabolites. A few algal species, for example, *Chlamydomonas reinhardtii*, are as of now perceived as conceivable modern biotechnological stages for the creation of an assortment of bioactive chemicals (Singh et al., 2010) and are presently dependent upon hereditary controls (Varfolomeev et al., 2011); the idea is to expand its biotechnological potentials.

Microalgae likewise offer much promise in the field of health-based diet because they create significant bioactive fixings, for example, carotenoids with definitely realized medical advantages (Rindi et al., 2007), additionally, other cancer prevention agents are also extracted from them, making the deal irrefutable (Lewis et al., 2002).

Algal nutraceuticals, as of yet, are in exceptionally high demand in the markets, generally. Although the extension of strains is exceptionally little, essential algal nutraceuticals cover a broad range, including nourishment supplements, dietary enhancements, processed foods with added value, and non-dietary enhancements encompassing soft gels and tablets (Chandra, 2014). Omega-3 polyunsaturated fatty acids (PUFA), β-carotene, astaxanthin, and other carotenoids are among the most significant items extracted from algae.

6.3.1 CAROTENOIDS

Micro-algae are by and large generally utilized to produce algal nutraceutical products. Various species of *Chlorella, Dunaliella, Haematococcus, Arthrospira, Aphanizomenon*, are broadly assessed by scientists for their latent capacity in this regard and are hence subjects of incessant research and speculation.

The validity of various algal species towards nutraceutical advancement is by and large dependent upon the protein percentage of its cells. Components extracted from *Chlorella* and *Arthrospira* have great cell reinforcing, inflammation mitigating, and tumor-suppressing properties. *Haematococcus* has numerous vitamins in it which makes it all the more intriguing towards algal-nutraceutical improvement. *Aphanizomenon*, on the other hand, assumes an immense job in cholesterol controlling, incitement of the capacities of the liver, and is a solution for some dermatological issues (Ajjane et al., 2015).

6.3.2 ASTAXANTHIN

Astaxanthins are pigments produced from carbon antecedents; they are highly dissolvable in lipids. Natural astaxanthins (nASX) are far better and much more preferable than the artificial synthetic versions. They are fundamentally anti-oxidizing agents with somewhat lesser action capacity yet a great free radical terminating of every carotenoid. Astaxanthins find immense utility in nourishment additives (Paul et al., 2013).

6.3.3   OMEGA-3 POLYUNSATURATED FATTY ACIDS (PUFA)

Polyunsaturated fatty acids (PUFA) play a significant role in body metabolism. Chlorophytes and Bryophytes are commonly utilized for the extraction of these unsaturated fatty acids, loaded with health benefits. One example health-rich unsaturated fatty acid is N-6 PUFA which is extracted from the aforementioned groups. The unsaturated fat percentage in algal species makes it a notable bioactive chemical substance that is valuable in the medical business (Aditya et al., 2018).

6.4   CONCLUSION

With the continual boost in the human populace, the interest for nutritive nourishment and well-being assuring items amplifies associatively. The wellsprings of healthy algal biomass that can fulfill this need are sought after wildly. Their wide assorted variety, quick growth rate, and different uses make them effortlessly acknowledged for business culture. Hence the need for algaculture is more than it ever was in the past. Not only is it an easy solution to an otherwise dire situation, but it is also a field with extensive research data associated with it. The advent of modern technologies such as photobioreactors (PBRs) has made the cultivation of algae more practical than it ever was in the past.

Micro-algae need comparatively fewer assets when contrasted with different yields. The need of algal species in ensuring human well-being and sustenance will persistently escalate with extra research being devoted to the territories of medical advantages and growing algal cultures. Algal species are at the moment being cultivated because their applications principally incorporate nourishment, nutritional additives, aquaculture, colorants, beauty care products, medicinal drugs, and nutraceuticals. Only a handful of species of algae are being developed for human use as of yet. Potentially, the uses of thousands of species are not known to us at the time being, but their capabilities might soon be discovered. Subsequently, the possibility of using algal species in the domains of nourishment utilization, well-being supplements, fuels, and energy; surely a lot more applications are probably going to surface shortly.

KEYWORDS

- biochemical
- cancer
- eukaryotic
- multicellular
- nutraceutical
- pharmaceutical
- unicellular

REFERENCES

Aditya, T., Bitu, G., & Mercy, E. G., (2018). The role of algae in pharmaceutical development. *Reviews on Pharmaceutics and Nanotechnology, 4*, 82–89.

Amouzgar, P., & Salamatinia, B., (2015). A short review on the presence of pharmaceuticals in water bodies and the potential of chitosan and chitosan derivatives for elimination of pharmaceuticals. *J. Mol. Genet. Med., S4*, 001.

Ana, V. B., Enrique, R. Z., Sara, P. C., Sofía, P. G., Luis, V. R. J., & Ricardo, G. C., (2016). Myxofibrosarcoma following chemotherapy and radiotherapy for Hodgkin's lymphoma: Case study and review. *J. Clin. Case Rep., 6*, 816.

Andersen, R. A., (2013). The microalgal cell. In: Richmond, A., & Hu, Q., (eds.), *Handbook of Microalgal Culture: Applied Phycology and Biotechnology* (2nd edn., pp. 1–20). Wiley, Oxford.

Ben-Amotz, A., & Avron, M., (1980). Glycerol, β-carotene, and dry algae meal production by commercial cultivation of *Dunaliella*. In: Shelef, G., & Soeded, S. J., (eds.), *Algae Biomass* (pp. 603–610). Elsevier North-Holland Biomedical Press, Oxford.

Blaga, M., Sava, M., Pedro, F., & Strahil, B., (2015). Microbial transformations of plant origin compounds as a step-in preparation of highly valuable pharmaceuticals. *J. Drug Metab. Toxicol., 7*(2), 1–11.

Borowitzka, L. J., & Brown, A. D., (1974). The salt relation of marine and halophilic species of *Dunaliella*: The role of glycerol as a compatible solute. *Arch Microbiol., 96*, 37–52.

Borowitzka, M. A., & Borowitzka, L. J., (1987). Vitamins and fine chemicals from micro-algae. In: Borowitzka, M. A., & Borowitzka, L. J., (eds.), *Micro-Algal Biotechnology*. Cambridge University Press, New York.

Borowitzka, M. A., (2013). High-value products from microalgae - their development and commercialization. *J. Appl. Phycol., 25*, 743–756.

Chandra, K. K., (2014). Growth, fruit yield and disease index of *Carica papaya* L. inoculated with *Pseudomonas straita* and inorganic fertilizers. *J. Biofertil. Biopestici., 5*, 146.

Costa, J. A. V., Colla, L. M., & Duarte, P., (2003). *Spirulina platensis* growth in open raceway ponds using freshwater supplemented with carbon, nitrogen, and metal ions. *Zeits Naturforsch, C-A J Biosci., 58*, 76–80.

EPA (US Environmental Protection Agency), (2013). *Introduction to In Situ Bioremediation of Groundwater.* https://www.epa.gov/remedytech/introduction-situ-bioremediation-groundwater (accessed on 2 July 2021).

EPA, (2011). *Green Remediation Best Management Practices: Sites with Leaking Underground Storage Tank Systems.* https://www.epa.gov/remedytech/green-remediation-best-management-practices-sites-leaking-underground-storage-tank (accessed on 2 July 2021).

FAO Fisheries and Aquaculture Secretariat, (2010). *The State of World Fisheries and Aquaculture-2010.* Food and Agriculture Organization of the United Nations, Rome, Italy. http://41.215.122.106/dspace/handle/0/210 (accessed on 2 July 2021).

Gors, M., Schumann, R., Hepperle, D., & Karsten, U., (2010). Quality analysis of commercial *Chlorella* products used as dietary supplement in human nutrition. *J. Appl. Phycol., 22,* 265–276.

Hochman, G., & Zilberman, D., (2014). Algae farming and its bio-products. In: Maureen, C. M., Marcos, S. B., & Nicholas, C. C., (eds.), *Plants and BioEnergy, Advances in Plant Biology* (Vol. 4, pp. 49–64). Springer Science+Business Media, New York.

Hossain, A. B. M., Salleh, A., Boyce, A. N., Chowdhury, P., & Naqiuddin, M., (2008). Biodiesel fuel production from algae as renewable energy. *Am. J. Biochem. Biotechnol., 4,* 250–254.

Jia, C. X., & Zhang, M. P., (2015). Study on the methods of β-carotene extraction of *Spirulina platensis. Research and Reviews: Journal of Botanical Sciences.* http://www.rroij.com/open-access/study-on-the-methods-of--carotene-extraction-of-spirulina-platensis.php?aid=57950 (accessed on 2 July 2021).

Khan, Z., Bhadouria, P., & Bisen, P. S., (2005). Nutritional and therapeutic potential of *Spirulina. Curr. Pharm. Biotechnol., 6,* 373–379.

Kosta, S. P., Kosta, Y. P., Bhatele, M., Pandya, K., Soni, D., Gour, A., Dubey, Y. M., et al., (2013). Biological human tissue-based electronic circuits-an alternate to drug therapy for sick-man (a perspective visionary concept). *J. Bioeng. Biomed. Sci., 3,* 127.

Lee, Y. K., (2001). Microalgal mass culture systems and methods: Their limitation and potential. *J. Appl. Phycol., 13,* 307–315.

Lewis, L. A., & Flechtner, V. R., (2002). Green algae (Chlorophyta) of desert microbiotic crusts: Diversity of North American taxa. *Taxon., 51,* 443–451.

Munoz, R., & Guieysse, B., (2006). Algal-bacterial processes for the treatment of hazardous contaminants: A review. *Water Res., 40,* 2799–2815.

Paul, N., Cruz, P. C., Aguilar, E. A., Badayos, R. B., & Hafele, S., (2013). Evaluation of biofertilizers in cultured rice. *J. Biofert. Biopest., 4,* 133.

Prakash, S. B., (2016). Nutritional therapy as a potent alternate to chemotherapy against cancer. *J. Cancer Sci. Ther., 8*(6), 168.

Pulz, O., & Gross, W., (2004). Valuable products from biotechnology of microalgae. *Appl. Microbiol. Biotechnol., 65,* 635–648.

Radakovits, R., Jinkerson, R. E., Darzins, A., & Posewitz, M. C., (2010). Genetic engineering of algae for enhanced biofuel production. *Eukaryote Cell, 9,* 486–501.

Ratledge, C., (2004). Fatty acid biosynthesis in microorganisms being used for single-cell oil production. *Rec. Adv. Lip. Metab. Relat. Disord., 86,* 807–815.

Richmond, A., (2004). Principles for attaining maximal microalgal productivity in photobio-reactors: An overview. *Hydrobiologia, 512,* 33–37.

Rindi, F., McIvor, L., Sherwood, A. R., Friedl, T., Guiry, M. D., & Sheath, R. G., (2007). Molecular phylogeny of the green algal order *Prasiolales* (Trebouxiophyceae, Chlorophyta). *Journal of Phycology, 43,* 811–822.

Rosenberg, J. N., Oyler, G. A., Wilkinson, L., & Betenbaugh, M. J., (2008). A green light for engineered algae: redirecting metabolism to fuel a biotechnology revolution. *Tissue Cell Pathw. Eng., 19*, 430–436.

Saleh, T. A., (2016). Nanomaterials for pharmaceuticals determination. *Bioenergetics, 5*(1), 1–6.

Sheehan, J., Dunahay, T., Benemann, J., & Roessler, P., (1998). *A Look Back at the US Department of Energy's Aquatic Species Program: Biodiesel from Algae*. National Renewable Energy Laboratory, Golden, NREL/TP-580-24190.

Singh, J., & Gu, S., (2010). Commercialization potential of microalgae for production of biofuels. *Renewable and Sustainable Energy Reviews, 14*, 2596–2610.

Siva, K. R. R., Madhu, G. M., Satyanarayana, S. V., Kalpana, P., Bindiya, P., & Subba, R. G., (2015). Equilibrium and kinetic studies of lead biosorption by three *Spirulina* (*Arthrospira*) species in open raceway ponds. *Journal of Biochemical Technology, 6*(1), 894–909.

Trentacoste, E. M., & Zenk, T., (2015). The place of algae in agriculture: Policies for algal biomass production. *Photosynth Res., 123*, 305–315.

Varfolomeev, S. D., & Wasserman, L. A., (2011). Microalgae as a source of biofuel, food, fodder, and medicine. *Applied Biochemistry and Microbiology, 47*, 789–807.

Vinale, F., (2015). Biopesticides and biofertilizers based on fungal secondary metabolites. *J. Biofertil. Biopestici., 5*, e119.

Visconti, G. L., Mazzoleni, L., Rusconi, C., Grazioli, V., Roda, G., Manini, G., & Gambaro, V., (2015). Determination by UPLC/MS-MS of Coenzyme Q10 (CoQ10) in plasma of healthy volunteers before and after oral intake of food supplements containing CoQ10. *J. Anal. Bioanal. Tech., S13*, 011.

Ward, O. P., & Singh, A., (2005). Omega-3/6 fatty acids: Alternative sources of production. *Process Biochem, 40*, 3627–3652.

Wen, Z. Y., & Chen, F., (2003). Heterotrophic production of eicosapentaenoic acid by microalgae. *Biotech. Adv., 21*, 273–294.

Williams, P. W., & Phillips, G. O., (2000). Agar is made from seaweed and it is attracted to bacteria. *Handbook of Hydrocolloids* (p. 91). Woodhead Publishing, Cambridge.

CHAPTER 7

BIOLOGICAL IMPORTANCE OF ALGAL METABOLITES

USMAN ALI CHAUDHRY,[1] ANEELA NAWAZ,[2] AMEET KUMAR,[2] FARIHA HASAN,[2] MALIK BADSHAH,[2] and SAMIULLAH KHAN[2]

[1]Infection Control and Disease Prevention Center, Ministry of Health, Tabuk, Kingdom of Saudi Arabia

[2]Department of Microbiology, Faculty of Biological Sciences, Quaid-i-Azam University, Islamabad–45320, Pakistan

ABSTRACT

Algae are the complex and diverse group of photosynthetic organisms that can be unicellular and multicellular. Algae being autotrophic organisms channelize sunlight to produce food and metabolites. Algae are the producers of many bioactive metabolites with biological activities. The diverse compounds produced by algae using different metabolic pathways are considered as promising source of agarose, alginic acid, carrageenan, carotenoids, fatty acids, halogenated compounds, lectins, mycosporine-like amino acids (MAAs), polysaccharides, polyketides, and steroids. Because of biochemical composition and therapeutic properties, algae have been used for medicinal purposes for many centuries. Algae are considered for medicinal purposes because of their antioxidant, anti-inflammatory, antimicrobial, antiviral, and anticancer properties. Algae are now considered as novel source of bioactive metabolites and have been reported as the richest source of curacin and apratoxins. Some compounds of algae exhibited interesting results and have reached to phase II and III of clinical trials. Algae can reduce the risk of development of degenerative diseases and can also possess health-promoting potentials. From last few decades, algae have received increasing interest in the pharmaceutical industry because of their convincing properties that make them stand out over synthetic drugs.

7.1 INTRODUCTION

Algae can be unicellular or multicellular, and on the basis of its size, it has been divided into microalgae and macroalgae. Microalgae are considered as the earliest form of life that started their existence 3 billion years ago from the oceans of earth (Russell, 2006). The diversity of algae is vast and has not been fully exposed yet. The efficiency of algae in the conversion of solar energy into chemical energy and fixation of atmospheric carbon dioxide is 10 times greater than plants (Sayre, 2010). Macroalgae are the source of renewable and abundant compounds that can be explored for nutraceutical and pharmaceutical applications. The algal biomass is considered as immense source of valuable compounds like phycotoxins, polyunsaturated fatty acids (PUFA), alkaloids, phenols, flavonoids, tannins, saponins, terpenes, vitamins, carotenoids, phycoerythrin (PE), polysaccharides, phycocyanin (PC), phycobilin, and phycobiliproteins that can be exploited for different biotechnological applications. Algae can produce toxic metabolites that are called phycotoxins that can be used in agriculture as pesticides (Sosa-Hernández et al., 2019).

The growing market of nutraceutical and pharmaceutical is attracting consumers, globally. The growing research on the pharmaceutical products is focusing on the exploration of bioactive compounds from algae for exploring the pharmaceutical potential and global marketing of its valuable compounds. Because of biochemical compositions of algae that have potential health benefits, are considered as the promising source of sustainable food. Research on exploration of pharmaceutical potential of algae is just started that involve the enhancement of the production of bioactive metabolites with nutraceutical and pharmaceutical value. Scientific studies have demonstrated the beneficial effect of natural compounds (low toxicities and high efficacies) as compared to synthetic, so the interest of researchers in the exploration of natural compounds is increasing. The products of algae that have been commercialized or under consideration are sterols, vitamins, fatty acids, polysaccharides, carotenoids, and phycobilin and some of these products have antimicrobial, antiviral, antifungal, antiplasmodial, anticancerous, immunosuppressive, and neuroprotective potential. The use of algae in the field of medicine is rising because of high pharmaceutical values of algal compounds (Skjånes et al., 2013). The algae use different metabolic pathways for the production of algal metabolites, in prokaryotes; isoprenoids are synthesized by mevalonate pathway while in algae non-mevalonate pathways are used (Lohr et al., 2012). The production of bioactive metabolites can be by primary metabolic pathways or by secondary metabolic pathways. Most of the algae accumulate bioactive compounds in biomass while some excrete them out of the cell.

During the last few decades, several compounds from algae with antimicrobial, anticancerous, and anti-inflammatory activity have been discovered, showing that these organisms have great biotechnological importance. Recently two anticancer compounds dragonamide C and dragonamide D have been discovered from cyanobacterium (Gunasekera et al., 2008). Dragonamide A, and E showed in-vitro anti-leishmaniasis activity. In 2010 almiramides A–C from *Lyngbya majuscula* have been identified with anti-leishmaniasis activity. Cryptomycin was extracted from *Nostoc* with anti-cancer activity, almost 26 different forms of cryptomycin are discovered yet and Crytomycin 52 is in phase II of clinical trials against lungs and ovarian cancer. Borophycin obtained from *Nostoc linckia* and *Nostoc spongiaeforme* also appeared as anticancer (Singh et al., 2005). β-carotene, a fat-soluble photosynthetic pigment, has been announced as an anticancer compound by National Cancer Institute (Tanaka et al., 2012). It has also been proved as effective in controlling heart diseases. Because of these properties, the demand of β-carotene is increasing in the market. *Arthrospira* is the important source of γ-linolenic acid (GLA) that is important in lowering blood pressure and lipid metabolism (De La Jara et al., 2018).

This chapter deals with the biochemical components of algae with pharmaceutical value, biological activities of algae and its derived bioactive compounds, and the importance of algae for the production of valuable bioactive compounds of pharmaceutical importance.

7.2   ALGAE AS A SOURCE OF BIOACTIVE COMPOUNDS

Algae have been given too much attention by researchers and companies for their application in different fields of life because of a high proportion of production of bioactive compounds. The applications of algae range from food and feed to the production of compounds of pharmaceutical values. Algae are becoming the promising source of novel products because of a vast diversity of algae and developments in genetic engineering. Algae use inorganic compounds and sunlight to produce compounds like proteins, polysaccharides, lipids, phycobilins, vitamins, and sterols of various biological activities. A huge number of bioactive metabolites has been identified with antimicrobial, antifungal, antiviral, anti-inflammatory, anticancer, antioxidant, and anti-enzymatic activity by extensive research on *Arthrospira*

(*Spirulina*), *Botryococcus braunii, Chlorella vulgaris, Dunaliella salina, Haematococcus pluvialis*, and *Nostoc* (Amaro et al., 2011).

Intracellular and extracellular metabolites of Cyanobacteria with the potential biological activities like antibacterial, antiviral, antifungal, anti-plasmodial, anti-inflammatory, immunosuppressant, and herbicidal are identified. *Spirulina* produces essential PUFA (10–20% w/w), proteins (50–70% w/w) of high value and digestibility, vitamins (B1, B2, B12, and E), pigments, minerals, oligo-elements (iron, potassium, magnesium, phosphorous, and calcium) and phenolic compounds. *Spirulina* because of production of active bio-compounds is studied, worldwide. Several studies reported the therapeutic importance of *Spirulina*, for example, in the treatment of cancer, AIDS, hyperlipidemia, obesity, hypertension, and diabetes (Shao et al., 2019). Biomass of *Nostoc* is used as a dietary supplement and also in pharmaceutical products. This alga is helpful in the treatment of cancers and fistula. The biomass of *Nostoc* helps in controlling blood pressure, treating infections and cancers, boosting the immune system and also in treating digestive problems. Cyanovirin is the protein produced by *Nostoc* showed positive effect against influenza A and human immunodeficiency virus (HIV). *Nostoc* produces essential fatty acids (EFAs) that are the precursors of prostaglandins. The nutritional and pharmaceutical importance of *Nostoc* is encouraging its cultivation on a large scale (Bhattacharjee, 2016). *Chlorella* was discovered by the Japanese and was used as food supplements because of production of high valued nutrients especially vitamin B12 that play an important role in the formation of blood cells. Bioactive substances produced by *Chlorella* showed anticoagulant, antioxidant, antitumor, antibacterial, and immunostimulatory activity. The most important bioactive compounds of *Spirulina* with free radicals scavenging property are ascorbic acid, α-carotene, β-carotene, and α-tocopherol. β-1,3-glucan produced by *Chlorella* is immune-stimulator, antioxidant, can reduce blood cholesterol, provides protection against gastric ulcer, and reduces constipation (Nakano et al., 2010). Biomass of *Dunaliella* has bioactivities like bronchodilatory, analgesic, antioxidant, hepatoprotective, antihypertensive, antiedemal, and antibacterial. Crude extract of *Dunaliella* showed antibacterial activity against *Staphylococcus aureus, Bacillus cereus, Bacillus subtilis, Enterobacter aerogenes,* and *Escherichia coli* and antifungal activity against *Candida albicans* and *Aspergillus niger*. In the USA, Australia, China, and Israel *Dunaliella* is cultivated for the production of β-carotene (Jena and Subudhi, 2019). Fucoxanthin produced by Phaeophyta and phycobilin, phycocyanin, and phycoerythrin produced by Rhodophyta are of pharmaceutical importance (Rasmussen and Morrissey, 2007). Fucoxanthin provides protection against cancer by acting as an antioxidant, antiproliferative, and by anti-angiogenic mechanisms. Dieckol that is phlorotannin produced by *Ecklonia cava* provides protection against type II diabetes (Lee et al., 2010). Seaweeds are used to treat goiter because of the presence of appreciable levels of iodine in their biomass. *Laminaria japonica* and *Sargassum fusiforme* had been used for the treatment of goiter, edema, and scrofula (Olasehinde et al., 2019). Phlorotannins extracted from brown algae have shown to increase the activity of alkaline phosphatase, protein, and collagen synthesis and mineralization in human osteosarcoma cells (Li et al., 2011).

7.3 BIOCHEMICAL NATURE OF BIOACTIVE COMPOUNDS OF ALGAE

7.3.1 POLYSACCHARIDES

Accumulation of carbohydrates in algae is in the form of glucose, starch, and other polysaccharides are in large fraction. The composition of carbohydrates in algae varies from species to species and is also dependent on environmental conditions. Polysaccharides are the most important components of the algal cell wall, help in the signaling pathways of algae, and regulate defense against the environmental changes. In recent years the biological effect of polysaccharides of seaweeds was investigated. Biological activity of polysaccharides is based on the bound sulfate to the hydroxyl group of sugar. Formation of sulfated polysaccharides like carrageenan and alginic acid help to reduce the absorption of cholesterol from the gut. Algal polysaccharides are now the focus of researchers because they are bioactive compounds that show various biological activities like antiviral, antitumor, antioxidant, anti-inflammatory, anticoagulant, and antiproliferative (Gupta and Abu-Ghannam, 2011). Agarose, highly purified agar is used in pharmaceutical products as laxatives, for the formation of capsule, anticoagulant, bulking agent, tablets, and suppositories. In vitro apoptotic activity of agar persuaded that it can be employed in cancer therapy.

7.3.2 FATTY ACIDS

PUFA like omega-3 fatty acids are the important component of eukaryotic cell membranes, principally of neuronal cells and algae. Algae are considered as the richest source of EFAs like eicosapentaenoic acid (EPA) and docosahexaenoic acid (DHA) (Ryan et al., 2009). The supplements of EPA and DHA prevent cardiovascular inflammatory diseases. Regular consumption of omega-3 fatty acids reduces the risk of thrombosis, hypertension, myocardial infarction, and cardiac arrhythmia by decreasing blood cholesterol level and low-density fatty acids. These fatty acids play an important role in the neural tube development of fetus. DHA plays a crucial role in the development and normal function of the brain. Severity of depression can also be reduced by the consumption of DHA (Mori et al., 2003). During the treatment of asthma, ulcerative colitis, Crohn's disease, lupus, rheumatoid arthritis, cystic fibrosis, and psoriasis immune-modulatory effect of omega-3 fatty acid has also been observed (Farooqui, 2009).

7.3.3 PROTEINS

The use of algae alternative to food started with Second World War for overcoming the dietary protein deficiency. Because of the high protein content of algae, it is considered as potential source of production of valuable proteins of therapeutic importance. Comprehensive and nutritional analysis of proteins of algae has revealed that the proteins produced by algae are of high quality compared to proteins of plants. Due to high protein content *Arthrospira, Cholorella*, and *D. salina* were utilized in human diet (Spolaore et al., 2006). Recombinant DNA technology and hybridoma made the large-scale production of protein possible that can be used in biopharmaceuticals. Phycobiliproteins (PBPs) from algae have been reported with antibacterial, anticancer, anti-inflammatory, and antioxidant activities (Kannaujiya et al., 2017). The production of proteins from algae is of low cost and can be used as therapeutic proteins, vaccines, and antibodies. Proteins contain a large part of the biomass of algae, but still little attention is paid to the algal proteins.

7.3.4 VITAMINS AND MINERALS

Algae are considered as the richest source of vitamins and minerals. Algae are reported for the production of vitamin E, inositol, biotin, cobalamin, and thiamine. The vitamins from algae are used for the production of vitamin supplements. Biomass of algae contains several kinds of minerals in different compositions that are used as food and feed supplements. The vitamins and minerals of algae are used for improving cosmeceutical products and also for curing skin diseases (Bishop and Zubeck, 2012).

7.3.5 GLYCEROL

Glycerol, an organic osmolyte is accumulated in algae due to external water activity. Glycerol has a wide range of application in textile industry, leather, paper, and pulp, tobacco, cosmetics, food, paint, and pharmaceutical industry. Halotolerant species of algae accumulate glycerol in large amounts for-example, *Dunaliella* accumulates 17% w/w of glycerol, and *C. reinhardtii*, can accumulate glycerol in large amounts (Demmig-Adams et al., 2017).

7.3.6 STEROLS

Sterols are organic molecules that are a subgroup of steroids. Wide varieties of phytosterols are produced by microalgae that are stigmasterol, rassicast-erol, and sistosterol. These sterols have wide applications in pharmaceutical products and functional foods. *D. tertiolecta* and *D. salina* produce phytos-terol in high quantity under different salt concentrations. The annual increase in the global market of sterols is 5–7% (Francavilla et al., 2010).

7.3.7 CAROTENOIDS

Carotenoids are the natural pigments produced by algae in high quantity, used for coloring natural matrices and are essential for healthy eyes. Carotenoids constitute of eight isoprenoid units. The production of carot-enoids from microalgae is a successful activity. β-carotene is the precursor molecule of vitamin A and is used in lotions, shampoos, hair conditioners and makeup products (Christaki et al., 2013). Astaxanthin is an important carotenoid from microalgae having antioxidant activity, and it is used for the protection of skin from the effect of ultraviolet (UV) radiations and treatment age-related damages. It is also used in the foods of athletes for increasing their power and decreasing the time of their muscle's recovery. Lutein and zeaxanthin have extensively been used in nutraceutical and pharmaceutical products. Lutein and zeaxanthin protect the macula from photo-oxidative damage that causes blindness and age-related macular degeneration because these carotenoids concentrate in macula luteal where they absorb blue light in human retina. Luteins scavenge free radicals and absorb high energy of blue light of sunlight and ultimately protect human skin (Shegokar and Mitri, 2012). Lycopene was found effective against prostate cancer.

7.3.8   LECTINS

Lectin have a role in medical sciences, and algae are considered as good source. Due to bioadhesive property of lectin, it is used for the delivery of vaccine across the mucosal surface (Baudner and O'Hagan, 2010).

7.3.9   PHENOLICS

More than 8,000 phenolic compounds with different chemical structure have been reported and are divided into different classes. Phenolic compounds are considered as natural antioxidants, and the mechanism they use is the transfer of single electron and hydrogen atom from the reactive free radicals. Phenolic compounds produced by algae belong to flavonoids. Flavonoids are the high valued natural products that are used in cosmetics, nutraceutical, and pharmaceutical products and have been reported for their anti-inflammatory, antioxidant, and anticancer activity (Panche et al., 2016).

7.3.10   ISOTOPIC COMPOUNDS

Algae are considered as a source of stable isotopic compounds for the production of high valued stable isotopic organic chemicals. These stable isotopic compounds are used for the elucidation of the molecular structure of chemicals and the diagnosis of breathing and gastrointestinal problems. Because of various uses of these isotopic compounds, the market value of these chemicals has increased to 13 million US Dollars (Milledge et al., 2011). Production of ^{13}C or N^{15} labeled biomass from *P. tricornutum* on commercial-scale will open a new era for the isotopes production from natural sources.

7.4   BIOLOGICAL ACTIVITIES OF ALGAE AND ITS BIOACTIVE COMPOUNDS

7.4.1   ANTIMICROBIAL ACTIVITY

It is a fact that algae are the source of bioactive compounds, but the utilization of algal compounds as antibacterial agents is still in its infancy. Constant use of currently available antibiotics in clinical practices has developed resistance in pathogens that has become a threatening issue. Discovery or the development of new antibiotics is the biggest need of this era to deal the resistant pathogens. From natural sources, different kinds of antimicrobial compounds like phenols, indols, terpenes, and acetogenins are reported. Algae produce different antimicrobial compounds to deal with the pathogenic bacteria and other microorganisms of their environment (Falaise et al., 2016). The metabolites produced by algae are either bactericidal or bacteriostatic. Algal bioactive compounds that show antibacterial activity belong to carbohydrates, fatty acids, sterols, terpenoids, acrylic acids, phenols, halogenated aliphatic compounds, and sulfur-containing heterocyclic compounds. Antimicrobial activity of algae is mostly because of lipid components like phytol, neophytadine, β-cyclocitral, α- and β-ionone (de Morais et al., 2015). The antimicrobial activity of algae reported against *Escherichia coli*, *Pseudomonas aeruginosa*, *Staphylococcus aureus*, and *Staphylococcus epidermidis* is because of arachidonic acid, DHA, EPA, GLA, hexadecatrienoic acid, lactic acid, lauric acid, oleic acid and palmitoleic acid. Various structures in microorganisms are affected by fatty acids and the most effected areas are cell membranes of the target organisms. The mechanisms of action of these antimicrobial compounds includes, damaging the membrane, cause loss of internal material of cell, entry of harmful substances into the cell, and inhibition of cellular respiration that ultimately causes cell death. The efficiency of fatty acids to disrupt bacterial cells is dependent on the chain length and level of saturation of fatty acids. Fatty acids that contain more than 10 carbons can induce bacterial lysis (Desbois and Smith, 2010). Polysaccharides of algae have also been reported for antimicrobial activity. Sulphated polysaccharides are the potent antibacterial compounds that are produced by algae and are secreted in the external environment of algae (Ibañez et al., 2012). The current challenge of the pharmaceutical industry is to develop antibiotics against multidrug resistance *S. aureus* that has even developed resistance against vancomycin (Frieri et al., 2017). Halogenated compound produced by *Laurencia* species appeared as effective against multi drug-resistant *S. aureus* (Santos et al., 2010). γ-lactone malyngolide extracted by dichloromethane from *Lyngbya majuscula* was effective against *Mycobacterium smegmatis* and *Streptococcus pyogenes* (Jena and Subudhi, 2019). Diphenyl ether from *Cladophora fascicularis* inhibited the growth of *Escherichia coli*, *Bacillus subtilis* and *Staphylococcus aureus*. Cycloeudesmol, a cyclopropane antibiotic extracted from *Chondria oppositiclada* was effective against *Staphylococcus aureus* and *Candida albicans* (Sims et al., 1975). Lyengaroside A from *Codium iyengarii* and green algae and extract of *Caulerpa prolifera* appeared as antibacterial. Long-chain fatty acids from *P. nureskis* have been reported with antibacterial activity against *C. jejuni*, *E. coli*, and *S. enterica*.

Fucus vesiculosus has been reported for the production of lanosol enol ether with antifungal and antibacterial activity (Kim, 2011). *Cystoseira tamariscifolia* has been reported for the production of methoxybifurcarenone which appeared as active against *Agrobacterium tumefaciens (*de Sousa et al., 2017).

7.4.2 ANTIFUNGAL ACTIVITY

Lyngbya majuscula produces Majuscuiamide C 16, a cyclic depsipeptide that can inhibit fungal plant pathogens (Lavakumar, 2012). Antifungal compound zonarol was extracted from *Dictyopteris zonaroides*.

7.4.3 ANTIPROTOZOAL AGENTS

Marine algae produce valuable antimalarial compounds of diverse chemistry that include alkaloids, cyclic peptides, linear peptides, and alkylated phenols. Gallinamide A-18 that resembles linear peptides dolastatin 10 and 15 had showed antimalarial activity. Recently novel antimalarial compounds viridamide A-14 and B-15 were isolated from marine algae. Viridamide A, almiramide A-17, gallinamide A-18 and dragonamide E produced by algae have been reported for anti-leishmanial activity. Symplostatin 4–19 has been reported as the most potent antimalarial compound produced by marine algae with an EC_{50} value of 74 nM against *Plasmodium falciparum*. Ascosalipyrrolidinones A 61 and B 62 produced by *Ascochyta salicorniae* has been reported for antimalarial activity (Torres et al., 2014).

7.4.4 ANTHELMINTIC ACTIVITY

Chondria atropurpurea produced Chondriamide C and 3-indolacrylamide that showed anthelmintic activity against *Nippostrongylus brasiliensis*. *Jania rubens* has been reported for the production of brominated diterpenes that were antihelmintic (El Gamal, 2010).

7.4.5 ANTIVIRAL ACTIVITY

Highly sulfated polysaccharides compounds produced by algae with antiviral activity mostly consist of glucose, galactose, and xylose. The sulfate group, chemical structure, type of sugar, molecular weight, degree of sulfation, stereochemistry, and conformation determine the characteristics of sulfated polysaccharides, higher the presence of sulfate group higher the antiviral activity. The antiviral activity of sulfated polysaccharides is directly due to interaction with the receptors of virus or interaction with the positive charges present on cell surfaces ultimately cause inhibition of virus binding site on host cell that stop entry of virus into the host cell (Talarico et al., 2005). *Spirulina*, *Chorella*, and *Dunaliella* produce sulfated polysaccharides with antiviral activity. Sulphated polysaccharides produced by microalgae can inhibit virus that include encephalomyocarditis virus (HIV), herpes simplex virus types 1 and 2, swine fever virus, and varicella virus (Grice and Mariottini, 2018). Sulphated polysaccharides produced by marine algae can inhibit the replication of enveloped virus-like navirus, herpes virus, orthopoxvirus, rhabdovirus, and togavirus. Complex sulfated polysaccharides produced by marine algae showed inhibition of HIV (Harden et al., 2009). Sulphated polysaccharide that is sulfated glucuronogalactan produced by *Schizymenia dubyi* showed anti-HIV activity by inhibiting the attachment of virus to host receptors ultimately inhibit the entry of virus in host cell (Pomin et al., 2017). *F. vesiculosus, Lobophora variegata, Dictyota mertensii* and *Spatoglossum schroederi* were reported for the production of sulfated fucans that were effective against HIV by inhibiting its reverse transcriptase. Human papilloma causes female genital infection that leads to cervical cancer need to be controlled. Natural compounds as well as their derivatives can inhibit human papillomavirus effectively with few side effects and are also cost-effective. Marine algae can produce sulfated polysaccharides that can inhibit the papillomavirus. Carrageenan, a sulfated polysaccharide, binds to the human papilloma virus and inhibits the adsorption of virus to host cell and can also inhibit the uncoating of virus. The antiviral activity of carrageenan against human papillomavirus is three times higher than heparin. Carrageenan by blocking the receptors on human papillomavirus block the entry of virus into host cell and can also control infection through a post attachment heparin sulfate-independent effect because of similarity of carrageenan with heparin sulfate that is recognized by human papillomavirus attachment factor. Carrageenan is considered as potential anti-human papillomavirus compound. There are numerous advantages that are associated with the utilization of carrageenan as antiviral agents, and those are low toxicity, low cost of production, novel mode of action and broad spectrum of activity (Wang et al., 2011).

Sulquinovosyl diacylglycerol is a sulfolipid that has been reported from *Gigartina tenella, Caulerpa racemosa* and *Ishige okamurai* is a potent inhibitor of eukaryotic DNA polymerase and HIV-1 reverse transcriptase. It has also shown selective toxicity against *herpes simplex* virus 2. *Symphyocladia latiuscula* produced 2,3,6-tribromo-4,5-dihydroxybenzyl methyl ether that was effective against herpes simplex virus (Paudel et al., 2019). *Laurencia venusta* produced venustatriol, thyrsiferol, and thyrsiferyl 23-acetate with potential antiviral activity against vesicular stomatitis virus (VSV) and herpes simplex virus type 1 (Nishiguchi, 2005). Peyssonol A and B are sesquiterpene hydroquinone produced by *Peyssonnelia* species appeared as inhibitor of HIV reverse transcriptase (Lane et al., 2010). *Halimeda tuna* produced habitual that is novel diterpene aldehyde appeared as effective against murine coronavirus A59, *in vitro* (Islam et al., 2020). *Ulva fasciata* has been reported for the production of sphingosine effective against semliki forest virus (Santos et al., 1999). *D. menstrualis* has been reported for the production of diterpenes that exhibited antiretroviral activity. The derivatives of phlorotannin 8,8'-bieckol and 8,4"-bieckol produced by *Ecklonia cava* appeared as inhibitor of reverse transcriptase and protease of HIV (Artran et al., 2008).

7.4.6 ANTI-INFLAMMATORY ACTION

Inflammation is the body response to the injury of cells and tissues caused by toxins or pathogens. Foreign agents are recognized by the immune system that tries to neutralize the foreign agent quickly, the process of neutralization causes inflammation. Ingestion of compounds with anti-inflammatory activity enhances immune response and also prevents disease and helps in the healing process. Algae produce compounds in their biomass with anti-inflammatory activity that can protect the body from inflammatory reactions. Because of the anti-inflammatory activity of algae, it is used in developing scaffolds in tissue engineering and for the reconstitution of tissues and organs, especially for burnt patients with completely lost skin. Compounds of algae used in tissue engineering are polyunsaturated long-chain fatty acids, pigments, and sulfurized polysaccharides (Morais et al., 2014).

Macrophages regulate innate immune response and secrete cytokines and chemokines that act as signal for the activation of inflammatory reactions. Polysaccharides produced by algae can modulate immune response by the activation of macrophages and induction of production of reactive oxygen species (ROS), nitric oxides, and other cytokines or chemokines (Schepetkin et al., 2008). Sulfated polysaccharides when applied on skin for the treatment of skin inhibit migration and adhesion of polymorphonuclear leukocytes (PMN). In the treatment of chronic inflammations such as rheumatism and skin diseases, PUFA, especially eicosapentaenoic and arachidonic acid have been applied for reducing inflammation. Fucoxanthin carotenoid is a pigment produced by diatoms has anti-inflammatory activity, and it can also induce apoptosis in cancer cells (Peng et al., 2011). Phycocyanin is a pigment that showed anti-inflammatory activity by means of inhibiting the release of histamine.

7.4.7 ANTIOXIDANT ACTIVITY

The demand of natural antioxidant compounds is increasing globally for the protection of living cells from oxidative damages and prevention of pharmaceutical and food items from deterioration. Oxidation of lipids, proteins, and nucleic acids by production and action of ROS lead to the development of chronic diseases like cancer, aging, heart diseases and atherosclerosis (Lee et al., 2004). Intake of antioxidants can reduce the chances of these diseases. Algae are considered as the richest source of antioxidants with their application in the food, pharmaceutical, and cosmetic industry. From the algal biomass lipids, polysaccharides, amino acids, and pigments have been identified with antioxidant activities, but the most powerful antioxidants with higher water solubility are vitamins, PBPs, and polyphenols. Dimethyl sulfoniopropionate and mycosporine amino acids are the antioxidants that block UV radiation (Bhattacharjee, 2016). β-carotene and astaxanthin are the pigments used as antioxidants in pharmaceutical products and in algae, they provide protection against light and oxidative damage in the photosynthetic tissues. *Dunaliella salina* is the major producer of β-carotene while *H. pluvialis* is the major producer of astaxanthin (Rammuni, 2019). The sulfated polysaccharides produced by algae help in the modulation of immune response, activation of macrophages and are antioxidant in nature (Wijesekara et al., 2011). Some species of algae, for example, *Dunaliella* accumulate vitamin C in high amount that is used as antioxidant, essential in the synthesis of neurotransmitter, collagen, and carnitine, provide protection from atherosclerosis and cancers and act as immunomodulator (Skjånes, 2013). *B. braunii* produces butylated hydroxytoluene, a lipophilic antioxidant, is used in the preservations of food. *Dunaliella* produce glutathione in high amount that is non-protein thiol antioxidant that scavenges ROS to prevent the cellular damage (Hossain et al., 2012). *Cystoseira crinita* has been reported for the production of several kinds of prenyl toluquinones that exhibited free radicals scavenging activity (Praveen, 2015). The sargachromanols A–P isolated from *Sargassum siliquastrum* are antioxidant in nature.

7.4.8 ANTICANCER AND CYTOTOXIC ACTIVITIES

Cancer is a serious disease that has affected the world more than AIDS, tuberculosis, and malaria. To deal with this deadly disease, researchers are focusing on its effective treatment. The currently available treatments for cancer are platinum compounds, alkylating agents, ionizing radiations, topoisomerase inhibitors, and hyperthermia, but these treatments also affect normal cells with actively proliferating cancer cells. Algae being the richest source of bioactive compounds have been explored for anticancer compounds too. The pigments of algae protect normal cells and exert a cytotoxic effect on cancer cells. Curacin A is a new type of anticancer drug that have been extracted from *Lyngbya majuscula*, it works by inhibiting the binding of cholchicine to tubulin and by inhibiting the assembly of microtubules (Mahindroo et al., 2006). *Lyngbya majuscula* has also been reported for the production of majusculamide D5 and deoxymajusculamide D6 that are pentapeptides and are cytotoxic in nature. Lyngbyatoxin-A from *Lyngbya majuscula* has been investigated for the treatment of severe erythematous papulovesicular dermatitis and has also appeared as cytotoxic against leukemia (P388). Cryptophycin from *Nostoc* appeared as effective against solid tumors and act by depolymerizing microtubules (Moore, 1996). Depsipeptide kahalalide F56 extracted from *Bryopsis* species appeared as effective against prostate cancer. Linear cytotoxic diterpene bifurcadiol extracted from *Bifurcaria bifurcata* showed cytotoxicity against human tumor cell lines (Culioli, 2004). Turbinaric acid extracted from *Turbinaria ornata* appeared as cytotoxic. Terpenoic C from *Styolpopdium zonale* showed cytotoxicity against HT-29, H-116 and A-549 cell lines. Strong cytotoxic compound, 4,18-dihydroxydictyolactone was extracted from *Dictyota* species. National Cancer Institute has considered Halmon (polyhalogenated monoterpene) a novel antitumor agent produced by *Portieria hornemannii* (Lavakumar, 2012). The discovery of laurinterol from *Laurencia okamurai* has appeared as effective in the inhibition of melanoma by the induction of apoptosis (Kim et al., 2008). Thysiferyl 23-acetate extracted from *Laurencia obtusa* showed remarkable cytotoxicity. Sulfur-containing polybromoindoles produced by *Laurenda brongniartii* were active against HT-29 and P388 cell lines (Kim, 2011). Bromophycolide A from *Callophyeus serratus* showed cytotoxicity against human tumor cell lines by inducing specific apoptosis (Lavakumar, 2012). Thyresenol A and B (polyether squalene derivatives) and dehydrothyrsiferol extracted from *Laurencia viridis* showed cytotoxicity against P388 cell lines (Mandrekar et al., 2019). Majuscule amide D-34 and deoxymajusuculamide D-35 are cytotoxic peptides produced by *Lyngbya majuscula.* Apratoxins are cyclic lipopeptide produced by cyanobacteria showed excellent cytotoxic effect on cancer cells. Cytotoxic compounds like antillatoxin-51, aplysiatoxin-52, and debromoaplysiatoxin-53 were reported from algae. Curacin A-38 and dolostatin 10–39, microtubule polymerization inhibitors are the anticancer compounds produced by algae in clinical trials.

7.4.9   ACTIVITY AGAINST CARDIOVASCULAR DISEASES

The main risk factor for the development of coronary heart disease is dyslipidemia. The disorders in lipids lead to the development of atherosclerosis those results in heart failure. Nutraceutical products help in the reduction in the progression of atherosclerosis and coronary heart disease. Carotenoids are the fats-soluble pigments produced by plants and algae help in reducing the incidences of cardiovascular diseases by scavenging free radicals and by anti-inflammatory action on lipoxygenase activity (Shebis et al., 2013). Sterols are the important structural component of cell membranes, and some of these sterols are cardiac glycosides, and these are used in the treatment of atrial arrhythmias and cardiac failure. EPA and DHA have been reported in improving cardiovascular system and are considered as cardioprotective too (Harris et al., 2008). They protect the cardiac system by modulating inflammation in vascular wall and by influencing the level of cholesterols and triglycerides (TG) in plasma.

7.4.10   ANTI-COAGULANT ACTIVITY

In abnormal vascular condition, the flow of blood from the injured vessels does not stop, and it leads to the exposure of blood to non-endothelial surfaces at the site of vascular injury. The coagulation of blood can be stopped or prolonged by anticoagulants (endogenous or exogenous) that act by interfering with the coagulation system. The bioactive peptides from marine echiuroid worm prolong thrombin and prothrombin time and ultimately prolong anticoagulant activity (Ge et al., 2018). Sulfated polysaccharides from marine algae are alternative and novel anti-coagulant agents. Two sulfated polysaccharides with high anticoagulant activity are carrageenan and fucoidans (Qiu et al., 2006). *Codium fragile* produce sulfated polysaccharide (xyloarabinogalactans) with anticoagulant activity and *Codium cylindricum* produces sulfated galactan with anticoagulant activity (Wijesekara et al., 2011). *Monostroma nitidum* produce sulfated polysaccharide with six folds higher anticoagulant activity when compared with heparin (Zhang et al., 2008). Specific and non-specific binding of sulfated polysaccharides with biologically active protein can be increased by the sulfate group. The content of sulfate, position of sulfate group and type of sugar determines the anticoagulant activity of sulfated polysaccharides. High molecular weight carrageenans with a high number of sulfates show a high level of anticoagulant activity compared to low molecular weight carrageenan with a low number of sulfate groups (Liang et al., 2014). The currently used anticoagulant drugs are low molecular weight and unfractionated sulfated polysaccharides. Sulphated polysaccharides derived from seaweeds show anticoagulant activity similar or higher than heparin. Sulphated polysaccharides from seaweeds are the promising anticoagulant agent (Gupta and Abu-Ghannam, 2011). *Sargassum thunbergii* produces phlorotannins with potential anticoagulant activity. Phloroglucinol can also be exploited as a novel anticoagulant agent.

7.4.11 ANTI-OBESITY ACTIVITY

Excessive accumulation of fats in adipose tissue causes obesity. Obesity is a multifactorial metabolic disorder that leads to many complications and diseases like diabetes, cancer, cardiovascular diseases, and aging. Adipogenesis lead to the overgrowth of adipose tissues and by controlling cells involved in adipogenesis lead to control of obesity. Fat lowering agents and anti-hyperlipidemic agents from natural sources can control obesity, and algae are considered as potential source of these agents. Free radicals have been found involved in the progression of obesity (Fernández-Sánchez et al., 2011). The use of antioxidants can control the accumulation of free radicals induced fats. Fucoxanthin and fucoxanthin down-regulate peroxisome proliferator-activated receptor-c that ultimately inhibits the differentiation of 3T3-L1 to adipocytes. For the differentiation of adipocytes, free allenic and hydroxyl groups are required, and neoxanthin and fucoxanthin inhibit the accumulation of fats by trapping these groups (Peng et al., 2011). *Cylindrotheca closterium* and *Phaeodactylum tricornutum* produce fucoxanthin.

7.4.12 ACTIVITY AGAINST POTENTIALITY OVER DEGENERATIVE DISEASES

In humans, the oxidation reaction is responsible for the damage of cellular proteins, lipids, nucleic acids, and introduction of mutation by the production of ROS that leads to several syndromes like aging, some cancers, and cardiovascular diseases. Oxidative stress and inflammations are responsible for the development of age-related diseases and neurodegenerative diseases such as Alzheimer's disease, dementia, multiple sclerosis, and Parkinson's disease. Neuroprotective properties of natural pigments produced by algae are explored for the treatment of neurodegenerative diseases (Barbalace et al., 2019). Vitamin E appeared as protective against atherosclerosis and heart disease as well as in the treatment of multiple sclerosis (Marchioli et al., 2011). Carotenoids, because of their potential benefits on human health can be used in the treatment of macular degeneration and cataract. Phenolic compounds produced by algae have the potential of scavenging free radicals and decrease the occurrence of degenerative diseases. *Dunaliella salina* produces β-carotene naturally, that can be used for reducing the risks of cancers and degenerative diseases in humans. Lutein has been found effective in the treatment of atherosclerosis at an early stage, cataract, and macular degeneration (Carpentier et al., 2009). In rats, β-carotene and lutein extracted from *Chlorella* has prevented the development of cognitive disability that led to Alzheimer disease and lutein has also decreased the incidence of cancer (Olasehinde et al., 2017). Growth of colon cancer was inhibited by carotenoids extracted from *Chlorella vulgaris* and *Chlorella ellipsoidea* (Cha et al., 2008). In mice, the proliferation of prostate cancer was significantly reduced by lycopene extracted from *Chlorella marina* (Renju et al., 2014). Lycopene has also been reported for the treatment of rheumatoid arthritis, reducing total blood cholesterol level and the level of low-density lipoproteins (LDLs) (Ried and Fakler, 2011). Increased tendency of myocardial infarction was observed when the level of lutein was low in blood, while decreased risk of stroke was observed with the high intake of lutein. The risk of angina pectoris was reduced with a high level of β-cryptoxanthin, β-carotene and α-carotene. Low consumption of lutein and zeaxanthin, prevent macular degeneration that led to irreversible vision loss (Ma and Lin, 2010). Astaxanthin can prevent age-related immune dysfunction, cancer, cognitive impairment, cardiovascular diseases, degenerative diseases, metabolic syndrome, ocular diseases, and skin damages (Ambati et al., 2014).

7.4.13   ANTIDIABETIC AND ANTIHYPERTENSIVE ACTIVITY

Fucosterol produced by *Pelvetia siliquosa* showed antidiabetic activity (Lee et al., 2004). Eckol, phlorofucofuroeckol A and dieckol produced by *Ecklonia stolonifera* showed inhibitory activity against angiotensin-converting enzyme (Jung, 2006).

7.4.14   HEALTH PROMOTING FUNCTION

Algae are important source of functional ingredients and are used as a source of valuable bioactive compounds with antioxidant, anticancer, neuroprotective, anti-obesity, anti-inflammatory, and antiangiogenic activity. Xanthigen is a fucoxanthin reported from algae is a promising dietary supplement that can be used for the treatment of obesity (Kim and Pangestuti, 2011). Fucoxanthin has been reported as an effective therapeutic agent in the treatment of diabetes, osteoporosis, and rheumatoid arthritis. Bioactive peptides produced by algae are of great interest because of their therapeutic potential in the treatment of various diseases, and can act as growth factors, immunomodulators, and can provide nutritional benefits too (Ibañez and Cifuentes, 2013). Because of the rich protein content of *Chlorella* and *Spirulina*, they are used in functional foods for the prevention of cellular and tissues damages. Immunoglobulin and certain lytic enzymes such as lysozymes have been recommended in patients with diseases like Crohn's disease because of their chemistry and the presence of free amino acids (Proctor et al., 1998). Protein can also increase the production of cholecystokinin that suppresses appetite and reduces the level of LDLs. Proteins from *Chlorella* have been reported in preventing oxidative stress-related diseases such as cancers, atherosclerosis, and coronary heart disease.

EFAs are important for the integrity of tissues. Linolenic acid slows down the process of skin aging by revitalizing the skin. Linoleic and linolenic acids help in strengthening the immune system and in the regeneration process of tissues. In the treatment of hyperplasia of skin linoleic acid (LNA) is being used. The most studied lipids of algae are PUFA because of their health promoting benefits. DHA and EPA reduce the lipid content in blood, reduce blood pressure, problems associated with stroke and arthritis and also act as anti-inflammatory agents (Song et al., 2016). DHA helps in proper development and the functioning of the nervous system. Arachidonic acid and EPA have showed chemostatic activity in neutrophils, and antiaggregation action on endothelium and are also help in vasodilation, vasoconstriction, and in the aggregation of platelets. Other lipids, like sterols of algae have interesting bioactive properties, sterols help in the inhibition of cholesterols absorption from intestines (Lopes et al., 2013). Polysaccharides are dietary fibers that give different physiological benefits like increasing satiety, improving laxation, and promote the movement of materials through the digestive system (Ibañez and Cifuentes, 2013). They are also considered as prebiotics because they promote the growth of microflora of the gastrointestinal tract (GIT).

7.4.15   METABOLITES IN SKIN PRODUCTS

Skin is a complex physical barrier that provides protection against water loss and several environmental factors that include chemicals, pathogens, and ultraviolet radiations. Skin performs physiological functions like first line of defense, thermoregulation, endocrine, and metabolic functions for maintaining optimal health and provides sensory input from mechanoreceptors. Several intrinsic and extrinsic factors can bring multiple changes in skin like dryness, wrinkles, and fragility of structural proteins that cause skin aging (Kammeyer and Luiten, 2015). Photoaging is the aging of skin caused by exposure to ultraviolet radiations. In addition to photoaging, ROS produced by cellular metabolism result in the degradation of skin connective tissues and collagen. To protect skin from photo and oxidative damages mycosporin like amino acids that are produced by a variety of organisms are used (Wada et al., 2015). Cyanobacteria and macroalgae has been reported for the production of mycosporin like amino acids in excess amount. Bioactive substances extracted from natural sources are given priority in the cosmetic industry. Natural products are rich in anti-oxidants with beneficial metabolites and pose low toxicities. Microalgae are one of the potential sources of metabolites beneficial in cosmetics that can accelerate healing of damaged skin and also have antiblemish effect (Ariede et al., 2017). *Nostoc*, *Spirulina*, and *Aphanizomenon* have been extensively studied for their utilization in skin and hair protection products because these species of algae contain calcium, phosphorous, and iron in substantial amount and can produce beta-carotene, biotin, and vitamin B12.

7.5   ADVANTAGES OF USING BIOACTIVE COMPOUNDS OF ALGAE

Algae are considered as important source of bioactive compounds. Metabolites extracted from algae have the potential health benefit and pharmaceutical value. Secondary metabolites such as pigments and vitamins accumulated in algae have been used in food, cosmetics, and pharmaceutical products. Algae live in complex habitat and if subjected to extreme environments can adapt rapidly in new environment, survive, and produce variety of new biologically active compounds (Oarga, 2009). The cultivation of algae is associated with the chemical composition, its diversity, its growth under controlled conditions, and its production of secondary metabolites induced by stress (de Morais et al., 2015). Algae use solar light and carbon dioxide and have a higher growth rate compared to higher plants. Algae can be grown on areas that are unsuitable for agriculture and do not compete with agricultural land. Manipulation of cultural conditions for controlling the production of certain bioactive metabolites is another advantage of the utilization of algae. Algae reduce the atmospheric carbon dioxide by biofixation. Carbon dioxide-rich gas can be used in the industrial cultivation of algae that cause the reduction of greenhouse effect and global warming and reduce the cost of carbon source required for growth. Cultivation of algae is not seasonal can be cultivated in wastewater for the effective removal of pollutants (Hena et al., 2015). Algae use solar energy for the production of biomass and the cultivation of algae for metabolites production is sustainable. Integration of production of bioactive metabolites in biorefinery is a sustainable way of production of high valued compounds, food products, and energy production (Chandra et al., 2019). Through integrated processes, fractionation of biomass of algae leads to several isolated products with pharmaceutical importance, importance in food and biofuel industry, and are economically viable (Laurens et al., 2017).

7.6   CONCLUSION

Algae produce bioactive compounds with applications in the nutraceutical and pharmaceutical industries that are yet to be explored. The metabolites of algae appeared as protective against various metabolic and infectious diseases. Production of bioactive compounds from algae is attractive because algae can be grown on a large scale under a controlled condition and on an area that is not suitable for agriculture. Production of bioactive compounds by integrating it with biofuel production can be economically viable.

KEYWORDS

- algae
- bioactive metabolites
- human immunodeficiency virus
- nutraceutical
- pharmaceutical
- polysaccharides

REFERENCES

Amaro, H. M., Guedes, A. C., & Malcata, F. X., (2011). Antimicrobial activities of microalgae: An invited review. In: Mendez-Vilas, A., (ed.), *Science Against Microbial Pathogens: Communicating Current Research and Technological Advances* (Vol. 2, pp. 1272–1284). Formatex Research Center, Badajoz, Spain.

Ambati, R. R., Phang, S. M., Ravi, S., & Aswathanarayana, R. G., (2014). Astaxanthin: Sources, extraction, stability, biological activities, and its commercial applications: A review. *Marine Drugs, 12*(1), 128–152.

Ariede, M. B., Candido, T. M., Jacome, A. L. M., Velasco, M. V. R., De Carvalho, J. C. M., & Baby, A. R., (2017). Cosmetic attributes of algae: A review. *Algal Research, 25*, 483–487.

Artan, M., Li, Y., Karadeniz, F., Lee, S. H., Kim, M. M., & Kim, S. K., (2008). Anti-HIV-1 activity of phloroglucinol derivative, 6, 6′-bieckol, from *Ecklonia cava*. *Bioorganic and Medicinal Chemistry, 16*(17), 7921–7926.

Barbalace, M. C., Malaguti, M., Giusti, L., Lucacchini, A., Hrelia, S., & Angeloni, C., (2019). Anti-inflammatory activities of marine algae in neurodegenerative diseases. *International Journal of Molecular Sciences, 20*(12), 3061.

Baudner, B. C., & O'Hagan, D. T., (2010). Bioadhesive delivery systems for mucosal vaccine delivery. *Journal of Drug Targeting, 18*(10), 752–770.

Bhattacharjee, M., (2016). Pharmaceutically valuable bioactive compounds of algae. *Asian Journal of Pharmaceutical and Clinical Research, 9*, 43–47.

Bishop, W. M., & Zubeck, H. M., (2012). Evaluation of microalgae for use as nutraceuticals and nutritional supplements. *Journal of Nutrition and Food Science, 2*(5), 1–6.

Carpentier, S., Knaus, M., & Suh, M., (2009). Associations between lutein, zeaxanthin, and age-related macular degeneration: An overview. *Critical Reviews in Food Science and Nutrition, 49*(4), 313–326.

Cha, K. H., Koo, S. Y., & Lee, D. U., (2008). Antiproliferative effects of carotenoids extracted from *Chlorella ellipsoidea* and *Chlorella vulgaris* on human colon cancer cells. *Journal of Agricultural and Food Chemistry, 56*(22), 10521–10526.

Chandra, R., Iqbal, H. M., Vishal, G., Lee, H. S., & Nagra, S., (2019). Algal biorefinery: A sustainable approach to valorize algal-based biomass towards multiple product recovery. *Bioresource Technology, 278*, 346–359.

Christaki, E., Bonos, E., Giannenas, I., & Florou-Paneri, P., (2013). Functional properties of carotenoids originating from algae. *Journal of the Science of Food and Agriculture, 93*(1), 5–11.

Culioli, G., Ortalo-Magné, A., Daoudi, M., Thomas-Guyon, H., Valls, R., & Piovetti, L., (2004). Trihydroxylated linear diterpenes from the brown alga *Bifurcaria bifurcata*. *Phytochemistry, 65*(14), 2063–2069.

De La Jara, A., Ruano-Rodriguez, C., Polifrone, M., Assunçao, P., Brito-Casillas, Y., Wägner, A. M., & Serra-Majem, L., (2018). Impact of dietary *Arthrospira* (*Spirulina*) biomass consumption on human health: Main health targets and systematic review. *Journal of Applied Phycology, 30*(4), 2403–2423.

De Morais, M. G., Vaz, B. D. S., De Morais, E. G., & Costa, J. A. V., (2015). Biologically active metabolites synthesized by microalgae. *BioMedical Research International, 2015*.

De Sousa, C. B., Gangadhar, K. N., Macridachis, J., Pavao, M., Morais, T. R., Campino, L., & Lago, J. H. G., (2017). Cystoseira algae (Fucaceae): Update on their chemical entities and biological activities. *Tetrahedron: Asymmetry, 28*(11), 1486–1505.

Demmig-Adams, B., Burch, T. A., Stewart, J. J., Savage, E. L., & Adams, III. W. W., (2017). Algal glycerol accumulation and release as a sink for photosynthetic electron transport. *Algal Research, 21*, 161–168.

Desbois, A. P., & Smith, V. J., (2010). Antibacterial free fatty acids: Activities, mechanisms of action and biotechnological potential. *Applied Microbiology and Biotechnology, 85*(6), 1629–1642.

El Gamal, A. A., (2010). Biological importance of marine algae. *Saudi Pharmaceutical Journal, 18*(1), 1–25.

Falaise, C., François, C., Travers, M. A., Morga, B., Haure, J., Tremblay, R., & Leignel, V., (2016). Antimicrobial compounds from eukaryotic microalgae against human pathogens and diseases in aquaculture. *Marine Drugs, 14*(9), 159.

Farooqui, A. A., (2009). Status and potential therapeutic importance of n–3 fatty acids in other neural and non-neural diseases. In: *Beneficial Effects of Fish Oil on Human Brain* (pp. 333–365). Springer, New York.

Fernández-Sánchez, A., Madrigal-Santillán, E., Bautista, M., Esquivel-Soto, J., Morales-González, Á., Esquivel-Chirino, C., & Morales-González, J. A., (2011). Inflammation, oxidative stress, and obesity. *International Journal of Molecular Sciences, 12*(5), 3117–3132.

Francavilla, M., Trotta, P., & Luque, R., (2010). Phytosterols from *Dunaliella tertiolecta* and *Dunaliella salina*: A potentially novel industrial application. *Bioresource Technology, 101*(11), 4144–4150.

Frieri, M., Kumar, K., & Boutin, A., (2017). Antibiotic resistance. *Journal of Infection and Public Health, 10*(4), 369–378.

Ge, Y. H., Chen, Y. Y., Zhou, G. S., Liu, X., Tang, Y. P., Liu, R., & Yue, S. J., (2018). A novel antithrombotic protease from marine worm *Sipunculus nudus*. *International Journal of Molecular Sciences, 19*(10), 3023.

Grice, I. D., & Mariottini, G. L., (2018). Glycans with antiviral activity from marine organisms. In: *Marine Organisms as Model Systems in Biology and Medicine* (pp. 439–475). Springer, Cham.

Gunasekera, S. P., Ross, C., Paul, V. J., Matthew, S., & Luesch, H., (2008). Dragonamides C and D, linear lipopeptides from the marine cyanobacterium brown *Lyngbya polychroa*. *Journal of Natural Products, 71*(5), 887–890.

Gupta, S., & Abu-Ghannam, N., (2011). Bioactive potential and possible health effects of edible brown seaweeds. *Trends in Food Science and Technology, 22*(6), 315–326.

Harden, E. A., Falshaw, R., Carnachan, S. M., Kern, E. R., & Prichard, M. N., (2009). Virucidal activity of polysaccharide extracts from four algal species against herpes simplex virus. *Antiviral Research, 83*(3), 282–289.

Harris, W. S., Kris-Etherton, P. M., & Harris, K. A., (2008). Intakes of long-chain omega-3 fatty acid associated with reduced risk for death from coronary heart disease in healthy adults. *Current Atherosclerosis Reports, 10*(6), 503–509.

Hena, S., Fatimah, S., & Tabassum, S., (2015). Cultivation of algae consortium in a dairy farm wastewater for biodiesel production. *Water Resources and Industry, 10*, 1–14.

Hossain, M. A., Piyatida, P., Da Silva, J. A. T., & Fujita, M., (2012). Molecular mechanism of heavy metal toxicity and tolerance in plants: Central role of glutathione in detoxification of reactive oxygen species and methylglyoxal and in heavy metal chelation. *Journal of Botany, 2012*, 872875. https://doi.org/10.1155/2012/872875.

Ibañez, E., & Cifuentes, A., (2013). Benefits of using algae as natural sources of functional ingredients. *Journal of the Science of Food and Agriculture, 93*(4), 703–709.

Ibañez, E., Herrero, M., Mendiola, J. A., & Castro-Puyana, M., (2012). Extraction and characterization of bioactive compounds with health benefits from marine resources: Macro and microalgae, cyanobacteria, and invertebrates. In: *Marine Bioactive Compounds* (pp. 55–98). Springer, Boston, MA.

Islam, M. T., Sarkar, C., El-Kersh, D. M., Jamaddar, S., Uddin, S. J., Shilpi, J. A., & Mubarak, M. S., (2020). Natural products and their derivatives against coronavirus: A review of the non-clinical and pre-clinical data. *Phytotherapy Research, 34*(10), 2471-2492. doi: 10.1002/ptr.6700.

Jena, J., & Subudhi, E., (2019). Microalgae: An untapped resource for natural antimicrobials. In: *The Role of Microalgae in Wastewater Treatment* (pp. 99–114). Springer, Singapore.

Jung, H. A., Hyun, S. K., Kim, H. R., & Choi, J. S., (2006). Angiotensin-converting enzyme I inhibitory activity of phlorotannins from *Ecklonia stolonifera*. *Fisheries Science, 72*(6), 1292–1299.

Kammeyer, A., & Luiten, R. M., (2015). Oxidation events and skin aging. *Ageing Research Reviews, 21*, 16–29.

Kannaujiya, V. K., Kumar, D., Richa, J. P., Sonker, A. S., Rajneesh, V. S., Sundaram, S., & Sinha, R. P., (2017). Recent advances in production and the biotechnological significance of phycobiliproteins. In: *New Approaches in Biological Research*. Nova Science Publisher.

Kim, M. M., Mendis, E., & Kim, S. K., (2008). *Laurencia okamurai* extract containing laurinterol induces apoptosis in melanoma cells. *Journal of Medicinal Food, 11*(2), 260–266.

Kim, S. K., & Pangestuti, R., (2011). Biological activities and potential health benefits of fucoxanthin derived from marine brown algae. In: *Advances in Food and Nutrition Research* (Vol. 64, pp. 111–128). Academic Press.

Kim, S. K., (2011). *Handbook of Marine Macroalgae: Biotechnology and Applied Phycology*. John Wiley & Sons.

Lane, A. L., Mular, L., Drenkard, E. J., Shearer, T. L., Engel, S., Fredericq, S., & Aalbersberg, W., (2010). Ecological leads for natural product discovery: Novel sesquiterpene hydroquinones from the red macroalga *Peyssonnelia* sp. *Tetrahedron, 66*(2), 455–461.

Laurens, L. M., Markham, J., Templeton, D. W., Christensen, E. D., Van, W. S., Vadelius, E. W., & Pienkos, P. T., (2017). Development of algae biorefinery concepts for biofuels and bioproducts; a perspective on process-compatible products and their impact on cost-reduction. *Energy and Environmental Science, 10*(8), 1716–1738.

Lavakumar, V., (2012). *Anticancer and Antiviral Activity of Marine Algae (Doctoral Dissertation, Vels College of Pharmacy, Chennai)*.

Lee, J., Koo, N., & Min, D. B., (2004). Reactive oxygen species, aging, and antioxidative nutraceuticals. *Comprehensive Reviews in Food Science and Food Safety, 3*(1), 21–33.

Lee, S. H., Han, J. S., Heo, S. J., Hwang, J. Y., & Jeon, Y. J., (2010). Protective effects of dieckol isolated from *Ecklonia cava* against high glucose-induced oxidative stress in human umbilical vein endothelial cells. *Toxicology In Vitro, 24*(2), 375–381.

Lee, Y. S., Shin, K. H., Kim, B. K., & Lee, S., (2004). Anti-diabetic activities of fucosterol from *Pelvetia siliquosa. Archives of Pharmaceutical Research, 27*(11), 1120–1122.

Li, Y. X., Wijesekara, I., Li, Y., & Kim, S. K., (2011). Phlorotannins as bioactive agents from brown algae. *Process Biochemistry, 46*(12), 2219–2224.

Liang, W., Mao, X., Peng, X., & Tang, S., (2014). Effects of sulfate group in red seaweed polysaccharides on anticoagulant activity and cytotoxicity. *Carbohydrate Polymers, 101*, 776–785.

Lohr, M., Schwender, J., & Polle, J. E., (2012). Isoprenoid biosynthesis in eukaryotic phototrophs: A spotlight on algae. *Plant Science, 185*, 9–22.

Lopes, G., Sousa, C., Valentao, P., & Andrade, P. B., (2013). Sterols in algae and health. In: Hernández-Ledesma, B. & Herrero, M. (eds). *Bioactive Compounds from Marine Foods*: *Plant and Animal Sources*, John Wiley & Sons Ltd., Chichester, UK. 173–191.

Ma, L., & Lin, X. M., (2010). Effects of lutein and zeaxanthin on aspects of eye health. *Journal of the Science of Food and Agriculture, 90*(1), 2–12.

Mahindroo, N., Liou, J. P., Chang, J. Y., & Hsieh, H. P., (2006). Antitubulin agents for the treatment of cancer: A medicinal chemistry update. *Expert Opinion on Therapeutic Patents, 16*(5), 647–691.

Mandrekar, V. K., Gawas, U. B., & Majik, M. S., (2019). Brominated molecules from marine algae and their pharmacological importance. In: *Studies in Natural Products Chemistry* (Vol. 61, pp. 461–490). Elsevier, Amsterdam, Netherlands.

Marchioli, R., Schweiger, C., Levantesi, G., Tavazzi, L., & Valagussa, F., (2001). Antioxidant vitamins and prevention of cardiovascular disease: Epidemiological and clinical trial data. *Lipids, 36*(S1), S53–S63.

Milledge, J. J., (2011). Commercial application of microalgae other than as biofuels: A brief review. *Reviews in Environmental Science and Bio/Technology, 10*(1), 31–41.

Moore, R. E., (1996). Cyclic peptides and depsipeptides from cyanobacteria: A review. *Journal of Industrial Microbiology, 16*(2), 134–143.

Morais, M. G. D., Vaz, B. D. S., Morais, E. G. D., & Costa, J. A. V., (2014). Biological effects of *Spirulina* (*Arthrospira*) biopolymers and biomass in the development of nanostructured scaffolds. *BioMed Research International, 2014*.

Mori, T. A., Woodman, R. J., Burke, V., Puddey, I. B., Croft, K. D., & Beilin, L. J., (2003). Effect of eicosapentaenoic acid and docosahexaenoic acid on oxidative stress and inflammatory markers in treated-hypertensive type 2 diabetic subjects. *Free Radical Biology and Medicine, 35*(7), 772–781.

Nakano, S., Takekoshi, H., & Nakano, M., (2010). *Chlorella pyrenoidosa* supplementation reduces the risk of anemia, proteinuria, and edema in pregnant women. *Plant Foods for Human Nutrition, 65*(1), 25–30.

Nishiguchi, G. A., (2005). *Application of a Titanium (III)-Mediated Coupling Reaction Toward the Total Synthesis of 7,11-Epi-Thyrsiferol and its Pharmacological Properties*. University of California, Santa Barbara.

Oarga, A., (2009). Life in extreme environments. *Revista de Biologia e Ciencias da Terra, 9*(1), 1–10.

Olasehinde, T. A., Olaniran, A. O., & Okoh, A. I., (2017). Therapeutic potentials of microalgae in the treatment of Alzheimer's disease. *Molecules, 22*(3), 480.

Olasehinde, T. A., Olaniran, A. O., & Okoh, A. I., (2019). Macroalgae as a valuable source of naturally occurring bioactive compounds for the treatment of Alzheimer's disease. *Marine Drugs, 17*(11), 609.

Panche, A. N., Diwan, A. D., & Chandra, S. R., (2016). Flavonoids: An overview. *Journal of Nutritional Science, 5.*

Paudel, P., Seong, S. H., Park, H. J., Jung, H. A., & Choi, J. S., (2019). Antidiabetic activity of 2,3,6-tribromo-4,5-dihydroxybenzyl derivatives from *Symphyocladia latiuscula* through PTP1B downregulation and α-glucosidase inhibition. *Marine Drugs, 17*(3), 166.

Peng, J., Yuan, J. P., Wu, C. F., & Wang, J. H., (2011). Fucoxanthin, a marine carotenoid present in brown seaweeds and diatoms: Metabolism and bioactivities relevant to human health. *Marine Drugs, 9*(10), 1806–1828.

Pomin, V. H., Bezerra, F. F., & Soares, P. A. G., (2017). Sulfated glycans in HIV infection and therapy. *Current Pharmaceutical Design, 23*(23), 3405–3414.

Praveen, N. K., (2015). *Isolation and Characterization of Secondary Metabolites with Antioxidant Activity from Seaweeds from Southeastern Coasts of India.* Doctoral Dissertation, ICAR-Central Marine Fisheries Research Institute.

Proctor, V. A., Cunningham, F. E., & Fung, D. Y., (1988). The chemistry of lysozyme and its use as a food preservative and a pharmaceutical. *Critical Reviews in Food Science and Nutrition, 26*(4), 359–395.

Qiu, X., Amarasekara, A., & Doctor, V., (2006). Effect of over sulfation on the chemical and biological properties of fucoidan. *Carbohydrate Polymers, 63*(2), 224–228.

Rammuni, M. N., Ariyadasa, T. U., Nimarshana, P. H. V., & Attalage, R. A., (2019). Comparative assessment on the extraction of carotenoids from microalgal sources: Astaxanthin from *H. pluvialis* and β-carotene from *D. salina. Food Chemistry, 277*, 128–134.

Rasmussen, R. S., & Morrissey, M. T., (2007). Marine biotechnology for production of food ingredients. *Advances in Food and Nutrition Research, 52*, 237–292.

Renju, G. L., Kurup, G. M., & Bandugula, V. R., (2014). Effect of lycopene isolated from *Chlorella marina* on proliferation and apoptosis in human prostate cancer cell line PC-3. *Tumor Biology, 35*(11), 10747–10758.

Ried, K., & Fakler, P., (2011). Protective effect of lycopene on serum cholesterol and blood pressure: Meta-analyses of intervention trials. *Maturitas, 68*(4), 299–310.

Russell, M., (2006). First life: Billions of years ago, deep under the ocean, the pores and pockets in minerals that surrounded warm, alkaline springs catalyzed the beginning of life. *American Scientist, 94*(1), 32–39.

Ryan, A. S., Keske, M. A., Hoffman, J. P., & Nelson, E. B., (2009). Clinical overview of algal-docosahexaenoic acid: Effects on triglyceride levels and other cardiovascular risk factors. *American Journal of Therapeutics, 16*(2), 183–192.

Santos, A. O. D., Veiga-Santos, P., Ueda-Nakamura, T., Sudatti, D. B., Bianco, É. M., Pereira, R. C., & Nakamura, C. V., (2010). Effect of elatol, isolated from red seaweed *Laurencia dendroidea*, on *Leishmania amazonensis. Marine Drugs, 8*(11), 2733–2743.

Santos, M. G. M., Lagrota, M. H. C., Miranda, M. M. F. S., Yoneshigue-Valentin, Y., & Wigg, M. D., (1999). A screening for the antiviral effect of extracts from Brazilian marine algae against acyclovir-resistant herpes simplex virus type 1. *Botanica Marina, 42*(3), 227–230.

Sayre, R., (2010). Microalgae: The potential for carbon capture. *Bioscience, 60*(9), 722–727.

Schepetkin, I. A., Xie, G., Kirpotina, L. N., Klein, R. A., Jutila, M. A., & Quinn, M. T., (2008). Macrophage immunomodulatory activity of polysaccharides isolated from *Opuntia polyacantha*. *International Immunopharmacology, 8*(10), 1455–1466.

Shao, W., Ebaid, R., El-Sheekh, M., Abomohra, A., & Eladel, H., (2019). Pharmaceutical applications and consequent environmental impacts of *Spirulina* (*Arthrospira*): An overview. *Grasas y Aceites, 70*(1), 292.

Shebis, Y., Iluz, D., Kinel-Tahan, Y., Dubinsky, Z., & Yehoshua, Y., (2013). Natural antioxidants: Function and sources. *Food and Nutrition Sciences, 4*(6), 643–649. doi: 10.4236/fns.2013.46083.

Shegokar, R., & Mitri, K., (2012). Carotenoid lutein: A promising candidate for pharmaceutical and nutraceutical applications. *Journal of Dietary Supplements, 9*(3), 183–210.

Sims, J. J., Donnell, M. S., Leary, J. V., & Lacy, G. H., (1975). Antimicrobial agents from marine algae. *Antimicrobial Agents and Chemotherapy, 7*(3), 320, 321.

Singh, S., Kate, B. N., & Banerjee, U. C., (2005). Bioactive compounds from cyanobacteria and microalgae: An overview. *Critical Reviews in Biotechnology, 25*(3), 73–95.

Skjånes, K., Rebours, C., & Lindblad, P., (2013). Potential for green microalgae to produce hydrogen, pharmaceuticals, and other high value products in a combined process. *Critical Reviews in Biotechnology, 33*(2), 172–215.

Song, C., Shieh, C. H., Wu, Y. S., Kalueff, A., Gaikwad, S., & Su, K. P., (2016). The role of omega-3 polyunsaturated fatty acids eicosapentaenoic and docosahexaenoic acids in the treatment of major depression and Alzheimer's disease: Acting separately or synergistically? *Progress in Lipid Research, 62*, 41–54.

Sosa-Hernández, J. E., Romero-Castillo, K. D., Parra-Arroyo, L., Aguilar-Aguila-Isaías, M. A., García-Reyes, I. E., Ahmed, I., & Iqbal, H., (2019). Mexican microalgae biodiversity and state-of-the-art extraction strategies to meet sustainable circular economy challenges: High-value compounds and their applied perspectives. *Marine Drugs, 17*(3), 174.

Spolaore, P., Joannis-Cassan, C., Duran, E., & Isambert, A., (2006). Commercial applications of microalgae. *Journal of Bioscience and Bioengineering, 101*(2), 87–96.

Talarico, L. B., Pujol, C. A., Zibetti, R. G. M., Faria, P. C. S., Noseda, M. D., Duarte, M. E. R., & Damonte, E. B., (2005). The antiviral activity of sulfated polysaccharides against dengue virus is dependent on virus serotype and host cell. *Antiviral Research, 66*(2, 3), 103–110.

Tanaka, T., Shnimizu, M., & Moriwaki, H., (2012). Cancer chemoprevention by carotenoids. *Molecules, 17*(3), 3202–3242.

Torres, F. A., Passalacqua, T. G., Velásquez, A. M., De Souza, R. A., Colepicolo, P., & Graminha, M. A., (2014). New drugs with antiprotozoal activity from marine algae: A review. *Revista Brasileira de Farmacognosia, 24*(3), 265–276.

Wada, N., Sakamoto, T., & Matsugo, S., (2015). Mycosporine-like amino acids and their derivatives as natural antioxidants. *Antioxidants, 4*(3), 603–646.

Wang, W., Zhang, P., Hao, C., Zhang, X. E., Cui, Z. Q., & Guan, H. S., (2011). *In vitro* inhibitory effect of carrageenan oligosaccharide on influenza A H1N1 virus. *Antiviral Research, 92*(2), 237–246.

Wijesekara, I., Pangestuti, R., & Kim, S. K., (2011). Biological activities and potential health benefits of sulfated polysaccharides derived from marine algae. *Carbohydrate Polymers, 84*(1), 14–21.

Zhang, H. J., Mao, W. J., Fang, F., Li, H. Y., Sun, H. H., Chen, Y., & Qi, X. H., (2008). Chemical characteristics and anticoagulant activities of a sulfated polysaccharide and its fragments from *Monostroma latissimum*. *Carbohydrate Polymers, 71*(3), 428–434.

CHAPTER 8

BIOLOGICAL PROPERTIES OF MICROALGAE-BASED PRODUCTS

TATIELE C. DO NASCIMENTO,[1] ANDRESSA M. BASEGGIO,[2] MÁRIO R. MARÓSTICA JR.,[2] EDUARDO JACOB-LOPES,[1] and LEILA Q. ZEPKA[1]

[1]Department of Food Science and Technology, Federal University of Santa Maria (UFSM), Santa Maria, RS, 97105-900, Brazil

[2]Department of Food and Nutrition, University of Campinas (UNICAMP), Campinas, SP 13083-862, Brazil

ABSTRACT

In recent decades, the exploitation of microalgae-based products has received increasing attention due to several biomolecules with essential biological activities. These bioactive compounds include carotenoids, chlorophylls, phycobiliproteins (PBP), phenolic compounds, polysaccharides, fatty acids, proteins (including amino acids and peptides), and some vitamins. The bioactivities have shown several biological properties, like antioxidant, antiobesity, antibacterial, anti-inflammatory, and anticancer. For this, the use of experimental models *in vivo* highlights the potential employment of these microorganisms in health purposes. However, some barriers associated with the economic viability of these biomolecules' production and toxicity is challenges to overcome for their larger use. In this sense, this chapter presents a comprehensive description of the biological evidence of microalgae bioactive compounds, including structural aspects of main compounds, and their related biological activities in some disease or health models, as well as the main existing challenges for extensive use aimed at promoting health applications.

8.1   INTRODUCTION

In recent years, a global trend towards healthy eating habits has considerably boosted the search for natural alternatives for health promotion (Holban and Grumezescu, 2018; Fanzo et al., 2020). In this circumstance, microalgae are considered a promising alternative, once they present a range of biocompounds capable of positively modulating human health, including the prevention, maintenance, and treatment of health conditions (Matos, 2017; Borowitzka, 2018; Jacob-Lopes et al., 2019; Lupette and Benning, 2020; Tang et al., 2020).

Microalgae are prokaryotic (cyanobacteria) and eukaryotic (other microalgae) photosynthetic organisms capable of growing in diverse aquatic environments tolerating extreme conditions of temperatures, salinities, pH, and light intensity (Khan et al., 2018). As a consequence of this metabolic versatility, they have special abilities to produce numerous biomolecules with bioactive potential, including fatty acids, peptides, amino acids, vitamins, polysaccharides, phenolic compounds, and bioactive pigments (Singh et al., 2020).

Initially, microalgae were exclusively considered as a source of unicellular protein (total dry biomass), especially the genera *Spirulina* and *Chlorella* (Borowitzka, 2018). The industrial insertion of microalgae-based products emerged as a commercial opportunity to gain market share in the segment of bioactive molecules (Jacob-Lopes et al., 2019). This insertion was only possible because, in general, microalgae bioproducts have advantages over traditional sources (Matos, 2017). Among these advantages is the plurality of its bioactive properties that allow for varied applications from food, pharmaceuticals, and cosmetics (Tang et al., 2020).

Among the microalgae bioproducts, total dry biomass (protein), carotenoids, PC, and polyunsaturated fatty acid have biological effects in the ascending process of exploration (Abdel-Aziem et al., 2018; Khalil et al., 2018; Memije-Lazaro et al., 2018; Xiong et al., 2018; Abu-Taweel et al., 2019; Nascimento et al., 2019, 2020; Silva et al., 2020). In contrast, chlorophylls, vitamins, and polysaccharides are considered as emerging proposals, and the investigation of its biological effects is limited and recent. The higher portion of the existing data is limited to identification and *in vitro* evaluations (Rodrigues et al., 2015; Fernandes et al., 2016; Bernaerts et al., 2018; Tarento et al., 2018; Edelmann et al., 2019).

The bioactive properties of microalgae were mostly investigated for exploratory purposes; however, experimental inconsistencies verified at *in vitro* and *in vivo* assays turns the application in physiological conditions necessary to provide further credibility to the results. To date, these activities have been established for the entire biomass, isolated biocomposites, and extracts. The most relevant experimental models comprise the use of animals, such as rats (Wistar, Sprague-Dawley), mice (ICR, BALB, C57BL6, Swiss), and rabbits (Tsai et al., 2012; Chen et al., 2014; Rao et al., 2015; Khalil et al., 2018; Nascimento, 2019, 2020; Aladaileh et al., 2020).

Among the biological activities evidenced by microalgae, antioxidants, anti-inflammatory, hepatoprotective, immunomodulatory, and antiobesity properties stand out (Dvir et al., 2020; Singh et al., 2020). Many of these properties are believed to be linked with the specific structural characteristics of some biomass constituents (Raposo et al., 2015; Jacob-Lopes et al., 2019; Zepka et al., 2019). Despite the bioactive potential, proven or expected, the high cost of production and downstream processing remain critical barriers to the use of microalgae-based products (García et al., 2017; Khan et al., 2018; Lafarga, 2019; Zanella and Vianello, 2020). Also, the toxins produced by some species and the acid nucleic (purine) content are barriers to the establishment of the biological purposes of these compounds (Matos, 2017).

Based on the above, this chapter presents a comprehensive description of the biological evidence for microalgae use, including main products, structural aspects, and their related *in vivo* biological activities, as well as the main existing challenges for extensive use aimed at promoting health applications.

8.2 MICROALGAE-BASED PRODUCTS

Currently, microalgae-based products are obtained mainly from the following classes: Cyanophyceae (*Spirulina platensis* or *Arthrospira platensis, Oscillatoria, Nostoc*), Chlorophyceae (*Chlamydomonas, Dunaliella salina, Chlorella*), Prymnesiophyceae (*Isochrysis galbana*), and Bacillariophyceae (*Thalassiosira weissflogii, Cyclotella cryptica*) (Singh et al., 2020). However, *Spirulina*, and *Chlorella* are the best-known species for several reasons, as they show a valuable nutritional composition, which turns possible their employment as food supplement, and are deemed reliable for human intake (they are Generally Recognized as Safe, status provided by the Food and Drug Administration (FDA)) (García et al., 2017; Jacob-Lopes et al., 2019; Koyande et al., 2019).

Biomass is the main product based on microalgae, it is predominantly marketed as a food supplement, in bulk powder, in capsules, moist (in paste), and occasionally found added in food products (Barkia and Saari, 2019; Lafarga, 2019). Besides to the entire biomass, other compounds based on microalgae stand out commercially and have been approved as food ingredients by the competent regulatory authorities, like β-carotene obtained from *Dunaliella salina*, docosahexaenoic acid (DHA), from *Crypthecodinium cohnii*, single-cell oils, from *Ulkenia* sp., *Schizochytrium* sp., and astaxanthin esters by *Haematococcus pluvialis* (Jacob-Lopes et al., 2019). Other bioproducts with potential bioactive activities such as phenolic compounds, polysaccharides, vitamins, and other pigments (carotenoids, chlorophylls, phycobiliproteins (PBPs)) have been verified, aiming the regulation and expansion of the microalgae product portfolio for various applications (Klejdus et al., 2009; Fernandez et al., 2017; Matos, 2017; Barkia and Saari, 2019).

Representative fractions of microalgae achieve levels above 50% protein (Becker, 2013). The protein percentage is one of the main reasons for considering these microorganisms an excellent food source, especially *Spirulina* (55%–70%) and *Chlorella* (42%–55%) (Matos, 2019). In general, the protein of microalgae origin is comparable to conventional sources, once they present all essential amino acids (EAAs) and considerable biological value (Becker, 2013). The efficiency, productivity, and sustainability turn the extraction of proteins from microalgae interesting: while the protein yield from microalgae reported at 4–15 tons/ha/year, the production for other plants crops (as soybean) reaches 1.1 tons/ha/year and requires a considerable amount of fresh water and arable land (Koyande et al., 2019). Thus, microalgae-based protein is considered a valuable option to replace animal meat, a promising and sustainable food innovation (Weinrich and Elshiewy, 2019). In addition to the recognized nutritional benefits, microalgae protein (including its peptides and amino acids) has biological functions associated with other benefits to human health, especially antiobesity, and antioxidant effects (Patias et al., 2018; Abu-Taweel et al., 2019; Nascimento et al., 2019).

Numerous species of microalgae produce significant volumes of lipids, including polyunsaturated fatty acids (PUFAs) ω-3, as α-linolenic acids (ALA), eicosapentaenoic (EPA) and (DHA) and ω-6, as linoleic acids (LNA), γ-linolenic (GLA) and arachidonic (ARA) (Jacob-Lopes et al., 2019; Lupette and Benning, 2020). Among the producing species, *C. cohnii* is predominantly rich in DHA, *Nannochloropsis oculata*, *Phaeodactylum tricornutum*, and *Porphyridium cruentum* are rich in EPA, while *Arthrospira* sp., *Isochrysis* sp., *Odontella* sp., *Pavlova* sp., and *Porphyridium* sp. produce large quantities of both PUFAs (DHA and EPA) (Jiang et al., 1999; Martins et al., 2013; Enzing et al., 2014).

Some species, including *Desmococcus olivaceus*, *Chlorella vulgaris*, *Chlorella pyrenoidosa*, *Spirulina platensis*, *Nannochloropsis salina*, and *Nannochloropsis limnetica* synthetize phenolic compounds (Machu et al., 2015). Most of the identified phenolic compounds are acidic (gallic, coumaric, ferulic, and caffeic) and some flavonoids (mainly quercetin) (Klejdus et al., 2009; Goiris et al., 2014; Scaglioni et al., 2018). In quantitative terms, microalgae phenolic compounds are not competitive compared to the plant sources but are identified as significant contributors to antioxidant action combined with other bioactive compounds (Zanella and Vianello, 2020).

Besides vitamin A (provided by some carotenoids), recently, microalgae cells have been considered sources of vitamins, especially B complex, and vitamin K, with beneficial effects on human health (Cezare-Gomes et al., 2019; Edelmann et al., 2019). Although vitamins B are not produced by the microalgae, they are absorbed from the culture medium. In parallel, interest in microalgae polysaccharides is also recent. These macromolecules are complex, heterogeneous, and difficult to characterize (Delattre et al., 2016; Michaud, 2018; Gaignard et al., 2019a, b; Pierre et al., 2019). However, these biopolymers call attention for the possibility of use as prebiotics, as antioxidants, and in the control of dyslipidemias (Raposo et al., 2014; García et al., 2017; Gaignard et al., 2019b). The main related species are marine microalgae of the genus *Porphyridium*, *Dixioniella*, and *Rhodella* (Raposo et al., 2013; Gaignard et al., 2019b).

Considering the microalgae pigment fraction, they can synthesize up to three classes, including carotenoids, chlorophylls, and PBPs (Jacob-Lopes et al., 2019). Of the 1,183 natural carotenoids categorized to date (Yabuzaki, 2017), about 200 have been found in microalgae (Egeland et al., 2011). Among them, β-carotene, lutein, zeaxanthin, α-carotene, violaxanthin, echinenone, mixoxantophyll, astaxanthin, and canthaxanthin are frequently reported (Rodrigues et al., 2014; Haque et al., 2016; Nascimento et al., 2019; Fernandes et al., 2020). Although a range of species has been investigated, *D. salina* and *H. pluvialis* stand out commercially (Borowitzka, 2013; Koller et al., 2014; Borowitzka, 2018). The primary application of microalgae carotenoids is as a food colorant (especially β-carotene and astaxanthin), however due to their bioactive activities new applications are being discovered (Jacob-Lopes et al., 2019). On the other hand, the exploration of the microalgae chlorophyll fraction is still timid, but have become the target of investigations in recent years (Fernandes et al., 2016, 2020; Chen and Roca, 2018; Sarkar et al., 2020). According to Zepka et al. (2019), in addition to their primary pigmentation function, chlorophylls are being sought due to their bioactive properties that enable health benefits. According to Sarkar et al. (2020), *Chlorella thermophila* stands out from other microalgae species for having a high concentration of chlorophyll (~7%). Unlike conventional pigments, PBPs are protein pigments and are commercially produced from *Spirulina, Porphyridium,* and *Rhodella* (Spolaore et al., 2006; Borowitzka, 2013). Among these pigments, phycocyanin (PC) is the only natural blue dye available for use (Chew et al., 2017).

Finally, referring to the potential bioactive effects, some products based on microalgae have advantageous structural characteristics over their conventional competitors. In some cases, there are no similar alternatives available, as can be seen in the next section.

8.3 STRUCTURAL ASPECTS AND RELATIONSHIP STRUCTURE-ACTIVITY

Among microalgae-based products, proteins, fatty acids, phenolic compounds, and vitamins, are structurally comparable to conventional sources; in contrast, some carotenoids, chlorophylls, polysaccharides, and PBPs have some specific structural characteristics. The structure knowledge of any compound is essential to predict its bioactivities. Although it is not entirely clear, most biocompounds have some structure-activity relationship previously established, at least for some type of property, mainly antioxidant activity, as will be described below.

Microalgae protein, as well as proteins from other sources, is known to consist of amino acid subunits that determine its nutritional quality in terms of proportion and availability (Matos, 2019). Based on previous reports, the amino acid profile (including essentials) of almost all algae coincides with that required by the Food and Agriculture Organization (FAO) (Becker, 2007, 2013). Regarding structure-activity relationships, amino acids have been indicated as accountable for the antioxidant capacity of proteins. According to Agyei et al. (2015), the imidazole group present in histidine help to chelate metals and eliminate reactive species, and the hydrophobic amino acids can improve the solubility of peptides in the fat phase and consequently inhibit peroxidation at the water-lipid interface. Still, amino acids of an acidic nature alter the redox cycle and improve metallic bonds, aromatics allow the chelation of pro-oxidant metal ions while cysteine acts indirectly as a precursor of glutathione synthesis through its grouping thiol. The three-dimensional structure also allows the amino acids of the ingested proteins to promote conformational changes structural in the enzymes, acting indirectly against tissue lipid peroxidation (Wang et al., 2017).

In terms of fatty acids, those of biological interest in microalgae biomass are the essential omega 6 (n-6) or 3 (n-3) (LA, GLA, ARA, ALA, EPA, DHA). Structurally, they are carboxylic acids with a long carbon aliphatic chain (>18 carbons), 2–6 double bonds numbered by the carbon position ($^\Delta$) starting from the carboxyl group (LA: $18:2^{\Delta 6,12}$ n-6; GLA: $18:3^{\Delta 6,9,12}$ n-6; ARA: $20:4^{\Delta 5,8,11,14}$ n-6; ALA: $18:3^{\Delta 9,12,15}$ n-3; EPA: $20:5^{\Delta 5,8,11,14,17}$ n-3; DHA: $22:6^{\Delta 4,7,10,13,16,19}$ n-3) (Gunstone, 2003; Aldred et al., 2009; Lupette and Benning, 2020). In general, these lipophilic structures, especially those of very-long-chain (EPA and DHA) are essential for the integrity and function-ality of all cell membranes. They are precursors of many signaling molecules involved in humans' inflammatory processes (Lupette and Benning, 2020).

The phenolic acids are structurally simpler than fatty acids, being constituted by an aromatic ring linked to a carboxylic acid and hydroxyl groups, and usually divided into hydroxybenzoic acids (e.g., gallic acid) and acids hydroxy-dynamics (e.g., coumaric acid) with basic structures C6-C1 and C6-C3, respectively (Robbins, 2003). On the other hand, according to Panche et al. (2016), flavonoids, such as quercetin, are characterized by having two aromatic rings connected by a central ring C3 (C6-C3-C6). A well-known structure-activity relationship is the capacity to donate hydrogen (from the hydroxyl group-OH) to free radicals. Additionally, flavonoids such as quercetin can also eliminate radicals via electron donation (through the resonance system generated by the conjugated double bonds (CDBs) present in its chemical structure). Due to the presence of OH in ring B (in positions 3' and 4'), they can also act as metal ion chelators inhibiting the oxidation of low-density lipoproteins (LDL) (Leopoldini et al., 2004; Santos-Sánchez et al., 2019).

Vitamins are divided into two broad groups according to the solubility characteristics provided by the chemical structure (Combs and McClung, 2017). Among the vitamins recently reported in microalgae, those of the B complex (including B1, B7, B9, and B12) are highly water-soluble while K1 is fat-soluble. Due to these attributes, the B vitamins are easily absorbed. In contrast to the absorption of K (as well as the other lipophilic ones), the presence of bile or pancreatic lipase is required, in addition to combined fat intake in the diet (Costa-Pinto and Gantner, 2020). Another fat-soluble vitamin provided by microalgae is vitamin A obtained through the supply of precursor carotenoids, especially β-carotene (Grune et al., 2010). In general, vitamin structures have many roles recognized in the intermediate and specialized metabolism of specific organs. Once absorbed, vitamins are usually converted into more complex molecules acting as co-enzymes, and as they cannot be synthesized by human metabolism should be ingested through feeding (Costa-Pinto and Gantner, 2020).

Unlike the molecules mentioned above, the microalgae polysaccharides, carotenoids, chlorophylls, and PBPs do not have equivalent in higher plants. Their structures are shown in Figure 8.1 (except polysaccharides that have not yet been fully elucidated).

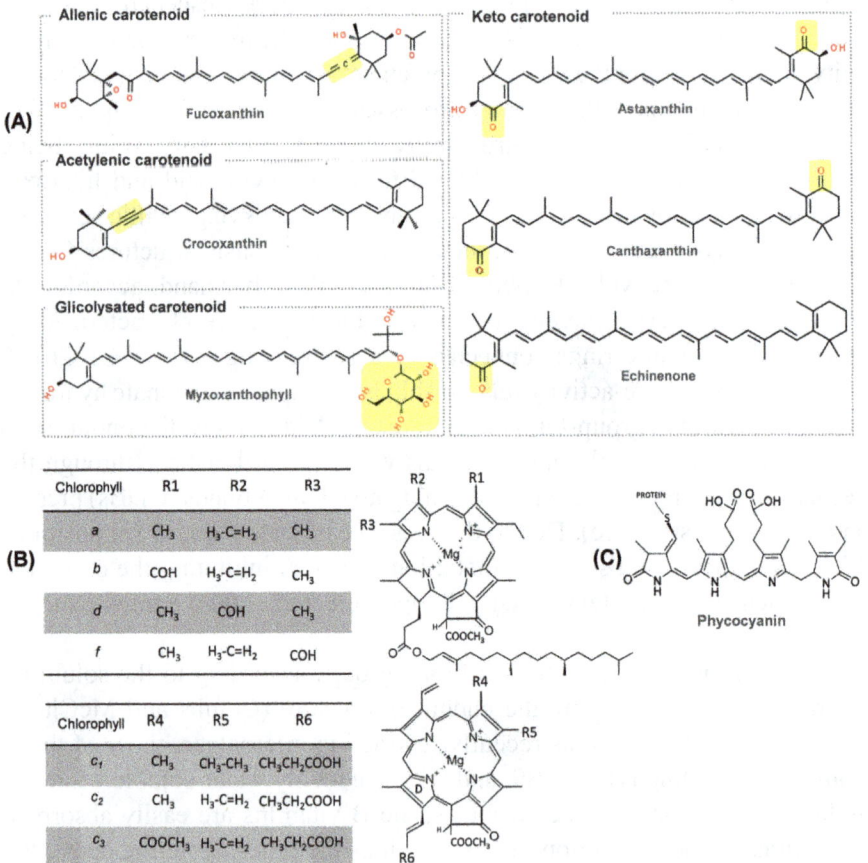

FIGURE 8.1 Chemical structures of microalgae compounds without equivalents in conventional sources (except chlorophyll α and β).

The merit for the potential biological functions identified in microalgae polysaccharides is quite recent, and the information is scarce (Delattre et al., 2016; Michaud, 2018; Pierre et al., 2019). Structurally, the microalgae bioactive polysaccharides are mostly secreted heteropolymers (exopolysaccharides: EPSs) in adverse situations, configured by a carbon chain formed from the polymerization of long chains of different simple sugars (mainly arabinose, fucose, fructose, galactose, galacturonic acid, glucose, mannose, ramose, ribose, xylose, and methylated sugars in varying proportions) joined by glycosidic bonds (Raposo et al., 2014; Delattre et al., 2016; Gaignard et al., 2019a).

Many of these microalgae polymers have different organic or inorganic substituents along their carbon chain (Xiao and Zheng, 2016). Among the substituents, the bioactivity attributed to these biopolymers is derived from the sulfate group. According to Senni et al. (2011), sulfated polysaccharides can bind to proteins at various specificity levels and are involved in multiple cellular activities, from cell development, differentiation, and adhesion. However, these biopolymers' structural complexity explains the lack of reports and limits the information needed to establish the relationships between structure and biological activity (Delattre et al., 2016; Michaud, 2018; Pierre et al., 2019).

Regarding carotenoids, their basic structure consists of a linear and symmetrical skeleton with a series of CDBs (Rodriguez-Amaya, 2001). These tetraterpenes are usually grouped into carotenes and xanthophylls, where carotenes are formed only by carbon and hydrogen (e.g., β-carotene) and xanthophylls contain oxygen (for example, lutein) (Fernandes et al., 2018).

Some microalgae xanthophylls are not synthesized in higher plants and are considered chemotaxonomic markers (Takaichi, 2011; Gee et al., 2018). These xanthophylls exhibit specific structural complexity (Figure 8.1(A)). They may contain allenic, acetylenic, glycosylated, and keto groups, such as fucoxanthin, crocoxanthin, myxoxanthophyll, canthaxanthin, astaxanthin, and echinenone (Takaichi and Mochimaru, 2007; Takaichi, 2011). The presence of these specific groups in the isoprenoid structure can result in the elongation of the chromophore (CDBs system), as observed in canthaxanthin (13 CDBs), astaxanthin (13 CDBs), myxoxanthophyll (12 CDBs) and echinenone (12 CDBs). Consequently, these pigments have their antioxidant capacity increased since the electron transfer capacity for the deactivation of reactive oxygen species (ROS) is directly proportional to the number of CDBs (Rodrigues et al., 2012).

Another important pigment is chlorophyll; whose basic structure consists of a tetrapyrrole replaced by a centrally coordinated magnesium atom that gives the greenish color of all algae, higher plants, and cyanobacteria (Zepka et al., 2019). Chlorophylls *a*, *b*, *c* (*c1*, *c2*, *c3*), *d*, and *f* are frequently reported in microalgae, except for the first two, the others are not found in higher plants (Zepka et al., 2019). Some structural peculiarities are responsible for the differentiation of these compounds (Figure 8.1(B)), for example, in R1 (C7) chlorophyll α has a methyl group and, chlorophyll *b* has an aldehyde (Fernandes et al., 2016). Chlorophyll *d* and *f* resemble chlorophyll α. However, the *d* has an aldehyde in R2 (C3) while the *f* in R3 (C2). In contrast, chlorophyll *c* has an unsaturated D ring and a characteristic propionic acid in R6 (C17), instead of a phytol group according to the other chlorophylls. Chlorophyll *c1* has an alkyl and *c2* a vinyl in R5 (C8), while chlorophyll *c3* has a methoxy group on R4 (C7) (Zepka et al., 2019). The number of CDBs and the different substituents on the pyrrole rings profoundly influences energy levels, color changes, and bioactive potential (Croce and Van Amerongen, 2014).

In contrast to carotenoids and chlorophyll, PBPs are highly water-soluble. They are characterized by a linear tetrapyrrole chromophore covalently linked to the protein skeleton through bonds with cysteine residues (Spolaore et al., 2006) (Figure 8.1(C)). The electron resonance system that makes up the structure of linear tetrapyrrole has been suggested to be responsible for its biological activity, especially PC (Bhat and Madyastha, 2000).

8.4 BIOLOGICAL EVIDENCE

The knowledge about microalgae usage as a food source is reported for hundreds of years, mainly as protein supply (García et al., 2017). Despite proteins, especially the presence of vitamins, lipids, sulfated polysaccharides, and secondary metabolites (as carotenoids and phenolic compounds) in microalgae biomass demonstrate their potential as functional food or ingredient. Among the biological activities already remitted, antioxidant, anti-inflammatory, hepatoprotective, immunomodulatory, and antiobesity properties are highlighted (Dvir et al., 2020), as summarized in Figure 8.2.

8.4.1 ANTIOXIDANT EFFECTS

The development of many chronic diseases is closely related to the increase in oxidative stress. This term refers to an imbalance between pro-oxidants chemical species, generates during own cell metabolism or introduced from external factors (as pollution, unhealthy eating habits, and irradiation, for example) and endogenous antioxidant cell defenses, represented by enzymes as superoxide-dismutase (SOD), catalase (CAT), glutathione peroxidase (GPx), glutathione reductase (GR), glutathione transferase (GST) and glutathione reduced (GSH) and oxidized (GSSG). The pro-oxidant species are reactive with many macromolecules with essential functions in the cell as lipids, proteins, and nucleic acids could lead to damages, such as lipid peroxidation, mitochondrial swelling, mutagenic actions, and posttranslational protein modifications (Barkia and Saari, 2019).

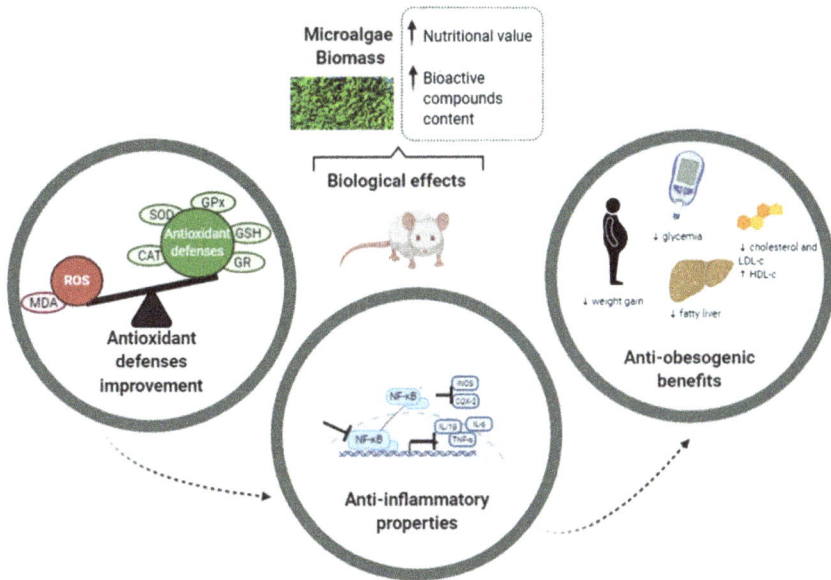

FIGURE 8.2   Biological properties already demonstrated by microalgae.

On the other hand, phytochemicals, vitamins, and minerals act as antioxidants *in vivo*, through to the endogenous antioxidant defenses improvement or scavenging of the reactive species, turning the employment of plants an interesting therapeutic approach against oxidative stress and diseases related (Barkia and Saari, 2019). Due to the several environmental conditions and phototropic growth mode, microalgae are generally exposed to high oxidative stress conditions: for self-protection, the synthesis of antioxidants and pigments, as chlorophylls, carotenoids, and PBPs are increased (Koyande et al., 2019). Thus, microalgae are considered sources of molecules with high antioxidant capacity that cannot be produced by mammals. Therefore, their consumption reduces the harmful effects of oxidative damage, as demonstrated in some studies presented in Table 8.1.

TABLE 8.1 Antioxidant Properties of Microalgae Species Oral Ingestion *In Vivo*, Using Different Experimental Models

Microalgae Species and Class	Antioxidant Compounds Identified	Experimental Model	Microalgae Dosage	Effects	References
Spirulina platensis or *Arthrospira platensis* Cyanophyceae	Chlorophyll Phycocyanin β-carotene Tocopherols Zinc Manganese	Rabbits submitted to high-fat diet consumption	1% and 5% added to the diet	↓ MDA-TBA in liver, plasma, heart, and kidney ↑ GSH level in liver and blood ↑ GR, GPx, and GST activities in liver and blood ↓ Oxidative damage of DNA in lymphocytes	Kim et al. (2010)
Spirulina maxima or *Arthospira maxima* Cyanophyceae	C-phycocyanin	Chronic renal failure induced in Wistar rats	1 g/kg/day	↓ ROS formation in kidney and heart ↓ lipid peroxidation in kidney ↑ GSH^2/GSSG ratio in heart	Memije-Lazaro et al. (2018)
Spirulina platensis or *Arthrospira platensis* Cyanophyceae	—	Hepatic impairment induced in Sprague-Dawley rats	300 mg/kg/day	↑ SOD and CAT activities in liver ↑ GSH content and total antioxidant capacity ↓ MDA, protein oxidation and DNA damage	Khalil et al. (2018)
Chlorella vulgaris Trebouxiophyceae	—	Hepatic impairment induced in ICR mice	100 mg/kg/day and 200 mg/kg/day	↓ MDA in liver ↑ SOD, CAT, and GST activities ↑ GSH content	Li et al. (2013)
Chlorella vulgaris Trebouxiophyceae	—	Ovary dysfunction induced in female mice	500 mg/kg/day	↑ GPx, Cu-Zn SOD and CAT gene expression ↑ SOD, CAT, and GR activities	Abdel-Aziem et al. (2018)

TABLE 8.1 *(Continued)*

Microalgae Species and Class	Antioxidant Compounds Identified	Experimental Model	Microalgae Dosage	Effects	References
Scenedesmus obliquus Chlorophyceae	Fatty acids (linolenic, linoleic, oleic) Amino acids (histidine, cysteine, glutamine) Carotenoids (mainly lutein and β-carotene)	Healthy BALB/ cAnUnib mice	400 mg/kg/day 800 mg/kg/day	↓ MDA, SOD, and CAT in heart, liver, kidneys, and spleen ↑ GPx in kidneys ↑ GR in liver	Nascimento et al. (2019)
Scenedesmus obliquus Chlorophyceae	Carotenoids (mainly all-*trans*-β-carotene, all-*trans*-lutein, and all-*trans*-zeaxanthin)	Healthy BALB/ cAnUnib mice	0.25 mg/kg/day 2.5 mg/kg/day of carotenoids	↑ MDA in liver ↓ MDA, GPx, and GR in spleen ↓ GR and CAT in liver ↑ GR and GPx in heart ↑ GPx in kidneys	Nascimento et al. (2020)
Dunaliella salina Chlorophyceae	Carotenoids (β-carotene, α-carotene, lutein, and lycopene)	Hepatic impairment induced in Wistar rats	125 µg/kg/day 250 µg/kg/day	↑ SOD, CAT, Peroxidase in liver ↓ MDA	Murthy et al. (2005)
Dunaliella salina Chlorophyceae	–	Corneal oxidative damage induced in ICR mice	123 mg/kg 615 mg/kg	↑ SOD, CAT, GPx, and GR activities in cornea ↑ GSH levels ↓ MDA	Tsai et al. (2012)
Hematococcus pluvialis Chlorophyceae	Carotenoids (Astaxanthin and astaxanthin esters)	Hepatic impairment induced in Wistar rats	100 µg/kg/day 250 µg/kg/day	↑ SOD, CAT, and GPx activities in liver ↓ MDA	Rao et al. (2015)

The antioxidant properties of microalgae consumption have been identified in some conditions, being that the main target organs are the liver,
the kidneys, the heart, the ovary, and the cornea. These tissues present
high affinity with bioactive compounds as carotenoids (Perera and Yen,
2007), important antioxidants found in microalgae biomass. In the organ,
carotenoids could take over the antioxidant enzyme's actions through their
free radical scavenging capacity; or still, can stimulate the activity of these
enzymes, enhancing the efficacy of detoxification (Babin et al., 2015). On
the other hand, polyphenolic metabolites and polysaccharides from microalgae activated the Nrf2 transcription, a factor responsible for the stimulus
of some antioxidant enzymes synthesis, like SOD and CAT (Fernando et
al., 2020).

Still, the most pronounced effect of supplementation in antioxidant
enzymes activity occurred in injured tissues, which could indicate their
action as a stimulator of defenses against oxidative stress. The majority of
studies evaluating the antioxidant microalgae effects defenses show this
trend mainly for the lipid peroxidation reduction.

8.4.2   ANTI-INFLAMMATORY EFFECTS

Once oxidative stress is a common feature underlying several inflammatory
pathways (Barkia and Saari, 2019), it is no surprise that the microalgae can
alleviate inflammation. The inflammation refers to the immune response
against injury, aiming to generate tissue repair and foreign agent elimination. This process is self-limited and shows initiation, progression, and
resolution. When occurring hyperactivation or persistence, the process
turns chronic, leading to organ commitment and systemic decompensation
(Mendes et al., 2018).

The chronic inflammation is signalized by constant pro-inflammatory
cytokines and chemokines synthesis and release [such as tumor necrosis
factor-α (TNF-α), interleukin-1 *beta* (IL-1β), IL-6, IL-8, and monocyte
chemoattractant protein-1 (MCP-1)], activation of inducible enzymes [as
cyclooxygenase-2 (COX-2) and inducible nitric oxide synthase (iNOS)],
and release of inflammatory mediators (as prostaglandins, leukotrienes,
and thromboxane), in a vicious circle. These pathologic events result
in structural remodeling in the affected tissue and systemic alterations
in homeostasis, associated with development and progression of most
chronic diseases, like diabetes mellitus type II, obesity, neurodegenerative

disturbances, cardiovascular diseases, and cancer (Wu et al., 2016; Mendes et al., 2018).

The anti-inflammatory properties reported to microalgae biomass occur mainly through the reduction of pro-inflammatory cytokines and eicosanoids transcription and synthesis, and by the modulation of the inducible activity enzymes responsible for inflammation propagation, mostly represented by COX-2, phospholipase A2, and iNOS. These effects imply the reduction in nuclear transcription factor *kappa-beta* (NF-κB) translocation to the nucleus and mitogen-activated protein kinase (MAPK) phosphorylation, a critical factor in the signaling for continuous synthesis of pro-inflammatory mediators. The properties are remitted mainly to the carotenoids, omega-3 fatty acids, PC, and modified carbohydrates present in the raw material (Barkia and Saari, 2019). Several studies report these and other anti-inflammatory effects for different microalgae species, which could be associated with their main bioactive compounds (Table 8.2).

A highlight in anti-inflammatory benefits associated with microalgae consumption is the reduction in chronic noninfectious airway inflammation associated with asthma, once the traditional medicines employed to control this chronic disease, the glucocorticoids, lead to many adverse effects in long-term use. Among the microalgae properties, the reduction in inflammatory cells recruited to the lung, as eosinophils, mast cells, macrophages, and Th2 lymphocytes, implies less immunoglobulin E (Ig-E) and pro-inflammatory cytokines, as IL-13, IL-4, and transforming growth factor-beta (TGF-β) release. In contrast, the products stimulated of lymphocytes T differentiation in Th1 form, responsible for interferon-gamma (IFN-γ), IL-10, and IL-12 production and secretion, that stimulated the inflammation resolution. The mechanism of action involved could be related, in part, by the resolvins biosynthesis precursors' presence, like omega-3 (Xiong et al., 2018; Capek et al., 2020).

8.4.3 EFFECTS IN OBESITY AND COMORBIDITIES

Obesity is defined as abnormal or excessive fat accumulation that may impair health. The prevalence of obesity tripled between 1975 and 2016 worldwide, and approximately 13% of the world's adult population was considered obese in 2016. Among the external factors involved with obesity development, dietary parameters as the increase in the high-energy food intake (rich in fats and sugars) combined to physical inactivity, leading to

TABLE 8.2 Some Pathways Involved in Anti-Inflammatory Activities Reported to Microalgae Species

Microalgae Species and Class	Bioactive Compounds Identified	Experimental Model	Microalgae Dosage	Effects	References
Spirulina platensis or *Arthrospira platensis* Cyanophyceae	–	Rabbits exposed to toxic agent lead acetate	0.5, 1, and 1.5 g/kg, oral administration	↑ IgG in blood ↓ NF-κB p65 immunoreactivity in liver, kidneys, and heart ↓ Inflammatory infiltrates	Aladaileh et al. (2020)
Spirulina platensis or *Arthrospira platensis* Cyanophyceae	Amino acids (glutamine, leucine, isoleucine, valine, and lysine) Polyphenols (not specified) Fatty acids (linolenic acid, oleic acid, palmitoleic acid)	Wistar rats induced of paw edema	200 mg/kg, oral administration	↓ TNF-α, IL-6, IL-1β, PGE-2 and nitrite levels in paw ↓ COX-2, iNOS, NF-κB p50 gene expression	Abu-Taweel et al. (2019)
Spirulina platensis or *Arthrospira platensis* Cyanophyceae	Fatty acids (linolenic, oleic, palmitoleic, and linoleic)	BALB/c challenge by OVA to allergic airway inflammation	0.4 g/kg, 0.8 g/kg and 1.6 g/kg of spirulina extract combined with fish oil, oral administration	↓ inflammatory score, represented by leukocytes infiltration, in lungs ↓ IL-13, IL-5, IL-4, and CCL-11 concentrations in bronchoalveolar lavage fluid (BALF) ↑ IFN-γ in BALF ↓ OVA-IgE levels in serum	Xiong et al. (2018)
Chlorella vulgaris Trebouxiophyceae	Exopolysaccharides (α-L-arabino-α-L-rhamno-α, β-D-galactan)	Guinea-pig sensitized to allergic asthma inflammation	50 mg/kg (purified exopolysaccharides), oral administration	↑ IL-12 and IFN-γ in BALF ↓ TGF-β in BALF	Capek et al. (2020)

TABLE 8.2 *(Continued)*

Microalgae Species and Class	Bioactive Compounds Identified	Experimental Model	Microalgae Dosage	Effects	References
Chlorella pyrenoidosa Trebouxiophyceae	Polysaccharides	C57BL/6 mice induced to Parkinson disease	200 mg/kg (polysaccharides isolated), oral administration	↓ TNF-α, IL-6 and IL-1β in serum ↑ IgA in small intestine	Chen et al. (2014)
Chorella pyrenoidosa Trebouxiophyceae	Polysaccharides	C57BL/6 mice LPS-stimulated	100 mg/kg (polysaccharides), intraperitoneal administration	↑ TNF-α and IL-1β in serum ↑ IL-1β concentration in macrophages, dependent to TLR-4 activation	Hsu et al. (2010)
Dunaliella salina Chlorophyceae	Carotenoids	Wistar rats induced to obesity-induced by high fat diet	150 mg/kg (carotenoid-rich fraction), oral administration	↓ C-reactive protein, adhesion molecules (ICAM and VCAM) and LOX activity in blood ↓ cardiac tissue damage	El-Baz et al. (2020)
Dunaliella salina Chlorophyceae	Fatty acids (linolenic, linoleic, palmitoleic) Carotenoids Chlorophyll	Wistar rats exposed to whole-body gamma irradiation	100 mg/kg and 200 mg/kg, oral administration	↓ TNF-α and IL-1β in jejunal tissue ↓ Jejunal tissue damage	Khayyal et al. (2019)

an energy imbalance between calories consumed and calories expended, favoring the fat accumulation mainly in white adipose tissue (WAT) (World Health Organization, 2020).

The visceral WAT remodeling and expansion, adipocyte hypertrophy and hyperplasia, inflammatory cells infiltration, and the constant release of pro-inflammatory cytokine led to many systemic changes, as oxidative stress. The WAT is a complex organ that secretes paracrine hormones for control of hunger and satiety mechanisms (leptin and adiponectin), and an insulin-dependent tissue. In obesity, many changes in hormone signaling occur: the hypothalamic response to leptin is disrupted, leading to resistance in hormone action and food intake dysregulation, and the diminution in insulin response in the own WAT and other insulin-sensitive tissues combined to impaired secretion of the hormone by the pancreas generates a condition termed insulin resistance. Insulin resistance is linked to an ample cluster of metabolic abnormalities and obesity-comorbidities, as the diabetes type II development, hepatic steatosis, dyslipidemia, hypertension, and coronary diseases (Longo et al., 2019).

On the other hand, the consumption of some compounds with biological activity from foods or in nutraceuticals could modulate, direct or indirectly, cell pathways involved in the pathophysiology of obesity and comorbidities. The incorporation of these functional compounds regularly in a balanced diet may contribute to the weight management and prevention of related diseases or could be therapeutic support during treatments (Konstantinidi and Koutelidakis, 2019). Once the microalgae are a rich source of bioactive substances, with antioxidant and anti-inflammatory properties, the literature also reported benefits of their consumption as a preventive agent against excessive weight gain and in metabolic changes promoted by overweight and obesity, acting in critical molecular mediators of obesity and comorbidities pathophysiology (Table 8.3).

The hypoglycemic and hypolipidemic activities observed in animals fed with microalgae biomass could be related to inhibitory effects in intestinal enzymes responsible for carbohydrates and fats digestion, like α-glucosidase, α-amylase, and lipase, promoted by bioactive compounds found in the matrix, like proteins (mainly PC), dietary fibers, polyphenols, and carotenoids. Therefore, some active compounds from microalgae can bind directly to cholesterol and bile acid metabolites reducing the solubility and increasing their fecal excretion. Molecular mechanisms related to insulin metabolism and glucose uptake in different tissues insulin-dependents, as skeletal muscle, and adipose tissue, can be modulated by microalgae

TABLE 8.3 *In Vivo* Effects of Microalgae Consumption in Obesity and Comorbidities

Microalgae Species and Class	Bioactive Compounds Identified	Experimental Model	Microalgae Dosage	Effects	References
Scenedesmus obliquus Chlorophyceae	Fatty acids (palmitoleic, oleic, linolenic, and linoleic) Dietary fibers Chlorophyll Carotenoids Phenolic compounds	Wistar rats fed with diets containing protein from *S. obliquus*	The casein in tested diets were replaced by 50 or 100% of the protein from microalgae	↓ Fasting glucose ↑ Total cholesterol/HDL-cholesterol ratio ↓ Total triglycerides	Silva et al. (2020)
Spirulina or *Arthospira* (species not identified) Cyanophyceae	–	ICR mice injected with monosodium glutamate subcutaneously to induced metabolic syndrome	Standard diet added 5% Spirulina	↓ Weight gain ↑ Visceral fat ↓ Total cholesterol, insulin, and leptin in serum ↓ Non-alcoholic fatty liver disease activity score, total triglycerides, and cholesterol in liver ↓ Macrophage infiltration and cytokines release in epididymal visceral fat	Fujimoto et al. (2012)
Spirulina or *Arthospira* (species not identified) Cyanophyceae	–	Wistar rats fed with high-fat diet	150 mg/kg/day orally	↓ Weight gain ↓ Triglycerides, cholesterol, and LDL-c in serum ↑ HDL-c ↓ Adipogenic genes expression in liver Modulation of gut microbiota	Chen et al. (2019)

TABLE 8.3 *(Continued)*

Microalgae Species and Class	Bioactive Compounds Identified	Experimental Model	Microalgae Dosage	Effects	References
Spirulina platensis Cyanophyceae *Chlorella pyrenoidosa* Chlorophyceae	Polyunsaturated fatty acids	Rats fed with high-fat high-sucrose diet	150 mg/kg/day of ethanolic or aqueous extracts	↑ Glucose tolerance ↓ Relation firmicutes/Bacteroidetes in gut microbiota	Wan et al. (2019)
Parachlorella beijerinckii Chlorophyceae	Dietary fiber, protein	C57BL/6J mice or rats fed with high-fat diet	5% of microalgae biomass (mice) or 1.5% of dried microalgae extract (rats) added to the diets	↑ Glucose and insulin tolerance (mice) ↓ Leptin, insulin, MCP-1 levels in serum (mice) ↓ Visceral adipocyte size (mice) ↓ Triglycerides in serum (rats) ↓ Epididymal fat depot (rats)	Noguchi et al. (2013)
Haematococcus pluvialis Chlorophyceae	Astaxanthin	Swiss mice fed with high-fat high-fructose diet	6 mg/kg (astaxanthin isolated)	↓ Weight gain and epididymal adipose tissue ↓ Glucose, insulin, TNF-α and IL-6 levels in serum ↑ Insulin signaling in skeletal muscle (IRS-FI3K-Akt pathway)	Arunkumar et al. (2012)
Dunaliella bardawil Chlorophyceae	β-carotene	ApoE−/− mice fed with low and high-fat diets	8% of whole *Dunaliella* powder or β-carotene deficient powder	↓ Plasma cholesterol (high-fat diet + whole microalgae) ↑ Plasma triglycerides (low-fat diet and *Dunaliella* β-carotene deficient powder) ↓ Atherosclerotic lesion area	Harari et al. (2013)

biomass, generating improvement in insulin resistance. Furthermore, bioactive substances from the raw material can increase glucose transporter type 4 (GLUT-4) receptors expression, which improves the glucose uptake in cells, and the phosphorylation of insulin receptor substrate (IRS), phosphoinositide 3-kinases-AKT (PI3K-AKT), and adenosine monophosphate-activated protein kinase (AMPK) in forms active, which is fundamental to insulin signaling (Fernando et al., 2020).

8.4.4   MICROALGAE AS VITAMINS SOURCES

The vitamins act as an organic growth factor that regulates microalgae phytoplankton, stimulating their faster growth and acting as defense mechanisms. In this sense, the algae require the exogenous supply of some vitamins, as B12 (cobalamin), B7 (biotin), and B1 (thiamine) in different proportions (Cezare-Gomes et al., 2019). These vitamins are obtained by culture medium enrichment or through the symbiotic relationship between algae and some bacteria. Interestingly, the supplemented vitamins could be incorporated into the biomass of some microalgae, turning this raw material a source of complex B vitamins, including B12 (Edelmann et al., 2019). This absorption capacity is particularly important for people who follow vegan diets, once the primary source of active B12 in the human diet is foods of animal origin.

Recently, Edelmann et al. (2019) demonstrated that *Chlorella* sp. supplements commercialized in Finland contain high amounts of active B12 vitamins, providing a significant percentage of the recommended daily intake in little consumed quantities. Despite this, the *Chlorella* sp. powder is a good source of folate (vitamin B9), when 5 g of the product (1 commercial tablet) could offer nearly a quarter of folate RDA for adults. This micronutrient is essential in red blood cell function, as well as in make and repair DNA. Moreover, folate availability for women in childbearing age is fundamental, once him deficiency can lead to fetal malformation.

Vitamin K1, a fat-soluble vitamin with essential roles in blood coagulation, neuroprotection, and bone maintenance, was already reported in high amounts in microalgae biomass. The *Anabaena cylindrica*, a blue-green alga of the same class of *Spirulina* (Cyanophyceae), is the most abundant source of this micronutrient, with around six times high amounts of the vitamin than other dietary sources as spinach (Tarento et al., 2018).

Besides the presence of vitamins, some carotenoids found in microalgae have pro-vitamin A activity, as β- and α-carotene. The β-carotene is the carotenoid with high pro-vitamin A potential, to which it is attributed 100% of activity after absorbed by the human body (Grune et al., 2010). Among the microalgae most studied and known as safe for human consumption, the *Dunaliella salina* presents the largest β-carotene amount, with 10–14% of its dry matter as β-carotene, accumulates in plastid-localized fat globules (Koyande et al., 2019). Under stress conditions, like high light and high salinity, the β-carotene synthesis is enlarged until 7-fold in these microalgae, as a defense mechanism. Recently, Lou et al. (2020) demonstrated that this response is mediated by miRNA, mainly the novel-m0533-3p, up-regulated about 5-fold in microalgae cultivated in high salt concentration and light incidence. The miRNA inhibits the malate dehydrogenase, leads to less acetyl-CoA use by tricarboxylic acid cycle, and more use in the geranylgeranyl pyrophosphate (GGPP) synthesis, and a carotenoid precursor.

8.5 CHALLENGES

The high cost of production and downstream processing, as well as the biogenic (synthesized), abiogenic (bioaccumulated) toxins, and the purine content accumulated in the biomass are some of the challenges that need to be overcome to make the utilization of microalgae-base products viable and safe (Matos, 2017; Khan et al., 2018; Jacob-Lopes et al., 2019). It is required to develop successful farming technologies to produce biomass to make the utilization of microalgae economically viable. Optimal cultivation conditions should be established considering the following factors: nutrients, temperature, pH, salinity, inorganic carbon, oxygen, luminous intensity, and CO_2; the variability of microalgae species is extensive, and they can be affected by both the species and the factors mentioned above (Khan et al., 2018).

From a toxicological point of view, some microalgae are potential sources of toxins dangerous to human health, such as hepatotoxins and neurotoxins (Zanella and Vianello, 2020). For example, *Anabaena* sp., *Microcystis* sp., *Dinophysis* sp. and *Pseudo-nitzschia*, are potent producers of saxitoxins, brevetoxin, domoic, and okadaic acids. Still, some algae can accumulate heavy metals from habitat by biosorption, and metal traces can be found in their derivatives (Matos, 2017). The nucleic acid content varies from 4 to 6% and is another limiter that needs to be evaluated before ingestion. The nucleic acids are sources of purines that are directly linked to severe kidney damage (Kelley and Andersson, 2014; Matos, 2017).

Considering that the existence of many unexplored biochemicals, possible biogenic, and abiogenic toxins in the entire microalgae biomass can cause severe health conditions, the purified compounds and extracts can be preferable. Besides safe, they have more and economic value than the entire biomass (Barkia and Saari, 2019). Although the production cost of the compounds is high, the technologies using the biorefinery concept can be a smart strategy to minimize production costs (Chew et al., 2017; Jacob-Lopes et al., 2019).

In summary, for any product based on microalgae to become viable, and obtain the approval of the regulatory bodies, it is essential to select the species and define the culture conditions, isolate, and purify to the target bioproduct and finally prove its biological activity (Khan et al., 2018; Silva et al., 2020). In addition, authorization for ingestion only occurs if the bioproduct meets all chemical and biological safety requirements imposed by current regulations (Zanella and Vianello, 2020).

8.6 CONCLUSION

Finally, microalgae have been proven to be sources of bioproducts with important bioactive properties, especially antioxidant, anti-inflammatory, and antiobesity effects. Despite the existing biological evidence, there is still a range of structures unexplored that could be associated to these properties. Although numerous species are known, biological exploration is not very comprehensive. In addition, strategies are needed to maximize production, reduce costs, and prove bioactivity and compliance with toxicological safety requirements for the scientific and technological advancement of microalgae-based products.

ACKNOWLEDGMENTS

This study was financed in part by the Coordenação de Aperfeiçoamento de Pessoal de Nível Superior-Brazil (CAPES)-Finance Code 001; CNPq (403328/2016-0; 301496/2019-6) and FAPESP (2015/50333-1; 2018/11069-5; 2015/13320-9). MRMJ acknowledges Red Iberomericana de Alimentos Autoctonos Subutilizados (ALSUB-CYTED, 118RT0543).

KEYWORDS

- anti-inflammatory
- antiobesity
- antioxidant
- bioactivity
- biological assay
- biomass
- bioproducts
- carotenoids
- chlorophylls
- fatty acid
- microalgae
- phycocyanin
- polysaccharides
- protein
- vitamins

REFERENCES

Abdel-Aziem, S. H., El-Kader, H. A. M. A., Ibrahim, F. M., Sharaf, H. A., & El Makawy, A. I., (2018). Evaluation of the alleviative role of *Chlorella vulgaris* and *Spirulina platensis* extract against ovarian dysfunctions induced by monosodium glutamate in mice. *Journal of Genetic Engineering and Biotechnology, 16*(2), 653–660. doi: 10.1016/j.jgeb.2018.05.001.

Abu-Taweel, G. M., Mohsen, G. A. M., Antonisamy, P., Arokiyaraj, S., Kim, H. J., Kim, S. J., Park, K. H., & Kim, Y. O., (2019). *Spirulina* consumption effectively reduces anti-inflammatory and pain related infectious diseases. *Journal of Infection and Public Health, 12*(6), 777–782. doi: https://doi.org/10.1016/j.jiph.2019.04.014.

Agyei, D., Danquah, M. K., Sarethy, I. P., & Pan, S., (2015). Antioxidative peptides derived from food proteins. In: Rani, V., & Yadav, U. C. S., (eds.), *Free Radicals in Human Health and Disease* (pp. 417–430). Springer, New Delhi.

Aladaileh, S. H., Khafaga, A. F., El-Hack, M. E. A., Al-Gabri, N. A., Abukhalil, M. H., Alfwuaires, M. A., Bin-Jumah, M., et al., (2020). *Spirulina platensis* ameliorates the subchronic toxicities of lead in rabbits via anti-oxidative, anti-inflammatory, and immune-stimulatory properties. *Science of the Total Environment, 701*, 1–15. doi: https://doi.org/10.1016/j.scitotenv.2019.134879.

Aldred, E., Buck, C., & Vall, K., (2009). Lipids. In: Aldred, E., Buck, C., & Vall, (eds.), *Pharmacology* (pp. 73–80). Churchill Livingstone, New York.

Arunkumar, E., Bhuvaneswari, S., & Anuradha, C. V., (2012). An intervention study in obese mice with astaxanthin, a marine carotenoid-effects on insulin signaling and pro-inflammatory cytokines. *Food and Function, 3*(2), 120–126. doi: 10.1039/c1fo10161g.

Babin, A., Saciat, C., Teixeira, M., Troussard, J. P., Motreuil, S., Moreau, J., & Moret, Y., (2015). Limiting immunopathology: Interaction between carotenoids and enzymatic antioxidant defenses. *Developmental and Comparative Immunology, 49*(2), 278–281. doi: https://doi.org/10.1016/j.dci.2014.12.007.

Barkia, I., & Saari, N., (2019). Microalgae for high-value products towards human health and nutrition. *Mar. Drugs, 17*(5), 304. doi: 10.3390/md17050304.

Becker, E. W., (2007). Micro-algae as a source of protein. *Biotechnology Advances, 25*(2), 207–210. doi: https://doi.org/10.1016/j.biotechadv.2006.11.002.

Becker, E. W., (2013). Microalgae for human and animal nutrition. In: Richmond, A., & Hu, Q., (eds.), *Handbook of Microalgal Culture* (pp. 461–503), John Wiley and Sons, Oxford.

Bernaerts, T. M. M., Kyomugasho, C., Van, L. N., Gheysen, L., Foubert, I., Hendrickx, M. E., & Van, L. A. M., (2018). Molecular and rheological characterization of different cell wall fractions of *Porphyridium cruentum*. *Carbohydrate Polymers, 195*, 542–550. doi: https://doi.org/10.1016/j.carbpol.2018.05.001.

Bhat, V. B., & Madyastha, K. M., (2000). C-Phycocyanin: A potent peroxyl radical scavenger *in vivo* and *in vitro*. *Biochemical and Biophysical Research Communications, 25*, 20–25. doi: https://doi.org/10.1006/bbrc.2000.3270.

Borowitzka, M. A., (2013). High-value products from microalgae - their development and commercialization. *Journal of Applied Phycology, 25*, 743–756. doi: https://doi.org/10.1007/s10811-013-9983-9.

Borowitzka, M. A., (2018). Microalgae in medicine and human health: A historical perspective. In: Levine, I., & Fleurence, J., (eds.), *Microalgae in Health and Disease Prevention* (pp. 195–210). Academic Press, Cambridge.

Capek, P., Matulová, M., Šutovská, M., Barboríková, J., Molitorisová, M., & Kazimi-erová, I., (2020). *Chlorella vulgaris* α-L-arabino-α-L-rhamno-α,β-D-galactan structure and mechanisms of its anti-inflammatory and anti-remodeling effects. *International Journal of Biological Macromolecules, 162*, 188–198. doi: https://doi.org/10.1016/j.ijbiomac.2020.06.151.

Cezare-Gomes, E. A., Mejia-Da-Silva, L. D. C., Pérez-Mora, L. S., Matsudo, M. C., Ferreira-Camargo, L. S., Singh, A. K., & De Carvalho, J. C. M., (2019). Potential of microalgae carotenoids for industrial application. *Applied Biochemistry and Biotechnology, 188*(3), 602–634. doi: 10.1007/s12010-018-02945-4.

Chen, H., Zeng, F., Li, S., Liu, Y., Gong, S., Lv, X. C., Zhang, J., & Liu, B., (2019). *Spirulina* active substance mediated gut microbes improve lipid metabolism in high-fat diet-fed rats. *Journal of Functional Foods, 59*, 215–222. doi: https://doi.org/10.1016/j.jff.2019.04.049.

Chen, K., & Roca, M., (2018). *In vitro* digestion of chlorophyll pigments from edible seaweeds. *Journal of Functional Foods, 40*, 400–407. doi: https://doi.org/10.1016/j.jff.2017.11.030.

Chen, P. B., Wang, H. C., Liu, Y. W., Lin, S. H., Chou, H. N., & Sheen, L. Y., (2014). Immunomodulatory activities of polysaccharides from *Chlorella pyrenoidosa* in a mouse model of Parkinson's disease. *Journal of Functional Foods, 11*, 103–113. doi: https://doi.org/10.1016/j.jff.2014.08.019.

Chew, K. W., Yap, J. Y., Show, P. L., Suan, N. H., Juan, J. C., Ling, T. C., Lee, D. J., & Chang, J. S., (2017). Microalgae biorefinery: High value products perspectives. *Bioresource Technology, 229*, 53–62. doi: https://doi.org/10.1016/j.biortech.2017.01.006.

Combs, Jr. G. F., & McClung, J. P., (2017). *The Vitamins: Fundamental Aspects in Nutrition and Health* (pp. 15–18). Academic Press, San Diego.

Costa-Pinto, R., & Gantner, D., (2020). Macronutrients, minerals, vitamins, and energy. *Anesthesia and Intensive Care Medicine, 21*(3), 157–161. doi: https://doi.org/10.1016/j.mpaic.2019.12.006.

Croce, R., & Van, A. H., (2014). Natural strategies for photosynthetic light-harvesting. *Nature Chemical Biology, 10*(7), 492–501. doi: https://doi.org/10.1038/nchembio.1555.

Dvir, I., Moppes, D. V., & Arad, S., (2020). Foodomics: To discover the health potential of microalgae. In: Smithers, G., & Trinetta, V., (eds.), *Reference Module in Food Science* (pp. 1–11). Elsevier, Oxford.

Edelmann, M., Aalto, S., Chamlagain, B., Kariluoto, S., & Piironen, V., (2019). Riboflavin, niacin, folate, and vitamin B12 in commercial microalgae powders. *Journal of Food Composition and Analysis, 82*, 103–226. doi: https://doi.org/10.1016/j.jfca. 2019.05.009.

Egeland, E. S., Garrido, J. L., Clementson, L., Andresen, K., Thomas, C. S., Zapata, M., Airs, R., et al., (2011). Datasheets aiding identification of phytoplankton carotenes and chlorophylls. In: Roy, S. T., & Llewellyn, C. A., (eds.), *Phytoplankton Pigments: Characterization, Chemotaxonomy, and Applications in Oceanography* (pp. 655–822). Cambridge University Press, Cambridge.

El-Baz, F. K., Aly, H. F., & Abd-Alla, H. I., (2020). The ameliorating effect of carotenoid-rich fraction extracted from *Dunaliella salina* microalga against inflammation- associated cardiac dysfunction in obese rats. *Toxicology Reports, 7*, 118–124. doi: https://doi.org/10.1016/j. toxrep.2019.12.008.

Enzing, C., Ploeg, M., Barbosa, M., Sijtsma, L., Vigani, M., Parisi, C., & Cerezo, E. R., (2014). *Microalgae-based Products for the Food and Feed Sector: An Outlook for Europe.* https://publications.jrc.ec.europa.eu/repository/bitstream/JRC85709/final%20version%20 online%20ipts%20jrc%2085709.pdf (accessed on 2 July 2021).

Fanzo, J., Covic, N., Dobermann, A., Henson, S., Herrero, M., Pingali, P., & Staal, S., (2020). A research vision for food systems in the 2020s: Defying the status quo. *Global Food Security, 26*, 100397. doi: https://doi.org/10.1016/j.gfs.2020.100397.

Fernandes, A. S., Nascimento, T. C. D., Jacob-Lopes, E., Rosso, V. V. D., & Zepka, L. Q., (2018). Carotenoids: A brief overview on its structure, biosynthesis, synthesis, and applications. In: Zepka, L. Q., Jacob-Lopes, E., & De Rosso, V., (eds.), *Progress in Carotenoid Research* (pp. 1–16). IntechOpen, London.

Fernandes, A. S., Nogara, G. P., Zepka, L. Q., Jacob-Lopes, E., Menezes, C. R., Cichoski, A. J., & Mercadante, A. Z., (2016). Identification of chlorophyll molecules with peroxyl radical scavenger capacity in microalgae *Phormidium autumnale* using ultrasound-assisted extraction. *Food Research International, 99*, 1036–1041. doi: https://doi.org/10.1016/j. foodres.2016.11.011.

Fernandes, A. S., Petry, F. C., Mercadante, A. Z., Jacob-Lopes, E., & Zepka, L. Q., (2020). HPLC-PDA-MS/MS as a strategy to characterize and quantify natural pigments from microalgae. *Current Research in Food Science, 3*, 100–112. doi: https://doi.org/10.1016/j. crfs.2020.03.009.

Fernandez, F. G. A., Sevilla, J. M. F., & Grima, E. M., (2017). Microalgae: The basis of mankind sustainability. In: Moya, M. L., Gracia, M. D. S., & Mazadiego, L. F., (eds.), *Case Study of Innovative Projects-Successful Real Cases* (pp. 123–140). IntechOpen, London.

Fernando, I. P. S., Ryu, B., Ahn, G., Yeo, I. K., & Jeon, Y. J., (2020). Therapeutic potential of algal natural products against metabolic syndrome: A review of recent developments. *Trends in Food Science and Technology, 97*, 286–299. doi: https://doi.org/10.1016/j. tifs.2020.01.020.

Fujimoto, M., Tsuneyama, K., Fujimoto, T., Selmi, C., Gershwin, M. E., & Shimada, Y., (2012). *Spirulina* improves non-alcoholic steatohepatitis, visceral fat macrophage aggregation, and serum leptin in a mouse model of metabolic syndrome. *Digestive and Liver Disease, 44*(9), 767–774. doi: https://doi.org/10.1016/j.dld.2012.02.002.

Gaignard, C., Gargouch, N., Dubessay, P., Delattre, C., Pierre, G., Laroche, C., Fendri, I., et al., (2019b). New horizons in culture and valorization of red microalgae. *Biotechnology Advances, 37*(1), 193–222. doi: https://doi.org/10.1016/j.biotechadv.2018.11.014.

Gaignard, C., Laroche, C., Pierre, G., Dubessay, P., Delattre, C., Gardarin, C., Gourvil, P., et al., (2019a). Screening of marine microalgae: Investigation of new exopolysaccharide producers. *Algal Research, 44*, 101711. doi: https://doi.org/10.1016/j.algal.2019.101711.

García, J. L., De Vicente, M., & Galán, B., (2017). Microalgae, old sustainable food, and fashion nutraceuticals. *Microbial Biotechnology, 10*(5), 1017–1024. doi: 10.1111/1751-7915.12800.

Gee, D., Archer, L., Paskuliakova, A., Mc Coy, G. R., Fleming, G. T. A., Gillespie, E., & Touzet, N., (2018). Rapid chemotaxonomic profiling for the identification of high-value carotenoids in microalgae. *Journal of Applied Phycology, 30*(1), 385–399. doi: https://doi.org/10.1007/s10811-017-1247-7.

Goiris, K., Muylaert, K., Voorspoels, S., Noten, B., De Paepe, D., Baart, G. J. E., & De Cooman, L., (2014). Detection of flavonoids in microalgae from different evolutionary lineages. *Journal of Phycology, 50*(3), 483–492. doi: https://doi.org/10.1111/jpy.12180.

Grune, T., Lietz, G., Palou, A., Ross, A. C., Stahl, W., Tang, G., Thurnham, D., Yin, S., & Biesalski, H. K., (2010). Beta-carotene is an important vitamin A source for humans. *The Journal of Nutrition, 140*(12), 2268S–2285S. doi: 10.3945/jn.109.119024.

Gunstone, F. D., (2003). Fatty acids: Gamma-linolenic acid. In: Caballero, B., (ed.), *Encyclopedia of Food Sciences and Nutrition* (pp. 2308–2311). Academic Press, London. doi: https://doi.org/10.1016/B0–12–227055-X/00448-X.

Haque, F., Dutta, A., Thimmanagari, M., & Chiang, Y. W., (2016). Intensified green production of astaxanthin from *Haematococcus pluvialis*. *Food and Bioproducts Processing, 99*, 1–11. doi: https://doi.org/10.1016/j.fbp.2016.03.002.

Harari, A., Abecassis, R., Relevi, N., Levi, Z., Ben-Amotz, A., Kamari, Y., Harats, D., & Shaish, A., (2013). Prevention of atherosclerosis progression by 9-*cis*-β-carotene-rich alga *Dunaliella* in ApoE-deficient mice. *BioMed Research International, 1695*–1697. doi: 10.1155/2013/169517.

Holban, A. M., & Grumezescu, A. M., (2018). *Alternative and Replacement Foods* (p. 494). Academic Press, San Diego.

Hsu, H. Y., Jeyashoke, N., Yeh, C. H., Song, Y. J., Hua, K. F., & Chao, L. K., (2010). Immunostimulatory bioactivity of algal polysaccharides from *Chlorella pyrenoidosa* activates macrophages via toll-like receptor 4. *Journal of Agricultural and Food Chemistry, 58*(2), 927–936. doi: 10.1021/jf902952z.

Jacob-Lopes, E., Maroneze, M. M., Deprá, M. C., Sartori, R. B., Dias, R. R., & Zepka, L. Q., (2019). Bioactive food compounds from microalgae: An innovative framework on industrial biorefineries. *Current Opinion in Food Science, 25*, 1–7. doi: https://doi.org/10.1016/j.cofs.2018.12.003.

Jiang, Y., Chen, F., & Liang, S. Z., (1999). Production potential of docosahexaenoic acid by the heterotrophic marine dinoflagellate *Crypthecodinium cohnii*. *Process Biochemistry, 34*(6, 7), 633–637. doi: https://doi.org/10.1016/S0032-9592(98)00134-4.

Kelley, R. E., & Andersson, H. C., (2014). Disorders of purines and pyrimidines. In: Biller, J., & Ferro, J. M., (eds.), *Handbook of Clinical Neurology* (pp. 827–838), Elsevier, Kensington. doi: https://doi.org/10.1016/B978-0-7020-4087-0.00055-3.

Khalil, S. R., Elhady, W. M., Elewa, Y. H. A., Abd El-Hameed, N. E. A., & Ali, S. A., (2018). Possible role of *Arthrospira platensis* in reversing oxidative stress-mediated liver damage in rats exposed to lead. *Biomedicine and Pharmacotherapy, 97*, 1259–1268. doi: https://doi.org/10.1016/j.biopha.2017.11.045.

Khan, M. I., Shin, J. H., & Kim, J. D., (2018). The promising future of microalgae: Current status, challenges, and optimization of a sustainable and renewable industry for biofuels, feed, and other products. *Microbial Cell Factories, 17*(1), 1–21. doi: https://doi.org/10.1186/s12934-018-0879-x.

Khayyal, M. T., El-Baz, F. K., Meselhy, M. R., Ali, G. H., & El-Hazek, R. M., (2019). Intestinal injury can be effectively prevented by *Dunaliella salina* in gamma-irradiated rats. *Heliyon, 5*(5), 1–5. doi: https://doi.org/10.1016/j.heliyon.2019.e01814.

Kim, M. Y., Cheong, S. H., Lee, J. H., Kim, M. J., Sok, D. E., & Kim, M. R., (2010). *Spirulina* improves antioxidant status by reducing oxidative stress in rabbits fed a high-cholesterol diet. *Journal of Medicinal Food, 13*(2), 420–426. doi: 10.1089/jmf. 2009.1215.

Klejdus, B., Kopecký, J., Benešová, L., & Vacek, J., (2009). Solid-phase/supercritical-fluid extraction for liquid chromatography of phenolic compounds in freshwater microalgae and selected cyanobacterial species. *Journal of Chromatography A, 1216*(5), 763–771. doi: https://doi.org/10.1016/j.chroma.2008.11.096.

Koller, M., Muhr, A., & Braunegg, G., (2014). Microalgae as versatile cellular factories for valued products. *Algal Research, 6*, 52–63. doi: https://doi.org/10.1016/j.algal. 2014.09.002.

Konstantinidi, M., & Koutelidakis, A. E., (2019). Functional foods and bioactive compounds: A review of its possible role on weight management and obesity's metabolic consequences. *Medicines (Basel), 6*(3), 94. doi: 10.3390/medicines6030094.

Koyande, A. K., Chew, K. W., Rambabu, K., Tao, Y., Chu, D. T., & Show, P. L., (2019). Microalgae: A potential alternative to health supplementation for humans. *Food Science and Human Wellness, 8*(1), 16–24. doi: https://doi.org/10.1016/j.fshw.2019.03.001.

Leopoldini, M., Marino, T., Russo, N., & Toscano, M., (2004). Antioxidant properties of phenolic compounds: H-atom versus electron transfer mechanism. *Journal of Physical Chemistry A, 108*(22), 4916–4922. doi: https://doi.org/10.1021/jp037247d.

Li, L., Li, W., Kim, Y. H., & Lee, Y. W., (2013). *Chlorella vulgaris* extract ameliorates carbon tetrachloride-induced acute hepatic injury in mice. *Experimental and Toxicologic Pathology, 65*(1), 73–80. doi: https://doi.org/10.1016/j.etp.2011.06.003.

Longo, M., Zatterale, F., Naderi, J., Parrillo, L., Formisano, P., Raciti, G. A., Beguinot, F., & Miele, C., (2019). Adipose tissue dysfunction as determinant of obesity-associated metabolic complications. *International Journal of Molecular Sciences, 20*(9), 2–23. doi: 10.3390/ijms20092358.

Lou, S., Zhu, X., Zeng, Z., Wang, H., Jia, B., Li, H., & Hu, Z., (2020). Identification of microRNAs response to high light and salinity that involved in beta-carotene accumulation in microalga *Dunaliella salina. Algal Research, 48*, 1019–1025. doi: https://doi.org/10.1016/j.algal.2020.101925.

Lupette, J., & Benning, C., (2020). Human health benefits of very-long-chain polyunsaturated fatty acids from microalgae. *Biochimie., 178*, 15–25.

Machu, L., Misurcova, L., Ambrozova, J. V., Orsavova, J., Mlcek, J., Sochor, J., & Jurikova, T., (2015). Phenolic content and antioxidant capacity in algal food products. *Molecules, 20*(1), 1118–1133. doi: https://doi.org/10.3390/molecules20011118.

Martins, D. A., Custódio, L., Barreira, L., Pereira, H., Ben-Hamadou, R., Varela, J., & Abu-Salah, K. M., (2013). Alternative sources of n-3 long-chain polyunsaturated fatty acids in marine microalgae. *Marine Drugs, 11*, 2259–2281. doi: https://doi.org/10.3390/md11072259.

Matos, Â. P., (2017). The impact of microalgae in food science and technology. *Journal of the American Oil Chemists' Society, 94*(11), 1333–1350. doi: https://doi.org/10.1007/s11746-017-3050-7.

Matos, Â. P., (2019). Microalgae as a potential source of proteins. In: Galanakis, C. M., (ed.), *Proteins: Sustainable Source, Processing and Applications*. Academic Press, Cambridge. doi: https://doi.org/10.1016/b978-0-12-816695-6.00003-9.

Memije-Lazaro, I. N., Blas-Valdivia, V., Franco-Colín, M., & Cano-Europa, E., (2018). *Arthrospira maxima* (*Spirulina*) and C-phycocyanin prevent the progression of chronic kidney disease and its cardiovascular complications. *Journal of Functional Foods, 43,* 37–43. doi: https://doi.org/10.1016/j.jff.2018.01.013.

Mendes, A. F., Cruz, M. T., & Gualillo, O., (2018). Editorial: The physiology of inflammation - the final common pathway to disease. *Frontiers in Physiology, 9,* 1–3. doi: 10.3389/fphys.2018.01741.

Michaud, P., (2018). Polysaccharides from microalgae, what's future? *Advances in Biotechnology and Microbiology, 8*(2), 1–2. doi: https://doi.org/10.19080/AIBM.2018.08.555732.

Murthy, K. N. C., Vanitha, A., Rajesha, J., Swamy, M. M., Sowmya, P. R., & Ravishankar, G. A., (2005). *In vivo* antioxidant activity of carotenoids from *Dunaliella salina* - a green microalga. *Life Sciences, 76*(12), 1381–1390. doi: https://doi.org/10.1016/j.lfs.2004.10.015.

Nascimento, T. C., Cazarin, C. B. B., Maróstica, M. R., Mercadante, A. Z., Jacob-Lopes, E., & Zepka, L. Q., (2020). Microalgae carotenoids intake: Influence on cholesterol levels, lipid peroxidation and antioxidant enzymes. *Food Research International, 128,* 1–9. doi: https://doi.org/10.1016/j.foodres.2019.108770.

Nascimento, T. C., Cazarin, C. B. B., Maróstica, M. R., Risso, É. M., Amaya-Farfan, J., Grimaldi, R., Mercadante, A. Z., et al., (2019). Microalgae biomass intake positively modulates serum lipid profile and antioxidant status. *Journal of Functional Foods, 58,* 11–20. doi: https://doi.org/10.1016/j.jff.2019.04.047.

Noguchi, N., Konishi, F., Kumamoto, S., Maruyama, I., Ando, Y., & Yanagita, T., (2013). Beneficial effects of *Chlorella* on glucose and lipid metabolism in obese rodents on a high-fat diet. *Obesity Research and Clinical Practice, 7*(2), e95–e105. doi: https://doi.org/10.1016/j.orcp.2013.01.002.

Panche, A. N., Diwan, A. D., & Chandra, S. R., (2016). Flavonoids: An overview. *Journal of Nutritional Science, 5,* 1–15. doi: https://doi.org/10.1017/jns.2016.41.

Patias, L. D., Maroneze, M. M., Siqueira, S. F., de Menezes, C. R., Zepka, L. Q., & Jacob-Lopes, E., (2018). Single-cell protein as a source of biologically active ingredients for the formulation of antiobesity foods. In: Holban, A. M., & Grumezescu, A. M., (eds.), *Alternative and Replacement Foods* (pp. 317–353). Academic Press, Cambridge.

Perera, C. O., & Yen, G. M., (2007). Functional properties of carotenoids in human health. *International Journal of Food Properties, 10*(2), 201–230. doi: 10.1080/109429 10601045271.

Pierre, G., Delattre, C., Dubessay, P., Jubeau, S., Vialleix, C., Cadoret, J. P., Probert, I., & Michaud, P., (2019). What is in store for EPS microalgae in the next decade? *Molecules, 24*(23), 1–25. doi: https://doi.org/10.3390/molecules24234296.

Rao, A. R., Sarada, R., Shylaja, M. D., & Ravishankar, G. A., (2015). Evaluation of hepatoprotective and antioxidant activity of astaxanthin and astaxanthin esters from microalga-*Haematococcus pluvialis*. *Journal of Food Science and Technology, 52*(10), 6703–6710. doi: 10.1007/s13197-015-1775-6.

Raposo, M. F. D. J., De Morais, A. M. M. B., & De Morais, R. M. S. C., (2014). Bioactivity and applications of polysaccharides from marine microalgae. In: Ramawat, K. G., & Mérillon, J. M., (eds.), *Polysaccharides* (pp. 1–38). Springer International Publishing, Cham. doi: https://doi.org/10.1007/978-3-319-03751-6.

Raposo, M. F. D. J., De Morais, A. M. M. B., & De Morais, R. M. S. C., (2015). Marine polysaccharides from algae with potential biomedical applications. *Marine Drugs, 13*(5), 2967–3028. doi: https://doi.org/10.3390/md13052967.

Raposo, M. F. D. J., De Morais, R. M. S. C., & De Morais, A. M. M. B., (2013). Bioactivity and applications of sulphated polysaccharides from marine microalgae. *Marine Drugs, 11*(1), 233–252. doi: https://doi.org/10.3390/md11010233.

Robbins, R. J., (2003). Phenolic acids in foods: An overview of analytical methodology. *Journal of Agricultural and Food Chemistry, 51*(10), 2866–2887. doi: https://doi.org/10.1021/jf026182t.

Rodrigues, D. B., Flores, É. M. M., Barin, J. S., Mercadante, A. Z., Jacob-Lopes, E., & Zepka, L. Q., (2014). Production of carotenoids from microalgae cultivated using agro-industrial wastes. *Food Research International, 65*, 144–148. doi: https://doi.org/10.1016/j.foodres.2014.06.037.

Rodrigues, D. B., Menezes, C. R., Mercadante, A. Z., Jacob-Lopes, E., & Zepka, L. Q., (2015). Bioactive pigments from microalgae *Phormidium autumnale*. *Food Research International, 77*, 273–279. doi: https://doi.org/10.1016/j.foodres.2015.04.027.

Rodrigues, E., Mariutti, L. R. B., Chisté, R. C., & Mercadante, A. Z., (2012). Development of a novel micro-assay for evaluation of peroxyl radical scavenger capacity: Application to carotenoids and structure-activity relationship. *Food Chemistry, 135*(3), 2103–2111. doi: https://doi.org/10.1016/j.foodchem.2012.06.074.

Rodriguez-Amaya, D. B., (2001). *A Guide to Carotenoid Analysis in Foods* (pp. 1–2). ILSI Press, Washington.

Santos-Sánchez, N. F., Salas-coronado, R., Villanueva-Cañongo, C., & Hernández-Carlos, B., (2019). Antioxidant compounds and their antioxidant mechanism. In: Shalaby, E., (ed.), *Antioxidants* (pp. 1–28). IntechOpen, London.

Sarkar, S., Manna, M. S., Bhowmick, T. K., & Gayen, K., (2020). Extraction of chlorophylls and carotenoids from dry and wet biomass of isolated *Chlorella thermophila*: Optimization of process parameters and modeling by artificial neural network. *Process Biochemistry, 96*, 58–72. doi: https://doi.org/10.1016/j.procbio.2020.05.025.

Scaglioni, P. T., Quadros, L., De Paula, M., Furlong, V. B., Abreu, P. C., & Badiale-Furlong, E., (2018). Inhibition of enzymatic and oxidative processes by phenolic extracts from *Spirulina* sp. and *Nannochloropsis* sp. *Food Technology and Biotechnology, 56*(3), 344–353. doi: https://doi.org/10.17113/ftb.56.03.18.5495.

Senni, K., Pereira, J., Gueniche, F., Delbarre-Ladrat, C., Sinquin, C., Ratiskol, J., Godeau, G., et al., (2011). Marine polysaccharides: A source of bioactive molecules for cell therapy and tissue engineering. *Marine Drugs, 9*(9), 1664–1681. doi: https://doi.org/10.3390/md9091664.

Silva, M. E. T. D., Correa, K. D. P., Martins, M. A., Da Matta, S. L. P., Martino, H. S. D., & Coimbra, J. S. D. R., (2020). Food safety, hypolipidemic and hypoglycemic activities, and *in vivo* protein quality of microalga *Scenedesmus obliquus* in Wistar rats. *Journal of Functional Foods, 65*, 103711. doi: https://doi.org/10.1016/j.jff.2019.103711.

Singh, S. K., Kaur, R., Bansal, A., Kapur, S., & Sundaram, S., (2020). Biotechnological exploitation of cyanobacteria and microalgae for bioactive compounds. In: Verma, M., & Chandel, A., (eds.), *Biotechnological Production of Bioactive Compounds* (pp. 221–259). Elsevier, Oxford.

Spolaore, P., Joannis-cassan, C., Duran, E., Isambert, A., Génie, L. D., & Paris, E. C., (2006). Commercial applications of microalgae. *Journal of Bioscience and Bioengineering, 101*(2), 87–96. doi: https://doi.org/10.1263/jbb.101.87.

Takaichi, S., & Mochimaru, M., (2007). Carotenoids and carotenogenesis in cyanobacteria: Unique ketocarotenoids and carotenoid glycosides. *Cellular and Molecular Life Sciences, 64*(19, 20), 2607–2619. doi: https://doi.org/10.1007/s00018-007-7190-z.

Takaichi, S., (2011). Carotenoids in algae: distributions, biosyntheses and functions. *Marine Drugs, 9*(6), 1101–1118. doi: https://doi.org/10.3390/md9061101.

Tang, D. Y. Y., Khoo, K. S., Chew, K. W., Tao, Y., Ho, S. H., & Show, P. L., (2020). Potential utilization of bioproducts from microalgae for the quality enhancement of natural products. *Bioresource Technology, 304*, 122997. doi: https://doi.org/10.1016/j.biortech.2020.122997.

Tarento, T. D. C., McClure, D. D., Vasiljevski, E., Schindeler, A., Dehghani, F., & Kavanagh, J. M., (2018). Microalgae as a source of vitamin K1. *Algal Research, 36*, 77–87. doi: https://doi.org/10.1016/j.algal.2018.10.008.

Tsai, C. F., Lu, F. J., & Hsu, Y. W., (2012). Protective effects of *Dunaliella salina* - a carotenoids-rich alga - against ultraviolet B-induced corneal oxidative damage in mice. *Molecular Vision, 18*, 1540–1547.

Wan, X. Z., Li, T. T., Zhong, R. T., Chen, H. B., Xia, X., Gao, L. Y., Gao, X. X., et al., (2019). Anti-diabetic activity of PUFAs-rich extracts of *Chlorella pyrenoidosa* and *Spirulina platensis* in rats. *Food and Chemical Toxicology, 128*, 233–239. doi: https://doi.org/10.1016/j.fct.2019.04.017.

Wang, W., Cang, L., Zhou, D. M., & Yu, Y. C., (2017). Exogenous amino acids increase antioxidant enzyme activities and tolerance of rice seedlings to cadmium stress. *Environmental Progress and Sustainable Energy, 36*(1), 155–161. doi: https://doi.org/10.1002/ep.12474.

Weinrich, R., & Elshiewy, O., (2019). Preference and willingness to pay for meat substitutes based on micro-algae. *Appetite, 142*(1), 1–11. doi: https://doi.org/10.1016/j.appet.2019.104353.

World Health Organization (WHO), (2020). *Fact Sheets: Obesity and Overweight*. https://www.who.int/news-room/fact-sheets/detail/obesity-and-overweight (accessed on 2 July 2021).

Wu, Q., Liu, L., Miron, A., Klímová, B., Wan, D., & Kuča, K., (2016). The antioxidant, immunomodulatory, and anti-inflammatory activities of *Spirulina*: An overview. *Archives of Toxicology, 90*(8), 1817–1840. doi: 10.1007/s00204-016-1744-5.

Xiao, R., & Zheng, Y., (2016). Overview of microalgal extracellular polymeric substances (EPS) and their applications. *Biotechnology Advances, 34*(7), 1225–1244. doi: https://doi.org/10.1016/j.biotechadv.2016.08.004.

Xiong, J., Liu, S., Pan, Y., Zhang, B., Chen, X., & Fan, L., (2018). Combination of fish oil and ethanol extracts from *Spirulina platensis* inhibits the airway inflammation induced by ovalbumin in mice. *Journal of Functional Foods, 40*, 707–714. doi: https://doi.org/10.1016/j.jff.2017.12.014.

Yabuzaki, J., (2017). Carotenoids database: Structures, chemical fingerprints and distribution among organisms. *Database, 1*, 1–11. doi: https://doi.org/10.1093/database/bax004.

Zanella, L., & Vianello, F., (2020). Microalgae of the genus *Nannochloropsis*: Chemical composition and functional implications for human nutrition. *Journal of Functional Foods, 68*, 103919. doi: https://doi.org/10.1016/j.jff.2020.103919.

Zepka, L. Q., Jacob-Lopes, E., & Roca, M., (2019). Catabolism and bioactive properties of chlorophylls. *Current Opinion in Food Science, 26*, 94–100. doi: https://doi.org/ 10.1016/j.cofs.2019.04.004.

CHAPTER 9

MICROALGAE: A MULTIFACETED TREASURE OF PHARMACEUTICALS AND NUTRACEUTICALS

MOHAMMED REHMANJI, SUKANNYA SURESH, ASHA ARUMUGAM NESAMMA, and PANNAGA PAVAN JUTUR

Omics of Algae Group, Industrial Biotechnology, International Center for Genetic Engineering and Biotechnology, Aruna Asaf Ali Marg, New Delhi–110067, India

ABSTRACT

In recent years, due to the ease of cultivation, metabolic versatility, and high nutritional value, microalgae have steadily attracted the attention of various industrial sectors. Microalgal bioactive compounds have been reinforced as a sustainable alternative by increasing demand and high production costs for biofuels. These products serve as favorable sources of vitamins, minerals, proteins, polyunsaturated fatty acids (PUFA)/functional lipids, and carotenoids. These biomolecules primarily have biological activities such as antioxidants, anticancer, antibacterial, antiviral, etc., making them suitable as pharmaceuticals and nutraceuticals for commercial use. These beneficial effects of microalgae derivatives on human health may be beneficial in the long run in combating or delaying the onset of different diseases. This chapter presents a detailed update on the status of algal pharmaceuticals and nutraceuticals, providing a brief overview of these bioactive compounds, barriers to their industrialization, new progress, vulnerability assessments, and restrictions on the use of these microalgae commodities. Overall, the various strategies and new techniques for the development of these molecules from microalgae advocate for the incorporation of technology into a paradigm that can make use of capital to accomplish new efforts in the nutraceutical and pharmaceutical industries.

9.1   INTRODUCTION

The current trends in people's needs reflect an increased consciousness towards their health, and hence there is an intensified usage of health-associated products. Enormous enthusiasm and expectation for health-promoting bio-compounds, among the global population, further assisted and led to the development of the pharmaceutical and nutraceutical market. An expanding fraction of the current pharmaceutical and nutraceutical research concentrates on the generation of robust bioactive molecules of microalgal origin due to their ease of cultivation, significant nutritional factors, and therapeutic values (Meenakshi, 2016).

Microalgae was one of the first forms of life present in fresh water and aquatic ecosystems on earth (Falkowski et al., 2004; Mobin and Alam, 2017). Microalgae cell factories are photosynthetic 'plant-like' microorganisms that are rich and diverse sources of pharmaceutically efficient molecules and nutraceuticals. They grow and dwell in complex habitats with extreme environments and in order to sustain the adverse environmental conditions, termed as 'stress,' microalgae develop defense mechanisms leading to activation or inhibition of various metabolic pathways thus generating molecules with considerable levels of structural and chemical heterogeneity (Barros et al., 2005; Cardozo et al., 2007). Henceforth, microalgae have possible origins of new structure and biologically active roles of secondary metabolites (Joshi and Rahimbhai, 2019). Many products from microalgae like omega fatty acids, carotenoids, polysaccharides, sterols, phycobilin, and vitamins have already been commercialized globally.

These photosynthetic microbes are a modern model organism for a number of biotechnological purposes such as biodiesel production (Demirbaş and Demirbas, 2011), bioremediation of wastewaters (Bwapwa et al., 2017), and in the processing of animal and human dietary supplements (Spolaore et al., 2006; Leu and Boussiba, 2014). According to recent financial feasibility estimates, production of biofuels from microalgae is not cost-effective (i.e., due to low productivity of biomass and the expense of commercial production) unless this method is coupled with manufacturing of higher value co-products (Richardson et al., 2014). In comparison, these high-value molecules show a significant improvement in market value owing to their higher demands and greater ability than their conventional equivalents (Barkia et al., 2019).

Extensive research on the functional molecules from well-studied microalgal forms like *Botryococcus braunii*, *Chlorella vulgaris*, *Dunaliella salina*, *Haematococcus pluvialis* led to the identification of intracellular or extracellular compounds like sulfated polysaccharides (Raposo et al., 2013), marennine (Pouvreau et al., 2008), various carotenoids, e.g., astaxanthin, fucoxanthin, etc., (Yuan et al., 2011; Gammone et al., 2015), omega-3 fatty acids (Haimeur et al., 2012), and polyphenols (Goiris et al., 2012) that possess 'anti-' like activities (similar to -microbial, -viral, -coagulant, -enzymatic, -oxidant, -fungal, -inflammatory or -cancer) (Plaza et al., 2010; Rath, 2012; Meenakshi, 2016) for prevention, implementation, and management of physiological aberrations, as well as the availability of viable natural resources.

Additionally, microalgae are indeed a fascinating medium for the manufacture of recombinant proteins and other inescapable natural commodities, such as food, biofuels, medicines, and nutritional supplements (Das et al., 2011; Abreu et al., 2012; Yusuf, 2013). Compounds synthesized by common microalgal processes include growth factors, blood coagulation proteins, immuno-boosters, hormones, monoclonal antibodies, virus vaccines, and enzymes. These microorganisms are playing a role in the biotechnological focal point for applications and commercialization in the pharmaceutical and nutraceutical sectors since they can generate valuable molecules. In this chapter, various insights into the pharmaceutical and nutraceutical relevance of the microalgae-based products, technical advancements, and production along with commercial applications, global status, and safety standards are discussed.

9.2 BIOACTIVE COMPOUNDS OF NUTRACEUTICAL AND PHARMACEUTICAL RELEVANCE IN MICROALGAE

Algae are beginning to emerge as one of the most renewable alternatives of sustainable bioactive compounds with potential health and economic benefits *ex situ* (Table 9.1). Nutraceutical and pharmaceuticals from algae are regarded as bioactive compounds, because of its physiological impact on the human body. Thus, "Nutraceutical" is the term assigned to the combination of "nutrition" and "pharmaceutical," which applies to nutritional substances, which have physiological benefits attributed to control and/ or defense against chronic diseases. Bioactive compounds with specific biological functions and attributes help in the survival of microalgal strains in stress conditions. These compounds may be directly extracted from primary metabolism in microalgae (proteins, carbohydrates, and lipids) or synthesized as secondary metabolites in a later stage of development (carotenoids, sterols, polyphenols, and vitamins). These compounds possess several activities like antioxidant, antifungal, antibacterial, antiviral, etc.,

TABLE 9.1 Pharmaceutical and Nutraceutical Activity of Various Molecules from Microalgae

Microalgae	Bioactive Compounds	Activities/Applications	References
Arthrospira platensis, Nannochloropsis flagelliforme, Chlorella ellipsoidea, Porphyridium cruentum	Polysaccharides (Chlorellan, Nostoflan, CaSp, EPS)	Pharmaceutics: Antiviral, anti-inflammatory, antioxidant, antitumor effects	Spolaore et al. (2006); Raposo et al. (2013); Berthon et al. (2017)
Chlorella pyrenoidosa, Nannochloropsis oculata, Arthrospira maxima, Tetraselmis suecica	Protein (peptides, EAA, MAA)	Pharmaceutics: Anti-inflammatory, antihypertensive, anticancer, antibacterial, antioxidant properties. a platform for recombinant proteins production Nutraceutics: Hypolipidemic, hypoglycemic, anorectic, anorexigenic	Becker (2007); Yaakob et al. (2014); Berthon et al. (2017); Mourelle et al. (2017)
Nannochloropsis oculata, Tetraselmis sp., Porphyridium cruentum, Nannochloropsis sp., Phaeodactylum tricornutum, Crythecodinium cohnii, Schizochytrium sp.	Lipids (EPA, DHA, DPA, MGDG, DGDG, AA, GLA, FFA)	Pharmaceutics: Acts against diabetes, arthritis, cardiovascular disease, and obesity), Reduce the level of cholesterol triglycerides, prevents Alzheimer's disease, psoriasis, and a certain type of cancer Nutraceutics: Enrich formula in omega-3	Odjadjare et al. (2015); Hamed (2016); Katiyar and Arora (2020)
Chlorella vulgaris, Dunaliella tertiolecta, Isochrysis galbana, Navicula incerta, Schizochytrium aggregatum	Phytosterols (Ergosterol, 7-Dehydroporiferasterol, Ergost-5-en-3β-ol, Stigmasterol, Campesterol, Lathosterol)	Pharmaceutics: Anticancer, immunomodulatory, Anti-inflammatory, neuromodulatory, antituberculosis, antioxidant, cholesterol-lowering activities	Kim and Kang (2011); Ahmed et al. (2015); Luo et al. (2015)
Dunaliella salina, Dunaliella bardawil, Haematococcus pluvialis, Chlorella vulgaris, Chlorella zofingiensis, Nannochloropsis gaditana, Phaeodactylum tricornutum	Pigments (chlorophyll, carotenoids: β-carotene, astaxanthin, lutein, lycopene, violaxanthin, canthaxanthin, zeaxanthin, fucoxanthin)	Pharmaceutics: Antioxidant, anti-inflammatory, antihypertensive, neuroprotective, anticancer properties, effective against atherosclerosis, ulcers, and cardiovascular diseases Nutraceutics: Antioxidant properties	Borowitzka (2013); Yaakob et al. (2014); Hamed (2016); Berthon et al. (2017)

TABLE 9.1 *(Continued)*

Microalgae	Bioactive Compounds	Activities/Applications	References
Arthrospira platensis, G. sulphuraria	Phycobiliproteins (phycoerythrin, phycocyanin, allophycocyanin, phycoerythrocyanin)	Pharmaceutics: Fluorescent properties, antioxidant, anti-inflammatory, neuroprotective, hepatoprotective properties	Milledge (2011); Yaakob et al. (2014); Odjadjare et al. (2015); Hamed (2016)
Dunaliella tertiolecta, Tetraselmis suecica, Chlorella sp., *Nannochloropsis gaditana*	Vitamins (A, E, B1, B2, folic acid)	Pharmaceutics: Antioxidant properties Nutraceutics: Precursor properties of some important enzyme cofactors	Galasso et al. (2019)
Spirulina maxima, Chlorella ellipsoidea, Nannochloropsis sp., *Chlorella vulgaris*	Polyphenols (dioxinodehydroeckol, phloroglucinol)	Pharmaceutics: Antioxidant, anti-inflammatory, anticancer, anti-allergic, anti-diabetes, antimicrobial, antifungal, antimycotoxigenic properties, improve cardiovascular-associated disorders	Li et al. (2007); Galasso et al. (2019)

Abbreviations: CaSp: calcium-free spirulina; EPS: exopolysaccharides; EAA: essential amino acids; MAA: mycosporine-like amino acids; EPA: eicosapentaenoic acid; DHA: docosahexaenoic acid; DPA: docosapentaenoic acid; MGDG: monogalactosyl diacylglycerol; DGDG: digalactosyl diacylglycerol; AA: arachidonic acid; GLA: γ-linolenic acid; FFA: free fatty acid.

suitable for commercial use as pharmaceuticals and nutraceuticals. Although the potential of large varieties of algae species as commercial candidates is yet to be explored (Sathasivam et al., 2019), the research in this area aims to improve the nutritional, dietary, and therapeutic quality of specific products within some algae species. Herein, we have shown various microalgal based compounds with pharmaceutical and nutraceutical relevance.

9.2.1 POLYSACCHARIDES

Carbohydrates (oligosaccharides and polysaccharides) are synthesized by various microalgae either in the cell or excreted as exopolysaccharides (EPSs). Polysaccharides from microalgae are studied extensively due to their functions, reflected by their conformation and structural biology. Structure-wise, EPSs, in most of the marine algae, are heteropolymers of galactose, xylose, and glucose (Yim et al., 2007). In CaSp (calcium-free spirulan) from *Arthrospira platensis*, the interconnections and oligosaccharides of the backbone system are typically 1,3-linked rhamnose and 1,2-linked 3-o-methyl-rhamnose (Lee et al., 1998). Some marine diatom species like *Navicula salinarum*, *Phaeodactylum tricornutum* stores polysaccharides like glucose, xylose, galactose, mannose, chrysolaminarin (a polymer of glucose), etc., as energy reserves (Staats et al., 1999; Guzmán et al., 2003). Polysaccharides like hetero-ramified polymer, made up of β-(1,3)-linked mannose was reported in *P. tricornutum* (Granum and Myklestad, 2002). *Isochrysis* sp., a golden-colored unicellular algae, stores leucosin, whereas members of the Chlorophyta group like *Chlorella* sp. and *B. braunii*, stores starch as their energy currency (de Jesus Raposo et al., 2014).

Some microalgae produce a significant number of polysaccharides in their stationary phase of growth, while some produce them in the exponential phase. Bergman reported that glyoxylate highly influences the production of EPSs from *Cyanospira capsulata*, and *Scenedesmus obliquus*, possibly due to metabolization of glyoxylate into glycine and further to serine, a process in photorespiration (Bergman, 1986). Nevertheless, nitrogen starvation was found to induce polysaccharide production via alternative pathways in algae (De Philippis et al., 1993).

Several studies highlighted that EPSs from microalgae possess antiviral bioactivities. Polysaccharides from *Arthrospira* sp. and *Porphyridium* sp. were one of the most studied polymers possessing antiviral activities (Aleksandar et al., 2010). EPS from *P. purpureum* and *P. cruentum* showed antiviral activity against *Orthopoxyvirus* and *Vesiculovirus* respectively

(Raposo et al., 2014). CaSp is a polysaccharide produced intracellularly in *A. platensis*, prevents several viruses by inhibiting the entrance of the virus into host cells (Hayashi et al., 1996). Additionally, polysaccharides also possess antibacterial properties, against both gram-positive and -negative bacteria (Challouf et al., 2011). The microalgal polysaccharides may also be used as nutraceuticals because of its high degree of fiber; the potential for acid-binding and exchange of cations (Ciferri, 1983).

9.2.2 PROTEINS AND PEPTIDES

Proteins are significant biomolecules which are liable for microalgal cell structural and functional stability. The protein content ranges between 6 and 70% of dry weight, depending on the microalgal species and the environmental conditions; the majority have protein levels about 50% (Becker, 2007). In addition, microalgae consist of a rich and complex array of amino acids with aspartate (Asp) and glutamate (Glu) as the most abundant amino acids (Conde et al., 2013). These molecules have various dietary, pharmaceutical, and biological functions and contain all essential amino acids (EAAs), which cannot be synthesized by mammals. Furthermore, microalgae's amino acid profiles are balanced and much like high-quality protein sources, such as lactoglobulin, egg albumin and soy (Williams and Laurens, 2010). *Chlorella* and *Spirulina* species comprise of around 70% protein of its mass (Bleakley and Hayes, 2017) and under the recommendations of WHO/FAO/UN, microalgae belonging to these species are suitable for human consumption (Chronakis and Madsen, 2011) and are thus considered an ideal ingredient for the nutraceutical industry.

Arthrospira is branded as a source for high-quality protein, phycocyanin (PC), and γ-linolenic acid (GLA) (Zaid et al., 2015). Some of the other peptide products involved in the control of mammal's blood pressure, such as the angiotensin I-converting enzyme (ACE), have been found to play a powerful role as a pharmaceutical agent and was reported in microalgae such as *C. vulgaris*, *C. ellipsoidea*, *A. platensis*, and *Nannochloropsis oculata* (Ejike et al., 2017). Presently, there are several products in the market containing protein extracts, like, Dermochlorella DG® a product developed from *C. vulgaris* extracts which contains oligopeptides (Martins et al., 2014).

Mycosporine-like amino acids (MAAs) are indeed a remarkable category of naturally occurring compounds, made up of an amino acid binding to chromophore, which is used to absorb light at low wavelengths. These amino acids protect the organism against UV radiation and are produced by *C. nivalis* and many other green algal species in substantial quantities.

A recent study in the cell line of lung cancer, utilized glycoproteins from dinoflagellate *Alexandrium minutum* and were found to have no harmful influences on normal human cells (Wi38) (Galasso et al., 2018). Microalgae often comprise lectins, lower molecular proteins, implicated in a variety of biological processes, such as host infections, cell contact, and the activation of apoptosis (Takebe et al., 2013). The human colon cancer lines (SW480) have also demonstrated anti-proliferative effects of *D. salina* peptides (Levasseur et al., 2020). One of the latest studies has investigated the oral toxicity of these bioactive peptides on Wistar rats (Barkia et al., 2019) and the examples described above demonstrate the stability and safety of these biologically active microalgal molecules.

9.2.3 LIPIDS

Fatty acids (FA), hydrocarbons, sterols, and waxes constitute algal lipid fraction (Halim et al., 2011). Lipid accumulation in microalgae is usually 20% to 50% of the overall biomass (Katiyar and Arora, 2020). Biogenesis commences via conversion of acetyl-CoA to malonyl-CoA by a multifaceted enzyme, namely acetyl-CoA carboxylase (ACCase). Based on their number of C=C bonds, they are further classified as saturated (SFA) and unsaturated fatty acids (mono-; MUFAs and poly-; PUFAs). The PUFAs contain a range of unique commercially important fatty acid groups called OMEGAs, namely α-linolenic acid (ALA, C18:3), eicosapentaenoic acid (EPA, C20:5), and docosahexaenoic acid (DHA, C22:6). In contrary to other oils, these microalgae (such as *N. gaditana*, *P. tricornutum* and *Crythecodinium cohnii*, etc.), have also demonstrated the ability to synthesize good yields of OMEGAs (Demirbaş and Demirbas, 2010). The consumer's attraction to health-conscious food improved EPA and DHA demand for nutraceutical commodities. The leading EPA and DHA suppliers primarily provide algae-based EPA and DHA formulations for infants (Pulz and Gross, 2004).

OMEGAs derived from microalgae significantly reduces irregular cardiac problems, for example, arrhythmias, elevated blood pressure, and stroke, besides curing other inflammatory diseases such as Crohn's disease, cystic fibrosis, psoriasis, and lupus (Steinrücken et al., 2017). In the case of pregnant women, minimum DHA and EPA levels are important for the efficient growth and development of the fetus brain (Bhandari et al., 2015). OmegaTech (United States) cultivates *Schizochytrium* sp. to manufacture high-quality DHA-rich algal oil branded as "DHA Gold" as food supplement. From a biological perspective, fatty acids have been reported to be anti-carcinogenic, anti-fungal, and antiviral (Burja et al., 2001; Singh et al., 2017). The number of animal and *in vitro* studies support the anti-inflammatory properties of microalgae oils. It was confirmed that there has been an inhibitory secretion of interleukin (IL)-1 beta and tumor necrosis factor-alpha (TNF-α) in human peripheral blood mononuclear cells by the application of docosapentaenoic acid (DPA), another omega-3 PUFA, from *Schizochytrium* sp. (Nauroth et al., 2010). Anti-inflammatory effects in macrophages from lipid extracts comprising monogalactosyldiacylglycerols have also been documented (Banskota et al., 2013). In addition, the use of microalgae oils in the diet (especially those rich in OMEGAs) has shown the demonstration of anti-carcinogenic effects (van Beelen et al., 2009).

9.2.4   PHYTOSTEROLS

Phytosterols are found in all eukaryotic cells, either through *de novo* synthesis or from the natural environment (Piironen et al., 2000). They are significant cell membrane structural components and perform broad roles in controlling fluidity and permeability of the membrane. In microalgae families, such as Chlorophyceae, Rhodophyceae, and Phaeophyceae, they are essential. Phytosterols range between 7 and 34 g/kg in 4 distinct extracts of microalgae (*Isochrysis galbana, N. gaditana, Nannochloropsis* sp. and *P. tricornutum*) (Luo et al., 2015). They have several beneficial effects on human nutrition, including anti-inflammatory (Yasukawa et al., 1996), anti-hypercholesterolemic (Rasmussen et al., 2008), antioxidant (Lee et al., 2003), anticancer (Sheu et al., 1999; Choi et al., 2007) and anti-diabetic effects (Lee et al., 2004). Phytosterols are also recommended to be consumed as dietary supplements to reduce cholesterol and decrease cardiovascular diseases. They also act as secondary messengers in the form of hormones that lead to relevant cellular processes (Francavilla et al., 2010). Because phytosterols are complex, the ability of distinct sterol types in the microalgae must be differentiated, including the mechanism of action, interaction with other compounds, and long-term effects. Henceforth, these bioactive molecules from microalgae have a high potential as drugs.

9.2.5 PIGMENTS

As microalgae are photosynthetic organisms, they are composed of pigments in their cellular composition, which are responsible for different coloration due to the presence of chlorophylls, carotenoids, and phycobiliproteins (PBP), respectively (Hamed, 2016). These compounds are widely used for functionally generated foods and pharmaceuticals with proven health benefits, such as antioxidant factors, vitamin substitutes, immune stimulators, and anti-inflammatory properties (Borowitzka, 1995; García et al., 2017).

9.2.5.1 CHLOROPHYLLS

Chlorophylls are green pigments, which are necessary for photosynthesis and are found in almost all photoautotrophic ecosystems, and chlorophylls becoming increasingly relevant in the business sector owing to excessive pigmentation and the increased competition for more natural resources (Yaakob et al., 2014; Odjadjare et al., 2017). The chlorophyll content varies widely in microalgal strains according to their cultivation conditions like *Chlorella* sp. and *Monoraphidium* sp. decrease their chlorophyll content from 14 mg/L to 2 mg/L and 11 mg/L to 3 mg/L respectively under high light conditions (i.e., 400 µmol photon/m^2/s) (Yaakob et al., 2014). Moreover, chlorophyll is a really strong antioxidant that can neutralize free radicals in the body and can inhibit the ingestion of chlorophyll rich foods. Chlorophyll often has a major role in slowing the development of cancer cells and is considered to suppress certain carcinogenic compounds. Chlorophyll may also lower triglyceride (TG) and cholesterol levels in the human body, according to studies performed on animals (Udayan et al., 2017).

9.2.5.2 CAROTENOIDS

Carotenoids are another type of pigments contained abundantly in microalgae which are rich in colors and are mostly regarded as antioxidants and dyeing agents (Berthon et al., 2017; Odjadjare et al., 2017). The most complex and commonly dispersed family of pigments is made up of more than 600 carotenoid classes. They are made up of an 18-carbon double bond chain with two hexacarbonyl rings per end with the same chemical structure (Odjadjare et al., 2017). Carotenes (β-carotene, lycopene) as well as xanthophylls (astaxanthin, violaxanthin, zeaxanthin, and lutein, etc.), are the major classes of microalgal carotenoids. These carotenoids on average account for 0.1% to 0.2% of microalgae's dry matter (Spolaore et al., 2006). At present, β-carotene and astaxanthin in the genera *Dunaliella* and *Haematococcus* respectively are two pigments with the largest demand in the global carotenoid industry (Berthon et al., 2017; Rammuni et al., 2019).

9.2.5.2.1   β-CAROTENE

β-carotene commonly produced by *S. almeriensis, D. bardawil* and *D. tertiolecta* (Borowitzka, 1995; Hamed, 2016; Berthon et al., 2017; Rammuni et al., 2019) is also the first commercially made high-value commodity of microalgae (Yaakob et al., 2014; Rammuni et al., 2019). The richest source of natural β-carotene is *D. salina*, which has an ability to express up to 98.5% β-carotene in total carotenoids which is equivalent to about 13% of its dry biomass (Molino et al., 2018; Rammuni et al., 2019). In mice and humans, algal β-carotene had beneficial effects against atherosclerosis, and β-carotene-rich *Dunaliella* blocked the degradation of low-density lipoproteins (LDLs) and affected plasma TGs, cholesterol, and high-density lipoproteins (Sathasivam et al., 2019). Few immunological studies have shown that people feeding on β-carotene-rich diets have preserved the normal blood levels and were least susceptible to many tumors and neurodegenerative disorders (León et al., 2003).

9.2.5.2.2   ASTAXANTHIN

The second most industrially used and exploited carotenoid is astaxanthin, a red xanthophyll pigment. The *H. pluvialis* naturally develops up to 80% to 99% of astaxanthin in its total carotenoids (~7% of its dry cell weight) (Molino et al., 2018). This microorganism is thus known as the most beneficial species in industrial processing of natural astaxanthin (nASX) (Raposo et al., 2014; Molino et al., 2018; Rammuni et al., 2019). The most powerful natural antioxidant is Astaxanthin which possesses specific antioxidant properties (Borowitzka, 1995; Berthon et al., 2017; Mourelle et al., 2017). Studies have found astaxanthin to be effective against disorders such as leukemia, autoimmune factors, metabolic disorders, asthma, neurodegenerative diseases, eye problems, and diabetic nephropathy (Yuan et al., 2011). A well-known active pro-carcinogenic and broad-spectrum anti-inflammatory agent with an excellent protection profile, algae-derived nASX is defensive against the cytokine vortex. The new research also indicates nASX may be paired with primary anti-virus treatment to boost the wellbeing and healing period of the COVID-19 patients (Talukdar, 2020).

9.2.5.2.3 CANTHAXANTHIN

Canthaxanthin is a form of carotenoid that is used as a food dye, which has been correlated with increased liver vitamin E content in egg yolks and chicken skin, which has been used as dietary additives to poultry (Surai et al., 2003). This secondary carotenoid also has antioxidant, anti-inflammatory, and anti-carcinogenic effects (Chan et al., 2009). Strains such as *Coelastrella striolata* and *C. zofingiensis* produce massive quantities of canthaxanthin under the conditions of salt overload and nitrogen deficiency (Pelah et al., 2004; Abe et al., 2007). Some microalgae, for example, *S. komareckii* (Hanagata and Dubinsky, 2002) and *D. salina* have also exhibited increased canthaxanthin accumulation (Sathasivam et al., 2019).

9.2.5.2.4 LUTEIN

Lutein is a yellow carotenoid present in microalgae that is used primarily in medicines and cosmetics and has a notoriety for shielding the lens and retina of the eyes against photo-induced injury (Roberts and Dennison, 2015). The development of lutein from microalgae, however, is growing as a consequence of improved productivity. Strains such as *Muriellopsis* sp., *C. protothecoides*, and *Chromochloris zofingiensis* have a potential of producing up to 0.5% lutein on a dry weight basis (Yaakob et al., 2014; Molino et al., 2018). Toxicological trials have demonstrated that lutein is an important agent for the prevention of a wide spectrum of human diseases. Lutein, in conjunction with zeaxanthin, defends tissues from free radicals and can prevent atherosclerosis, cataract, hypertensive retinopathy, and age-related retinal deterioration (Seddon et al., 1994; Rasmussen and Johnson, 2013). Lutein also has efficacy against cancer and protects *in vitro* endothelial cells (Liu et al., 2017).

Lycopene, fucoxanthin, violaxanthin, and zeaxanthin are other commercially-based carotenoid pigments. Fucoxanthin has a range of features such as antioxidant, anti-inflammatory, anti-diabetic, and anti-carcinogenic impact. In addition, in various experimental models, fucoxanthin has demonstrated efficacy against cancer in different forms, particularly colon and leukemia (Takahashi et al., 2015).

9.2.5.3   PHYCOBILIPROTEINS (PBPs)

The pigments present in cyanobacteria, rhodophyta, and in certain cryp-tophytes or glaucophytes are PBPs (Arad and Yaron, 1992). The greatest amount of PBP, 40 mg/g dry weight in low light condition, has been found in *Porphyridium marinum* (Gargouch et al., 2018). They were classified in four distinct PBP groups: phycoerythrin (PE), phycoerythrocyanin, PC, and allophycocyanin (APC). In biotechnological applications, PBPs are used, for instance, in food processing or in cosmetics. The well-known and incorpo-rated health advantage of *Spirulina* depends on higher PC quality (Rojas-Franco et al., 2018). In fact, PC have been demonstrated as an antioxidant, anti-inflammatory, cardio-protective, respiratory depressant, free radical spray-based agent and anti-atherosclerotic, making it a novel candidate of pharmaceutical interest (Gantar et al., 2012; Manirafasha et al., 2016). PC was also evaluated in human models as an anti-inflammatory molecule, when an enzyme involved in the inflammation process was assessed for its capacity to effectively inhibit cyclooxygenase-2 activity (Gantar et al., 2012; Manira-fasha et al., 2016). These proteins have also been observed in human chronic leukemia cell lines (K562) for the anticancer effect (Ghosh et al., 2015).

9.2.6   VITAMINS

Apart from various bioactive compounds in microalgae, vitamins function as an antioxidant for the protection of algal cells. Algal diets produce high levels of vitamins, and in microalgae, there are increased concentrations of vitamins, namely provitamin A, vitamin E, vitamin B1 and folic acid, in comparison to traditional food sources. Vitamin B12 (cobalamin), vitamin B2 (riboflavin), vitamin E (tocopherol), and provitamin A (β-carotene) are synthesized by *D. tertiolecta*. *Tetraselmis suecica* is an outstanding source of vitamin B1 (thiamin), vitamin B3 (nicotinic acid), vitamin B5 (pantothenic acid), vitamin B6 (pyridoxine), and vitamin C (ascorbic acid) (Watanabe et al., 2002). Vitamin B7 (biotin) is reported to be highly enriched in *Chlo-rella* sp. The important source of vitamins was obtained from microalgae; commercially accessible powders from microalgae intended for use as a source of vitamins B2, B3, B9, and B12 were also investigated (Edelmann et al., 2019). The Spirulina powder content of vitamin B2 was identified as 40.9 µg/g. In contrast, *Chlorella*, and *N. gaditana* powders are a great source of vitamin B9, with a maximum reported content of 25.9 µg/g and 20.8 µg/g respectively. These levels equate to around a quarter of the permissible regular intake (400 µg) for 5 g microalgal powder consumption (Edelmann et al., 2019). Furthermore, findings have shown that algal vitamins are better than other herbal commodities.

Microalgal medicinal products and nutraceutical raw materials have an immense ability to slow down the incidence of malnutrition and health problems in developing countries. A broad algal-based food industry for the sale of nutritious and functional foods can be established through the availability of proteins and other important nutrients in microalgae. These functional foods include many bioactive ingredients serving as anti-carcinogenic, antioxidant, anti-hypertensive, and cardioprotective. However, the high costs associated with microalgae processing and product extraction are not used to their fullest. Incitement for microalgae dependent food processing, combined with a lack of knowledge on its health benefits are the main obstacle to the growth of microalgae-based related nutrition. The integration into microalgae of the pharmaceutical and nutraceutical industry improves the health and well-being of humans until these criteria and concerns are answered. It would also meet the food requirements of the increasing global population.

9.3   COMMERCIAL STATUS OF MICROALGAE-BASED NUTRACEUTICALS AND PHARMACEUTICALS

The overall demand of algal products and nutraceuticals in the global market is very high, even though their presence is very limited in the baseline algae strain. According to the report, "Algae Products-Global Market Outlook (2017–2026)" by Markets and Research, the global algae products market was estimated to be approximately $3.40 billion in 2017 and is anticipated to reach $6.09 billion by 2026 growing at a CAGR (compound annual growth rate) of 6.7%; driven by nutritional value-based market, food, drug sector including bio-fuels. The pharmaceutical and nutraceutical (nutritional supplements and functional foods) applications of algae currently accounts for more than 72% of the market volume, with a major share of more than 59% of the global volume by the nutritional supplements alone (Borowitzka, 2013). The global market value of various high value compounds is listed in Table 9.2. Aspects like increasing demand for algae-based functional molecules, rise in global population, and customization of the food portfolios are driving the market. A major microalgae increase in these industries is also attributed to reduced time and strong microalgae productivity relative to earth plants and other supplies, particularly in cheap substrates (Borowitzka, 2013). The impact of climate factors on algae production, increased health concerns, a lack of social understanding of the goods, rigid policies and regulations inhibit business development.

TABLE 9.2   Current and Forecasted Global Market Value of Various High Value Compounds

Bioactive Molecule	Current Market Value (USD)*	Estimated Market Value (USD)*	CAGR for the Forecast Period*
β-carotene	425 million (2015)	500 million (2023)	3.0% (2016–2023)
Astaxanthin	600 million (2018)	880 billion (2023)	3.5% (2019–2025)
Lutein	274.6 million (2018)	454.8 million (2026)	6.4% (2016–2026)
Phycocyanin	18 million (2018)	30 million (2025)	8.0% (2019–2025)
Omega-fatty acids (EPA and DHA)	4.1 billion (2019)	8.5 billion (2025)	13.1% (2019–2025)
Phytosterols	709.7 million (2019)	–	8.7% (2019–2027)
Phycobiliproteins	36 million (2019)	79 million (2024)	22.1% (2019–2024)
Vitamins	5.18 billion (2018)	7.35 billion (2023)	7.3% (2018–2023)

*As reported in "algae products market by type, application, source, form, and region-global forecast to 2023" (https://www.marketsandmarkets.com/Market-Reports/algae-product-market-250538721.html) and "algae products-global market outlook (2017–2026)" (https://www.researchandmarkets.com/reports/4753171/algae-products-global-market-out-look-2017–2026).

9.3.1   MICROALGAE AS NUTRACEUTICALS

Increasing health consciousness and concerns about the animal welfare among the millennials has led to the introduction of new diet patterns; vegan diets being the most popular, accounting for approximately 7 million global inhabitant followers (Nicoletti, 2016). A vegan diet encourages the use of non-animal sources of food and is thus considered nutrient-deficient due to lack of proteins, vitamins, and minerals, as the daily requirement of these nutrients could not be fulfilled by plant sources alone. This nutrient deficiency is tackled by the introduction of microalgae-based food supplements and whole foods, rich in vitamin B12, vitamin D, selenium, and long-chain OMEGAs (Nicoletti, 2016). Microalgae's essential nutraceuticals include physiologically active food and nutritional supplements, medicinal goods, refined foods, and non-food supplementations.

The commercial nutraceutical sector is powered by microalgae belonging to two genus, *Spirulina*, and *Chlorella* (Nicoletti, 2016). The large-scale cultivation of *Spirulina* is followed in certain countries like China and Mexico for the last 30 years, and this practice is spreading to other parts of the world at a rapid rate, due to the ease of cultivation involved. USA, Thailand, India, Taiwan, China, Bangladesh, Pakistan, Myanmar, Greece, and Chile are the principal trade Spirulina manufacturer, whereas Taiwan (Taiwan Chlorella Manufacturing Company, 2000 tons/year) (Andrade et al., 2018) is reported as the largest producer of GRAS (generally regarded as safe) *Chlorella* followed by US (Nicoletti, 2016). Ease of cultivation, high productivity, inexpensive substrate, conventional techniques, and nutritional values are the major factors that rank these microalgae as the best nutraceutical sources. *Chlorella* and *Spirulina* have been commonly used as a provider of vitamins and antioxidants for the manufacture of pills, capsules, and formulations. The key components classified in biomass of *Spirulina* and *Chlorella* are PC and β-1,3-glucan. These compounds are considered to activate the immune system and reduce the plasma cholesterol content (Bhalamurugan et al., 2018).

β-Carotene was the first commodity made of a microalgae, *D. salina* in the 1980s by four different manufacturers around the world (Borowitzka, 2013). The current market trends also show a rising demand for β-carotene as well as astaxanthin; astaxanthin forecasted to show more than 50% increase in its usage by 2025. In 2017, brown algae were the leading producer of algae products in the global algae market, followed by blue-green algae, as a result of its composition of high carrageenan and protein and is expected to continue its dominance for the next 10 years (Nicoletti, 2016). The products manufactured along with the basic techniques used by some of these companies are listed in Table 9.3.

9.3.2 MICROALGAE AS PHARMACEUTICALS

The global market for medications is increasingly rising, and algae are regarded as an incentive to satisfy the rising demands of customers. A wide range of products are obtained from microalgae with distinct pharmacologically active molecules like antimicrobials, antivirals, therapeutic proteins, drugs, antifungals, and many more. Various features, namely, high protein folding accuracy, minor differences in glycosylation patterns of microalgae when compared to animal cells, high product quality and protein yield, low overall cost and communication risk, higher safety, low storage cost and very easy distribution and reproduction AIDS the utilization of microalgae in the pharmaceutical sector (Basaran and Rodríguez-Cerezo, 2008; Yao et al., 2015). Microalgae such as *S. maxima*, *Synechococcus* sp., *S. obliquus*, *P. cruentum*, *D. salina*, *C. vulgaris*, *C. reinhardtii*, and *Anabaena cylindrica*

TABLE 9.3 Nutraceutical Products Produced from Microalgae

Company	Product	Microalgae Source	Functions/Benefits	Manufacturing/ Extraction Process	Certifications
Cyanotech Corporation (Hawaii) https://www.cyanotech.com/	BioAstin Hawaiian Astaxanthin	*Haematococcus pluvalis* (grown on Hawaiian coast)	Eye and brain health Joint and tendon health Skin and cardiovascular health	High pressure supercritical CO_2 extraction followed by blending with safflower oil to form oleoresins	Non-GMO project verified Vegan friendly Gluten-free
	Hawaiian Spirulina	*Arthrospira platensis*	Boosts immune system Eye and brain health Cardiovascular health	High pressure water rinsing followed by drying to minimize oxidative damage	
Earthrise Nutritional LLC (USA) https://www.earthrise.com/	Spirulina Natural® Spirulina Gold Plus® (Increased phycocyanin content and Vitamin C from Acerola berry extracts)	*Arthrospira platensis*	Healthy immune system Eye health (high β-carotene content) Cardiovascular health (antioxidant protection by phycocyanin)	Automated cell harvesting followed by flash evaporation of moisture Vacuum preservation and packaging	ISO 9001:2015 certified facility Non-GMO project verified Keto and Vegan friendly Prop65 certification (safe drinking water and toxic enforcement act) Pesticide-free Diary and gluten-free GRAS-FDA (generally regarded as safe) GMP (good manufacturing practices) and HACCP (hazards analysis and critical control points) certified
Euglena Co. Ltd. (Japan) https://www.euglena.jp/	Euglena green powder	*Euglena gracilis*	Dietary supplement (59 kinds of nutrients like vitamins, minerals, amino acids, and unsaturated fatty acids)	—	Certified by Japan Food Analysis Center Halal and Kosher verified
	Euglena plus	—		—	—

TABLE 9.3 (Continued)

Company	Product	Microalgae Source	Functions/Benefits	Manufacturing/ Extraction Process	Certifications
	Paramylon 580	–	Paramylon (β-glucan) acts as a dietary fiber		
	Euglena Green Koji	–	Enhances digestion/ absorption/metabolic cycle	–	–
Algatech (Israel) https://www. algatech.com/	AstaPure® Astaxanthin	*Haematococcus pluvalis*	Skin and eye health Improves immunity and fertility Cardiovascular and brain health	3-month cultivation process by mimicking nature using patented, eco-friendly, and closed systems	ISO 9001:2015 and ISO 22000:2018 certified facility USDA Organic, Halal, and Kosher verified Non-GMO Project verified EFSA (European Food Safety Authority) GRAS (generally regarded as safe) GMP (good manufacturing practices) certified
	AstaPure® Arava	–	The whole algae, whole-food rich, all-natural supplement (antioxidants, polysaccharides, dietary fibers, EFA, amino acids and minerals)		
	AstaPure® EyeQ	–	Eye and brain health	2% Astaxanthin immobilization using patented encapsulation technology with a unique formula of biodegradable proteins for increased bioavailability in eyes and brain	–

TABLE 9.3 (Continued)

Company	Product	Microalgae Source	Functions/Benefits	Manufacturing/ Extraction Process	Certifications
	FucoVital™ (only microalgae derived fucoxanthin product in the market)	*Phaeodactylum tricornutum*	Liver health Control of metabolic syndromes Glucose management	Year-round cultivation in closed and controlled systems fully exposed to sunlight	ISO 9001:2015 and ISO 22000:2018 certified facility Kosher verified GMP (good manufacturing practices) certified Non-GMO project verified NDIN (new dietary ingredient notification) from US-FDA
Cellana Inc. (Hawaii) http://cellana.com/	ReNew™ Algae	Marine microalgae	Whole algae enriched with omegas and antioxidants for food applications	Algae strain selection and method optimization Algae growth in closed contamination-free photobioreactor	Non-GMO project verified
	ReNew™ omega-3 (ReNew™ EPA and ReNew™ DHA)		PUFAs such as omega-3s boosts brain function, improves skin and hair condition, weight loss and countering depression	Growth in open ponds (ALDUO™) Harvest and processing of algae	
AlgiSys (Ohio, USA) https://www.algisys.com/	AlgiSys EPA	Heterotrophic microalgae	Cardiovascular health Macular degeneration Infant brain and eye development	Heterotrophic EPA producing microalgae are grown by fermentation technology	—

TABLE 9.3 *(Continued)*

Company	Product	Microalgae Source	Functions/Benefits	Manufacturing/ Extraction Process	Certifications
Polaris (France) https://www. polaris.fr/ english/	Omegavie® algae oils	Marine microalgae	Rich in the PUFAs (EPA and DHA); 30–80%	Enzymatic synthesis of active ingredients; purified and concentrated using high vacuum molecular distillation technology	US-FDA approved GRAS certified SEDEX approved (sustainable and ethical supply chain certification)

are the important strains in the commercial market. In addition to the utilization of the naturally occurring molecules, these organisms have also been
engineered for the increased expression of the indigenous genes or for the
production of recombinant proteins (He, 2007; Das et al., 2011). Certain
companies deriving pharmaceutical molecules from microalgae are Rincon
Pharmaceuticals, Piramal Healthcare, Idec Pharmaceuticals Labprocure
India Pvt. Ltd., among various others.

9.3.2.1   SUBUNIT VACCINES

It is anticipated that microalgae would become the next greatest plant
nominee in recombinant subunit manufacturing as they deliver many advantages compared with land-based plant systems such as versatile and regulated
expansion, quick, and efficient transition, easily achievable steady cell lines
and steady transgenic expression. Microalgae have been shown to produce
and specifically fold several different vaccine antigens, and attempts in the
production of recombinant algal fusion proteins that can enhance the anti-
genicity for orally compelling vaccines have been ongoing. These methods
are able to revolutionize the way vaccines from the pricey distribution of
distilled protein to a cheap built-in microalgae tablet, which has active oral
mucosa and area of cognitive reactions (Specht and Mayfield, 2014).

The chloroplasts of the microalgae provide a unique confined compartment that facilitates folding of the protein due to the presence of a sophis-
ticated cellular folding machinery similar to other eukaryotes like yeast
(Chebolu and Daniell, 2009), reduced gene silencing and high amounts of
the transgene products, up to 10% of total soluble protein (TSP) have been
reported to accumulate in the chloroplasts (Manuell et al., 2007; Surzycki
et al., 2009). Numerous human and animal therapeutically significant
proteins like the signaling molecules (Rasala et al., 2010), full-length human
antibodies (Tran et al., 2009) and structural proteins (Rasala et al., 2010)
have been produced using the microalga model organism *Chlamydomonas
reinhardtii* as the host. Some of the existing microalgae produced vaccines
are listed in Table 9.4.

9.3.2.2   OTHERS

Monoclonal antibodies are used for the treatment of multiple diseases of
humans. The first mammalian defense developed by the *Chlamydomonas*

TABLE 9.4 Subunit Vaccines Produced from Microalga

Microalgae	Antigen	Product Content/ Activity	Expression Location	References
Chlamydomonas sp.	E2 protein, antigen for classical swine flu virus (CSFV)	1.5–2.0% of TSP	Chloroplast	He et al. (2007)
Chlamydomonas reinhardtii	VP1 protein (with Cholera toxin B subunit, CTB adjuvant), antigen for foot and mouth disease virus (FMDV)	3.0% of TSP	Chloroplast	Sun et al. (2003)
Dunaliella salina	Hepatitis B surface antigen (HBsAg)	1.6–3.1 ng/mg of TP	Nucleus	Geng et al. (2003)
Dunaliella salina, Chlamydomonas reinhardtii	VP28 protein (viral envelop protein of white spot syndrome virus: WSSV)	21% of TSP 3 ng/mg of TP (0.0003% of TSP)	Nuclear Chloroplast	Surzycki et al. (2009); Feng et al. (2014)
Chlamydomonas sp.	D2 fibronectin-binding domain of Staphylococcus aureus and CTB mucosal adjuvant	0.7% of TSP	Chloroplast	Dreesen et al. (2010)
Chlamydomonas reinhardtii	E7 oncoprotein of Human papillomavirus (HPV-16)	0.12% of TSP	Chloroplast	Demirbaş and Demirbas (2011)
Chlamydomonas reinhardtii	Glutamic acid decarboxylase-65 (GAD65); autoantigen for human type 1 diabetes	0.25–0.30% of TSP	Chloroplast	Wang et al. (2008)
Chlamydomonas reinhardtii	GBSS (granule-bound starch synthase) with malarial antigen, i.e., P. berghei apical major antigen (AMA1) or P. falciparum major surface protein (MSP1) and P. berghei MSP1	0.2 to 1.0 μg of protein per mg of purified starch	Chloroplast	Dauvillée et al. (2010)
Chlamydomonas reinhardtii	p210 epitope of ApoB100 fused with CTB mucosal adjuvant (vaccine against atherosclerosis)	60ug/g of algae weight	Chloroplast	Barahimipour et al. (2016); Barkia et al. (2019)
Chlamydomonas reinhardtii	Angiotensin II fused to Hepatitis B virus capsid antigen (HBcAg)	0.05% of TSP	Nuclear	Soria-Guerra et al. (2014)

reinhardtii chloroplast against the glycoprotein D from herpes simplex virus (HSV) was a monoclonal antibody; a fusion protein for the heavy chain of IgA and a variable area of the light chain (Mayfield et al., 2003). In microalgal chloroplast, single, and divalent immunotoxins were found to inhibit tumor development significantly. These immunotoxins are a combination of toxin molecule and antibody domain that possess the ability to bind to the target cells and inhibit their further proliferation; thus, has been used in cancer treatment therapies (Tran et al., 2013).

Antibacterial property has also been induced in microalgae by virtue of transformation with Defensin, small cationic peptide, a potent alternative to antibiotics. *C. ellipsoidea* was transformed with NP-1 (mature rabbit neutrophil peptide 1), a defensin with broad spectrum of activity against gram-negative as well as positive bacteria, viruses, and other pathogenic fungi (Bai et al., 2013). Crude extract of *Dunaliella*, rich source of vitamins and enzymes, is also progressing as an effective antimicrobial agent (Meenakshi, 2016).

Glycerol is a widely employed *Chlamydomonas reinhardtii* product in pharmaceuticals where the microorganism is sulfur depleted (Skjånes et al., 2013). *Nostoc* is also gaining momentum in the medicinal sector due to production of a bioactive molecule, Cyanovirin, which exhibited a potential to be used to treat symptoms of HIV as well as influenza A (H1N1) (Smee et al., 2008). Additionally, Vitamin B12, α-carotene, β-carotene, lutein, and α-tocopherol extracted from the biomass of *Chlorella* has been exploited to be used as supplements for anticancer drugs and also for prevention of macular degeneration (Zhao and Sweet, 2008; Plaza et al., 2010). PC have been provided the GRAS status by the US-FDA (United States Food and Drug Administration) to be used for treatment of cancers in the form of photosensitizers in photodynamic therapy (Hu et al., 2008; Borowitzka, 2013). Various pharmaceutically relevant molecules extracted from microalgae are depicted in Table 9.5.

9.3.3 CHALLENGES FOR COMMERCIAL PRODUCTION

The progress of high-value goods from microalgae relies on the market's abundance and the capacity to manufacture and remove the commodity at a fair price that is liable for the normal supply and demand guidance. Numerous hindrances to the large-scale commercial production of various biomolecules exist. These involve lower cell densities and biomass productivity; inefficient and cost-ineffective cultivation, requirements of

TABLE 9.5 Pharmaceuticals Produced from Microalgae

Microalgae	Bioactive Molecule	Yield/Effect	References
		Anticancer Agents	
Arthrospira platensis	Extracellular polysaccharide	Cytotoxic against kidney and colon cancer cell line	Rafika et al. (2011)
Isochrysis galbana	(1–3, 1–6)-β-D-glucan sulfated exopolysaccharide	Cytotoxicity against lymphoma cells	Sadovskaya et al. (2014)
Chlorella pyrenoidosa	CPAP (*C. pyreniodosa* antitumor polypeptide)	Inhibitory activity on human liver cancer cell HepG2	Wang and Zhang (2013)
		Anti-Inflammatory Agents	
Tetraselmis suecica	Sulfated polysaccharide	Inhibition of NO, TNF-α, IL-6	Jo et al. (2010)
Nannochloropsis oculata	Docosapentaenoic acid (DPA)	Inhibition of pro-inflammatory prostaglandin E2 (PGE 2)	Nauroth et al. (2010)
Tetraselmis sp.	Docosahexaenoic acid (DHA)	Inhibition of IL-6, IL-β	Richardson et al. (2014)
Phaeodactylum tricornutum	Sulfated extracellular polysaccharide	Immuno stimulant	Guzmán et al. (2003)
Porphyridium sp.	Sulfated polysaccharides	Inhibition of the migration of polymorphonuclear leukocytes (PMN)	Matsui et al. (2003)
		Antiviral Agents	
Haematococcus pluvialis, Dunaliella salina	Pressurized liquid extraction against herpes simplex virus type 1	IC50 = 189.58±3.18 mg/ml 163.81±5.25 mg/ml	Santoyo et al. (2012)
Navicula directa	Polysaccharide against HSV 1 and 2 and influenza virus	IC50 = 240±42 mg/ml	Lee et al. (2006)

TABLE 9.5 *(Continued)*

Microalgae	Bioactive Molecule	Yield/Effect	References
		Antibacterial Agents	
Porphyridium cruentum	Sulfated exopolysaccharide against HSV virus, vaccinia virus, vesicular somatitis virus	—	Raposo et al. (2013)

Note: IC50: concentration causing 50% inhibition of the desired activity.

complex drying and extraction technologies; and deficient contaminant, parasite, and predator control; some of which are discussed below.

9.3.3.1   SYNTHETIC ALTERNATIVES

The competencies of several microalgae goods, in particular the carotenoids, whose huge mass is chemically generated, are threatened by the synthetic molecules arising from chemical processing. These chemical replacements threaten the microalgae that must either perform at cost or become optimized on the marketplace to be commercialized at higher prices. In the case of astaxanthin production, due to the big differences existing between the natural and synthetic versions, in terms of quality and selling price, Natural Algae Astaxanthin Association (NAXA), the group formed by industrial manufacturers of astaxanthin, an authentication program to confirm the credibility of product derived from *H. pluvialis* has been established. The program aims at reducing the potential threats from any attempts made for the promotion and sale of synthetic astaxanthin as 'natural;' to obtain higher profits due to increased selling prices. The reduced quality of the synthetic forms may be due to the free form in the synthetic source as compared to 95% esterified in algal-sourced product. Top market players in the algal derived astaxanthin industry are Cyanotech Corporation, Valensa International, Algatechnologies, Fuji Chemical Industry Co, BGG, Algaetech International and Parry Nutraceuticals. A description of current and potential products and alternate sources of high value from microalgae is listed in Table 9.6.

9.3.3.2   LARGE SCALE CULTIVATION

The safety and sustainability issues surrounding the use of fish stocks for LC-PUFA (EPA and DHA) led to the need for an alternative solution. The exploitation of microalgae for commercial fatty acid production is decelerated due to many factors, namely, high cost of production and requirement of efficient and complex extraction techniques. According to Qualitas Health, the Texas-based algal EPA producers, the primary obstacle for EPA production from an EPA-rich strain *Nannochloropsis* is the stable large-scale biomass production. They have been following several practices like crop protection, fertilizer management, integrated pest management and judicious water utilization in order to stabilize their open pond cultivation systems. Although microalgae reflect various impressive traits, due to numerous

TABLE 9.6 Summary of Existing and Potential High-Value Products from Microalgae and Their Alternative Sources

Microalgae Source	Product	Alternate Producers	References
Dunaliella salina	β-Carotene	*Blakesleya trispora*, synthetic	Borowitzka and Borowitzka (1989)
Chlorella ellipsoidea; *Dunaliella salina* (mutant)	Zeaxanthin	Paprika (*Capsicum annuum*); *Tagetes erecta*, synthetic	Jin et al. (2003); Koo et al. (2012)
Chlorella spp., other green algae	Canthaxanthin	*Dietzia natronolimnaea*, synthetic	Nasrabadi and Razavi (2010)
Haematococcus pluvialis, *Chlorella zofingiensis*	Astaxanthin	*Xanthophyllomyces dendrorhous*, synthetic	Cysewski (2007); Schmidt et al. (2010)
Scenedesmus sp., *Muriellopsis* sp., other green algae	Lutein	*Tagetes* sp., *Blakesleya trispora*	Piccaglia et al. (1998); Sánchez et al. (2008)
Dunaliella	Phytoene	Tomato (*Solanum lycopersicum*)	Oppen-Bezalel and Shaish (2009)
Parietochloris incisa	Arachidonic acid	*Mortierella* spp.	Solovchenko et al. (2008); Streekstra (2010)
Nannochloropsis sp., *Phaeodactylum tricornutum*, *Monodus subterraneus*	Eicosapentaenoic acid	Fish oil	Lu et al. (2001)
Crypthecodinium cohnii, *Schizochytrium* sp., *Ulkenia* sp.	Docosahexaenoic acid	Fish oil	Barclay et al. (2010)
Porphyridium sp., *Rhodella* sp., various cyanobacteria	Polysaccharides	Guar gum, xanthan	Arad and Levy-Ontman (2010)
Many species	Sterols	Various plants	Volkman (2003)

restraints involved; continuous light requirement, temperature fluctuations, need for expensive substrates, mixing equipment, continuous monitoring devices, harvest, and extraction protocols; many pharmaceutical and nutraceutical manufacturers are not yet attracted to microalgae producers.

9.3.3.3 CONTAMINATION

Heterotrophic mode of nutrition is also an area of focus, which involves a use of a variety of carbon sources, primarily for DHA production. Many studies have been conducted to optimize the omega acid production by utilizing agronomic wastes like sugarcane molasses, crude glycerol (by-product of the biofuel industry), industrial dairy wastes, bagasse, etc., that helps in reducing the finances involved in raw material when compared to use of refined substrates like glucose and acetate. However, additional costs are involved in the precautionary measures adopted to avoid contamination associated with the use of external carbon sources. Use of antibiotics and closed cultivation systems are generally encouraged to inhibit unwanted microbial growth.

BioReal (Sweden) was the first industrial producer of Astaxanthin (30 tons/acre) using the mixotrophy cultivation in indoor closed photobioreactors (PBRs) (Del Campo et al., 2007). *Chlorella* mixotrophic cultivation has been widely studied with acetic acid as a source of carbon for production and development, as well as biomass up to 1.5–2 times higher than photoautotrophic cultivation methods has been obtained. Bacterial degradation is therefore one of the biggest threats of the *Chlorella* producers in these processes (Iwamoto, 2007; Hudek et al., 2014). Contamination issues are very common in the open cultivation of freshwater species of microalgae, but similar instances are easily tackled for marine microalgae systems by the use of high brine concentrations, as in the case of β-Carotene production using *Dunaliella* (Ben-Amotz, 2007). The most frequent pathogens are bacteria and other lower metazoans, whose growth may exceed algal growth by fighting for nutrients and can release certain toxins, which can inhibit algal production. Instead, cases have also been recorded of certain bacteria with growth stimulants (Kim et al., 2015).

9.3.3.4 ECONOMIC FEASIBILITY

There is currently no industrial development of microalgae of carotenoids such as lutein and zeaxanthin. Various obstacles that challenge their commercial viability like low carotenoid content of the microalgae, harvesting expenses, and the significant energy requirements for cell disruption and extraction steps of the process needs to be addressed. However, efforts are being made to increase the commercial practicality of these production processes by making them cost-effective. The microalgae derived production facilities are linked to a CO_2 emission abatement project, and even the strains used for large scale cultivation are acclimatized to easily achievable and inexpensive environments. Microalgae with commercial production potential for lutein like *Dunaliella* sp. (especially *D. salina* and *D. tertiolecta*), *Microcystis aeruginosa* and *Nannochloropsis* are grown at hypersaline conditions to reduce the water usage costs by utilizing seawater. Recent studies have made a breakthrough in improving zeaxanthin yield of *D. tertiolecta* by almost 10–15% via random chemical mutagenesis using ethyl methane sulphonate (EMS) as a mutagen (Kim et al., 2017). In spite of technological progresses in optimization of yield and cultivation protocols, the establishment of a commercially feasible largescale manufacturing facility for the two carotenoids lutein and zeaxanthin remains evasive.

The total method expense is raised further by the advanced methods used in biomass processing and chemical extraction. The limited size and existence of sturdy cell walls of the algal source preclude its handling with clear standard techniques. The need for consistent lighting often raises costs, because installations relying on light sources face problems such as regular or seasonal light intensity variations. Expensive machinery for mixing also becomes necessary in scaled up cultivation in order to maintain homogeneity (Jha et al., 2017).

9.4 RECENT ADVANCEMENTS OF PHARMACEUTICALS AND NUTRACEUTICALS PRODUCTION FROM MICROALGAE

Microalgae have been extensively exploited for their commercially valuable pharma and nutraceuticals products modified from carotenoids, vitamins, polysaccharides, long-chain fatty acids, single-cell proteins, sterols, etc. The development of these bioactive compounds has been documented to increase in microalgal cell factories under different stress conditions (Pulz and Gross, 2004; Wijffels et al., 2010). Hence, the ability of microalgal cells to produce similar molecules can be controlled by various biotic and abiotic factors like nutrients, pH, temperature, and light, among numerous others. Below are some of the main techniques adopted for improved bioproducts from microalgae. The chemical constraints of metabolic precursors, phytohormones, chemical inducers, and other molecular methods are suggested for the development of these products from microalgae, such as metabolic modification, transcriptional, and genetic engineering.

9.4.1 MICROALGAE CULTIVATION

One of the major problems with the development of bioactive compounds like pharmaceuticals or nutraceuticals from microalgae is the fact that the supply of algal biomass is inadequate (Olaizola, 2003). Over the last decade, major advancements made in algae cultivation systems mainly focuses on techniques that can be employed for enhancing biomass yield or productivity. Improvement in designs and engineering of PBRs led to a progress in the biomass productivity and hence the economically relevant development of microalgae bioactive compounds. Outdoor cultivation of microalgae is practiced in open as well as closed ponds. Open systems are best off because of low investments and cost of maintenance but suffer from unregulated microbial growth parameters (enlightenment, temperature, pH, nutrient levels) and degradation (Stephenson et al., 2010). Small wetlands, lakes, reservoirs, and large dams are part of these networks. Closed devices, i.e., PBRs (flat plate, tubular column type), are fitted with controllable machines for modifying the growth parameters and are also provided with steps to resolve contamination. It was reported that in a closed system, yearly biomass production of *C. vulgaris* was 25–30 tons biomass/ha, i.e., 7.68 g/m^{2}d productivity (Pruvost et al., 2016). Additionally, developing an open culture cascade method with large cell densities, that enabled high biomass production up to 10 g/L and productivity of 25 g dry biomass/m^{2}d in the *Chlorella* cultivation process. Thus, it was successfully demonstrated that traditional designs of closed and open cultivation systems greatly impact the accumulation of biomass in the cultivation process of microalgae.

Microalgal cultivation, in particular photoautotrophic, heterotrophic, and mixotrophic, often rely primarily on the mode of cultivation (Fernandes et al., 2015). *D. salina* commercially develops β-carotene for phototrophic production. Betaten and Aquacaroten Australian companies develop microalgae in open ponds, which results in an output of β-carotene of about 13 tons/annum (approx. 510 hectares of arable cropland) (Del Campo et al., 2007). Nature Beta Technologies, Israel, recorded 3 tons/annum β-carotene output annually through the cultivation of raceway pond microalgae. The first enterprise to use mixotrophy indoors in closed PBRs for Astaxanthin (30 tons/annum) in industrial development was BioReal, Sweden (Del Campo et al., 2007).

The production of biofilm with liquid suspensions in open pond and PBR systems has been recently documented. The growth of *Pseudochlorococcum* sp. of microalgae was found to support the buildup on the hydrophilic substratum in an attached biofilm rather than on the hydrophobic substratum (Ji et al., 2014). In order to increase biomass production, microalgae may also be immobilized in gel percolations. Cheirsilp's analysis revealed a biomass yield of 1 g/L and lipid yield of 0.356 g/L for the culture of *Nannochloropsis* sp. in alginated gel beads (Cheirsilp et al., 2017). Lastly, hybrid cultivation was often used on a commercial scale as well as a two-stage growing method incorporating various step development in the PBR and open pond systems. In a PBR, microalgae development is controlled for the initial step, while in the second phase, the necessary substance is synthesized from microalgae cells under suitable conditions (Brennan and Owende, 2010).

9.4.2   ENHANCEMENT OF BIOACTIVE COMPOUNDS IN MICROALGAE

In previous research, influences on the development of metabolites from microalgae were extensively assessed and checked on various stresses (like light exposure, temperature, nutrients, and salinity) or various conditions of cultivation (Paliwal et al., 2017). Various strategies for the enhancement in the production of bioactive compounds with pharmaceutical and nutraceutical importance have been employed. The latest developments in these strategies involve optimization of cultivation conditions by means of abiotic stress, environmental stress (addition of metabolic precursors, phytohormones, environmental inhibitors and chemicals causing oxidative stress), and molecular methods (mainly RNA-based knockdown and CRISPR/Cas9) such as biochemical, transcriptional, and gene dysfunction methodologies. It involves studies on multi-omics to help classify important genes, transcription factors and control hubs involved in bioactive compound biosynthesis (Guarnieri and Pienkos, 2015).

9.4.2.1   ABIOTIC FACTORS

For neutral lipids and/or carotenoids, high light intensity serves as a stimulant. The lipid content of *Chlamydomonas* sp. JSC4 has been identified was considerably higher than with low light-sensitive (30 μmol photons/m²s) and a lipid productivity of 312 mg/L/d (optimal, 300 μmol photons/m²s) (Ho et al., 2015). A research reveals that the astaxanthin yields of *H. pluvialis* mutant cells were 1.7 times greater than those of a wild strain under high light (250 μmol photons/m²/s) and 15% CO_2 (Cheng et al., 2017). Low-temperatures typically contribute to polar lipids (i.e., phospholipid) aggregation in most microalgae, with high-temperature resulting in enhanced non-polar lipids (i.e., neutral lipid) (Paliwal et al., 2017). The *C. reinhardtii* lipidomic analysis under heat stress (60 min at 42°C) showed a substantial decrease in unique polyunsaturated membrane lipids, induced by a growing amount of polyunsaturated TAGs and DAGs (Légeret et al., 2016). The DHA contents of *Aurantiochytrium* sp. SD116 can be considerably increased under cold stress in a recent study (Ma et al., 2017).

9.4.2.2   MIXOTROPHY

Even organic carbon substrates such as glucose, acetate, and glycerol can be used by microalgae along with inorganic carbon as sources of CO_2 (Cheng et al., 2017). In the case of *C. vulgaris*, the use of sodium acetate as carbon source resulted in high lipid content (42.5% dcw) with nitrogen deficiency (Abedini Najafabadi et al., 2015). When used as the main carbon source, glycerol (the major byproduct of biodiesel production), resulted in a high oil content (50% dcw) in semi-contained *C. protothecoides*, and there also existed substantial quantities of other essential bioproducts such as vitamin A, riboflavin, and lutein. In addition, higher concentrations of indole-3-butyric acid (IBA) caused the buildup of PC, APC, and PE in the treated *Nostoc linckia*, which had a relaxing effect on chlorophyll a and carotenoids accumulation (Mansouri and Talebizadeh, 2017). In response to the antagonist of squalene epoxidase (SQE), the amount of sterol was reduced, but the carotenoid levels and production of the fatty acids were dramatically increased in *Nannochloropsis oceanica* (Lu et al., 2014).

9.4.2.3   RECOMBINANT PROTEIN PRODUCTION

Chlamydomonas reinhardtii is considered a bioreactor for the manufacture of recombinant protein for pharmaceuticals and is used for the synthesis of antibodies, protein therapeutics, vaccinations, enzymes, and additives, such as viral proteins, phytase, etc. Also, chloroplast can produce the antigen E2, which constitutes an antigen for classical swine fever (CSFV). The gene which codes the antigen surface virus of hepatitis B (HBsAg) has been successfully electroporated to *D. salina*, and a stable nuclear transformants were achieved (Geng et al., 2003). Western blot analysis found that HBsAg was produced with a total protein concentration of 1.6–3.1 ng/mg in various transformants with varied insertion sites. Nevertheless, process for accumulation of the flounder growth hormone (FGH) was developed in protoplasts of *C. ellipsoidea* (Kim et al., 2002).

9.4.2.4 METABOLIC ENGINEERING

Metabolic engineering of microalgae also aided in overexpression of genes and knockout studies are undertaken for enhancement of a particular molecule or compound. Overexpression of GPAT (glycerol-3-phosphate acyltransferases) in the model diatom *P. tricornutum* enhanced TAG concentration by almost 2-fold, and cell morphology examination shows increased lipid droplets in cytosols and TAG-rich plastoglobuli in their chloroplasts (Balamurugan et al., 2017). An overexpression of PSY from *C. zofingiensis* (*CzPSY*) in *C. reinhardtii*, by nuclear transformation resulted in an improvement in the CzPSY transcript level, and also in violaxanthin and lutein contents that were 2.0 and 2.2 times greater than in wild-type cells (Cordero et al., 2011). A previous study found that the gene expression of the BKT (beta-carotene ketolase) gene in *H. pluvialis* resulted in ~3-fold greater carotenoid and astaxanthin content than those in control cells and ~10 times more of intermediary products such as echinenone and canthaxanthin (Kathiresan et al., 2015).

Overexpression of the Δ5-elongase gene (OtElo5) of *Ostreococcus tauri* resulted in an eight-fold increase in DHA levels of the *P. tricornutum*, and a further improvement in PUFAs by co-expressing an acyl-CoA-dependent Δ6-desaturase (OtD6Pt) resulted in an increase in overall DHA levels (Hamilton et al., 2014). By delivering preassembled Cas9 protein-gRNAs, zeaxanthin yields have been enhanced in two-gene knockout (*ΔZEP/ΔCpFTSY*) mutants and also photosynthetic changes were also well illustrated in *C. reinhardtii* (Baek et al., 2016). All above-mentioned techniques for improved competitiveness of bioactive compounds are efficient strategies but would definitely help to resolve obstacles and limits in the processing of new bioactive constituents derived from microalgae.

9.4.3   PROCESSING OF BIOACTIVE COMPOUNDS

The algal cells consist of solid, complicated, and multi-layered barriers that create difficulties during intracellular molecular extraction. The solid cell wall avoids the interaction of the solvent with intracellular substances and prevents mass movement of molecules during the extraction process. The goal is to increase extraction production before biomass treatment. Consequently, cell breakage is a crucial step in eliminating multiple bioactive intracellular compounds. Microalgal cells may be administered individually or in multiple combinations, via chemical, enzymatic/biological and mechanical/physical means for disruption (Molino et al., 2020). For chemical pretreatment of biomass of microalgae to boost bioactive compounds recovery, various chemicals such as organic solvents, acids, and bases, hydrogen peroxide, ozone, and ionic liquids may be used (Kim et al., 2016). In Thraustochytrids, acetone-based astaxanthin extraction was used employing different chemicals for pretreatment. Furthermore, biological hydrolysis is an excellent pretreatment, which can protect thermolabile carotenoids from physiological loss (Singh et al., 2015). Enzymatic/biological pretreatment consists of the application of cell-wall reducing enzymes, such as cellulose and amylase, which help in cellular interference, and intracellular recovery (Gerken et al., 2013). This approach has many benefits, including a decreased energy consumption, the absence of harmful solvents and molecular regeneration without thermal degradation and a downstream cost reduction. In conclusion, the strategies of mechanical/physical cell destruction are successful and quicker than those used during the scaling process. Mechanical pretreatment may be used to break the microalgae cell membrane, thereby facilitating the recovery of carotenoids and Fas-based compounds. This can be done with multiple methods, such as ball milling, high-pressure homogenization (HPH), ultrasonication, and steam explosion (Kim et al., 2016). Some studies on pretreatment process in *Thraustochytrium* sp. resulted in a 5-fold increase in astaxanthin yields as compared to the non-treated biomass, i.e., from 26.77±1.23 μg/g to 138.53±4.27 μg/g (Singh et al., 2015). Sonicated of cells (namely *Botryococcus* sp., *C. vulgaris*, and *Scenedesmus* sp.) at the intensity of 600 W for 5 min resulted in reduced enzyme release, which was around 10-fold lower than HPH pretreatment (Roh et al., 2008). The lipid recovery from *N. gaditana* significantly improved yields from 0.3% to 3.6% (at 120°C) and to 8.8% (at 180°C) (Nurra et al., 2014) and such enhancement in lipid extraction efficiency is due to the disruption of the cell wall caused by the steam explosion.

The next major feature of the metabolite production includes removing bioactive molecules predominantly by traditional solvent recovery and hydrolysis from the algal biomass. Extraction in Soxhlet equipment, solid-liquid extraction, and liquid-liquid extraction are also standard methods. Chemical solvents are used in conventional extraction processes. Alternatively, novel techniques featured super-critical fluid (SFE), microwave-assisted (MAE), ultrasound-assisted (UAE) and enzyme-assisted (EAE) extraction methods are employed (Kadam et al., 2013). In supercritical fluid extraction, CO_2 has physicochemical characteristics that are both a gas and a liquid, but have comparable viscosity, mean diffusions and high density that improve both microalgae biomass penetration and solubility of intracellular compounds (Goto et al., 2015). Krichnavaruk et al. (2008) optimized the effect of extraction time on astaxanthin recovery from *H. pluvialis* using SCF-CO_2 at 40 MPa and 70°C and shown greater selectivity for carotenoids, and the other molecules that could be directly supplemented for nutraceuticals (Krichnavaruk et al., 2008). Additionally, MAE, and UAE can work simultaneously, as in a method developed by Cravotto et al. (2008) in which the above methods were used to derive oil from the *Crypthecodinium cohnii*. These novel technologies contribute to the development of the next generation of sustainable functional foods and pharmaceutical products having a vast plethora of bioactive compounds from microalgae.

9.4.4 RECENT PHARMACEUTICALS AND NUTRACEUTICALS FROM MICROALGAE

The use of microalgae bioactive compounds as pharmaceuticals as well as nutraceuticals is discussed broadly in this chapter. In recent years, several new methods such as the development of the HIV microalgae vaccine have been explored. In this process, the release of transgenic material into the environment was inhibited via a dysfunctional cell technology. As an edible vaccine, microalgae may be used. The vaccine was developed to avoid the development of HIV in host cells through components including sulfated polysaccharides and bioactive peptides. Some diseases like smallpox, influenza, tuberculosis, and others should be found in the same treatment (Vo and Kim, 2010). A new mechanism of customized drug processing is recently studied in microalgal cells. The manufacture of medicines in algae will cause designer drugs to be produced in days and not months. Monoclonal antibodies and single-chain antibodies were documented to be generated using similar techniques in microalgae.

Fucoidan is an algae-containing bioactive compound. The defense of the algae against marine microbes, environmental stress and UV degradation in nature plays a vital role through fucoidan. Analysis has shown that fucoidan provides defensive health benefits, such as inflammatory suppression, balance of gut microbiota, viruses anti-proliferative, and the activation of immune cells. The component derived from microalgae uses green synthesis methods to preserve the cell's bioactivities (www.marinova.com. au). Although innovative techniques and the latest technology exist, there are still many obstacles to their effects on human health in the commercial processing of these drugs and nutraceuticals. Despite all these barriers, microalgae development provides an efficient means of making it possible for the organism to thrive.

9.5 RISK ASSESSMENT AND SAFETY ASSOCIATED WITH MICROALGAE-BASED PRODUCTS

In the coming decades, new protein resources from microalgae are expected to conquer the global market as a potent alternative to the existing animal-derived proteins. Although there are abundant opportunities existent for increasing productivity and efficient harvesting of the microalgae bioactives, their safety is still not well-documented (van der Spiegel et al., 2013). The safety aspects of such resources are a primary concern, especially when these products are to be used for human applications.

9.5.1 SAFETY OF CULTIVATION AND CONSUMPTION OF MICROALGAE BIOMASS

For centuries, microalgae have been utilized for the human consumption to sustain their extreme nutritional requirements (Mobin and Alam, 2017). The potential risks of their usage may include a high level of nucleic acid and rare structure of amino acid as well as the existence of many contaminants such as heavy metal, mycotoxins, pesticide residues and pathogenic substances (van der Spiegel et al., 2013; Navarro et al., 2016). Thus, the nutraceutical and pharmaceutical molecules from microalgae need to survive the socio-ethnological and toxicity-related examinations, to be accepted for consumers as safer food ingredients.

Although the nutritional and therapeutic value of various microalgae molecules are well-cited, certain chaos still exists regarding specific biomass qualities like digestibility, taste, odor, and texture (García et al., 2017) which may differ when a culture cultivated under controlled laboratory environment is grown in natural habitats. Also, high nucleic acid content of these microorganisms poses a serious threat to health, as they are metabolized into uric acid that may cause kidney stones or gout on consumption (Gantar and Svirčev, 2008). Even higher dosages of amino acids from red algae induce neurotoxic effects (Mouritsen et al., 2013). There are various other factors that may also impose a serious threat on the use of microalgae in nutraceutical and pharmaceutical sectors that are described below.

9.5.1.1   TOXICITY

Microalgae must be assessed for toxicity and the possible naturally occurring toxins (Draaisma et al., 2013). Traces of potent hepatotoxins and neurotoxins have been identified in some cyanobacteria and dinoflagellates, thus necessitating the need for periodic analysis during their manufacture to detect the presence of biological and nonbiological contaminants (Grobbelaar, 2003). Toxic microcystins, causative of pansteatitis disease, have been detected in *Aphanizomenon flos-aquae* used in nutraceutical industry (Spolaore et al., 2006). No similar toxins have been identified in algal species but polycyclic aromatic hydrocarbons (PAHs) levels were found to exceed, which signifies the need for monitoring at every production stage (Mouritsen et al., 2013; Jha et al., 2017). Certain human poisoning episodes were also witnessed with the consumption of wild-harvested *Spirulina* because of their contamination by *Microcystis* and other freshwater cyanobacteria as neurotoxins (Gellenbeck, 2011).

9.5.1.2   ALLERGENS

Microalgal nutraceuticals are rare in allergic studies. Allergenicity has been expressed by airborne cyanobacteria such as *Phormidium fragile*, *Nostoc muscorum* (Sharma and Rai, 2008), and certain green algae like *Chlorella* (Tiberg and Einarsson, 1989). Algalin, a lipid product composed of dried milled *C. protothecoides* showed very low or no tendency to impart allergies (Szabo et al., 2012). Reports have been published regarding the intake of *Spirulina* tablets resulting in anaphylaxis (Le et al., 2014). Development of allergic symptoms due to photosensibilization has also been reported in some humans on consumption of *Chlorella* due to pheophorbide formation, a product of chlorophyll degradation, which when utilized in a judicious way can act as a therapeutic for cancer treatment.

9.5.1.3   PATHOGENS

The emergence of pathogenic microorganisms is another factor of protection that typically contaminates the biomass obtained from outdoor/open pond cultures. In *Spirulina* and *Chlorella* grown in open basins, microbiological pollution of birds, insects, and rodents are well documented. Despite the recent advances in long-term microalgae growing processes, undesirable pathogens on a large scale are neither economical nor feasible to fully capture (Cooper and Smith, 2015). Additionally, *Pseudobodo* exhibited a broad predatory spectrum and negatively influenced the growth of *D. salina* (Chen et al., 2014). A great reduction in the algal cell densities has also been induced by rotifer grazers and ciliates (Moreno-Garrido and Canavate, 2001; Sarma et al., 2001). Also, fungi are well known to involve in parasitic relationships and often lead to malignant infections in algal cultures (Hoffman et al., 2008). So far, reports of fungal infections in *Chlamydomonas* (Shin et al., 2011), *H. pluvialis* (Hoffman et al., 2008) and *Scenedesmus* (Carney and Lane, 2014) are been well documented.

9.5.1.4   NONBIOLOGICAL CONTAMINATION

Microalgae shows a tendency of heavy metal accumulation and, being the first trophic level, possess the ability to act as a vector for the passage of these contaminants to the higher levels in the food pyramid of the aquatic environments (Souza et al., 2012). Sludge-grown microalgae is composed of significant levels of heavy metals which can be transferred to the consumers via processed biomass and also through the extracted micro/macromolecules (Hung et al., 1996; Wong et al., 1996; van der Spiegel et al., 2013). *Spirulina* also absorbs heavy metals at rates higher than *Chlorella*, but some of the researches conducted could detect no exceedance of their levels above the legal limits (Marles et al., 2011). Apart from heavy metals, microalgae like *Pseudokirchneriella subcapitata* and *Chlorococcum* spp. have exhibited efficient transformation and accumulation of harmful pesticide fenamiphos and its metabolites (Cáceres et al., 2008). As a consequence, contamination of the natural environments proved to imply undesirable impacts on the food chain and therefore on the individuals who consume these cells in their native or refined forms.

9.5.1.5   HAZARDS FROM HANDLING

Numerous non-GM (genetic modification) strategies like, media engineering, mixotrophy, and optimization of the abiotic factors have been fruitfully employed at the laboratory scale with a successful industrial scale-up for improving the quality and quantity of microalgae and derived products. The protection elements of microalgae products remain connected with algae or the product itself, but due to manufacturing procedures and process parameters which need to be examined and addressed during the development of the product several possible hazards can indeed be identified (García et al., 2017; Nethravathy et al., 2019). A pathogenic or chemical pollution can exist because it may include heavy metals, pesticides, or antimicrobials, radio-isotopes, poisonous synthetic chemicals, endocrine disruptors, or microbial loads, like coliform contaminants (Henley et al., 2013; García et al., 2017). Strict guidelines have been enforced by the pollution control boards in accordance with the respective governments regarding the quality of the water usage for biomass production and further downstream processing. Therefore, the producers of microalgal biomass and associated products must consider and overcome the possible dangers expected during production and also during product packaging operations.

9.5.2   RISK ASSESSMENT OF GM MICROALGAE

There is tremendous potential for genetic engineering of microalgae for the improved development of market-driven metabolites. However, their protection to human health and the atmosphere is a function of the performance of commercial operation. Because of some frequent and unexpected problems in the outdoor cultivation of GM microalgae, few progress reports are available (Henley et al., 2013). Potential threat is genetic pollution/ interbreeding with the wild or sexually competitive kin, natural species competitiveness, the targeted and non-target organism selection strain, effect on the environment, inefficient screening mechanism, horizontal transfers on recombinant genes to other micro-organisms, adverse human health impacts (Prakash et al., 2011; Nethravathy et al., 2019). The intended or accidental release (spills) of the non-native microalgae into the local environment also involves risks which needs to be carefully monitored and assessed (Prakash et al., 2011). The threats associated with spill include weakening of the vigor of invasive non-native algae and of the resilience of the local algal populations and rivalry between native phytoplanktons and invasive organisms (Henley et al., 2013). Although the existing policies are not adequate to carry out a comprehensive and reliable risk assessment, detailed know-how should be placed and applied to ensure the healthy use of GM microalgae within various biotech industry.

9.5.3 REGULATIONS REGARDING THE USE OF MICROALGAE AND ITS PRODUCTS

Many countries have implemented strict laws and regulations for algae-derived food products. In the EU (European Union), two EU regulations which affect the production and marketing of food and feed based on microalgae (Vigani et al., 2015) are prevalent. An independent agency, the European Food Safety Authority (EFSA) has also been established, that is responsible for scientific advice and support. The products approved from microalgae in the EU has also been graded safe (García et al., 2017). In United States, the Food and Drug Administration (FDA) regulates food safety, including the algal products, and is graded as "other dietary supplement." The consumption of processed biomass or products like protein powders or oils from certain microalgae are included as the GRAS. Various biotechnological firms ensure a stable supply of large quantities of bioproducts to the nutraceutical markets (Jha et al., 2017). Other products like oils from *Schizochytrium* and *Ulkenia*, and lipid as well as protein powder from whole cells of *Chlorella* are also categorized as GRAS by FDA (García et al., 2017).

Laws have also been enforced in the United States Environment Protection Agency (USEPA), necessitating the production and processing industries to address all the issues related to the release of the waste algae biomass and wastewater into the environment. Other developed countries like Australia and New Zealand have specific bodies like Food Standards Australia New Zealand (FSANZ) to regulate the use of novel functional foods (García et al., 2017). In Canada, the use of cyanobacterial biomass except *Spirulina* is restricted by Health Canada, the department responsible for federal health policy (García et al., 2017). Numerous guidelines have also been issued to the manufactures in many countries regarding the product labels; making the display of the ingredients and allergy warnings mandatory. A detailed awareness of the growing value of the algal industry, its technological advances, and the possible risks associated with its use, has prompted many governments to enact or change their current regulations to guarantee improved health and environmental protection.

9.6   CONCLUSIONS AND FUTURE PERSPECTIVES

As a potential for the development of biorefineries for commercial industrial applications, microalgae-based nutraceuticals and drugs are emerging. In comparison to animal and plant capital, microalgae tend to be a suitable supply of many bioactive compounds for achieving higher yield potentials. Though there are continuing challenges such as high cost and poor product yields after processing, refining of the algal biomass using productive greener technology has come to rescue nutraceuticals and pharmaceuticals from microalgae for cost-effective extraction. In spite of these concerns, demand for foods, supplements, and possible therapeutic components for microalgal is continuing to grow. Therefore, more development and effort are required to pursue cost-effective solutions such that high quality goods can be produced in large quantities. In this context, however, even further research is required to define the biochemical content of candidate microalgae in order to fully understand their advantages and potential issues. Moreover, with the complete processing and analysis of only a few development strains of microalgae, there are enormous possibilities for finding new bioactive metabolites with potential health benefits.

KEYWORDS

- **anti-carcinogenic activity**
- **antioxidant activity**
- **antiviral activity**
- **bioactive compounds**
- **metabolic engineering**
- **microalgae**
- **nutraceuticals**
- **omegas**
- **omics**
- **pharmaceuticals**
- **production technologies**

REFERENCES

Abe, K., Hattori, H., & Hirano, M., (2007). Accumulation and antioxidant activity of secondary carotenoids in the aerial microalga *Coelastrella striolata* var. *multistriata*. *Food Chemistry, 100*(2), 656–661. doi: https://doi.org/10.1016/j.foodchem.2005.10.026.

Abedini, N. H., Malekzadeh, M., Jalilian, F., Vossoughi, M., & Pazuki, G., (2015). Effect of various carbon sources on biomass and lipid production of *Chlorella vulgaris* during nutrient sufficient and nitrogen starvation conditions. *Bioresource Technology, 180*, 311–317. doi: 10.1016/j.biortech.2014.12.076.

Abreu, A. P., Fernandes, B., Vicente, A. A., Teixeira, J., & Dragone, G., (2012). Mixotrophic cultivation of *Chlorella vulgaris* using industrial dairy waste as organic carbon source. *Bioresource Technology, 118*, 61–66. doi: https://doi.org/10.1016/j.biortech.2012.05.055.

Ahmed, F., Zhou, W., & Schenk, P., (2015). *Pavlova lutheri* is a high-level producer of phytosterols. *Algal Research, 10*. doi: 10.1016/j.algal.2015.05.013.

Aleksandar, R., Stefanie, T., Achenbach, J., Andreas, K., Anna, V., Tanja, K., Christian, W., Kurt, P., & Nitsche, A., (2010). Anionic polysaccharides from phototrophic microorganisms exhibit antiviral activities to *Vaccinia virus. Journal of Antivirals and Antiretrovirals, 02*. doi: 10.4172/jaa.1000023.

Andrade, L., De Andrade, C. J., Dias, M., Nascimento, C., & Mendes, M., (2018). *Chlorella* and *Spirulina* microalgae as sources of functional foods, nutraceuticals, and food supplements; an overview. *MOJ Food Processing and Technology, 6*, 00144.

Arad, S., & Levy-Ontman, O., (2010). Red microalgal cell-wall polysaccharides: Biotechnological aspects. *Current Opinion in Biotechnology, 21*(3), 358–364.

Arad, S., & Yaron, A., (1992). Natural pigments from red microalgae for use in foods and cosmetics. *Trends in Food Science and Technology, 3*, 92–97. doi: https://doi.org/10.1016/0924-2244(92)90145-M.

Baek, K., Kim, D. H., Jeong, J., Sim, S. J., Melis, A., Kim, J. S., Jin, E., & Bae, S., (2016). DNA-free two-gene knockout in *Chlamydomonas reinhardtii* via CRISPR-Cas9 ribonucleoproteins. *Sci Rep., 6*, 30620. doi: 10.1038/srep30620.

Bai, L. L., Yin, W. B., Chen, Y. H., Niu, L. L., Sun, Y. R., Zhao, S. M., Yang, F. Q., et al., (2013). A new strategy to produce a defensin: Stable production of mutated NP-1 in nitrate reductase-deficient *Chlorella ellipsoidea. PLoS One, 8*(1), e54966. doi: 10.1371/journal.pone.0054966.

Balamurugan, S., Wang, X., Wang, H. L., An, C. J., Li, H., Li, D. W., Yang, W. D., et al., (2017). Occurrence of plastidial triacylglycerol synthesis and the potential regulatory role of AGPAT in the model diatom *Phaeodactylum tricornutum. Biotechnology for Biofuels, 10*, 97. doi: 10.1186/s13068-017-0786-0.

Banskota, A. H., Gallant, P., Stefanova, R., Melanson, R., & O'Leary, S. J. B., (2013). Monogalactosyldiacylglycerols, potent nitric oxide inhibitors from the marine microalga *Tetraselmis chui. Natural Product Research, 27*(12), 1084–1090. doi: 10.1080/14786419.2012.717285.

Barahimipour, R., Neupert, J., & Bock, R., (2016). Efficient expression of nuclear transgenes in the green alga *Chlamydomonas*: Synthesis of an HIV antigen and development of a new selectable marker. *Plant Molecular Biology, 90*(4), 403–418. doi: 10.1007/s11103-015-0425-8.

Barclay, W., Weaver, C., Metz, J., & Hansen, J., (2010). Development of a docosahexaenoic acid production technology using *Schizochytrium*: Historical perspective and update. In: Cohen, Z., & Ratledge, C., (eds.), *Single Cell Oils* (pp. 75–96). AOCS Press, Champaign.

Barkia, I., Ketata, B. H., Sellami, B. T., Aleya, L., Gargouri, A. F., & Saari, N., (2019). Acute oral toxicity study on Wistar rats fed microalgal protein hydrolysates from *Bellerochea malleus*. *Environmental Science and Pollution Research*. doi: 10.1007/s11356-018-4007-6.

Barkia, I., Saari, N., & Manning, S. R., (2019). Microalgae for high-value products towards human health and nutrition. *Marine Drugs, 17*(5), 304.

Barros, M. P., Pinto, E., Sigaud-Kutner, T. C. S., Cardozo, K. H. M., & Colepicolo, P., (2005). Rhythmicity and oxidative/nitrosative stress in algae. *Biological Rhythm Research, 36*(1, 2), 67–82. doi: 10.1080/09291010400028666.

Basaran, P., & Rodríguez-Cerezo, E., (2008). Plant molecular farming: Opportunities and challenges. *Critical Reviews in Biotechnology, 28*(3), 153–172. doi:10.1080/073885 50802046624.

Becker, E. W., (2007). Micro-algae as a source of protein. *Biotechnology Advances, 25*(2), 207–210. doi: https://doi.org/10.1016/j.biotechadv.2006.11.002.

Ben-Amotz, A., (2007). Industrial production of microalgal cell - mass and secondary products-major industrial species: *Dunaliella*. In: Richmond, A., (ed.), *Handbook of Microalgal Culture: Biotechnology and Applied Phycology* (pp. 273–280). Blackwell Publishing Ltd.

Bergman, B., (1986). Glyoxylate induced changes in the carbon and nitrogen metabolism of the cyanobacterium *Anabaena cylindrica*. *Plant Physiology, 80*(3), 698. doi: 10.1104/pp.80.3.698.

Berthon, J. Y., Nachat-Kappes, R., Bey, M., Cadoret, J. P., Renimel, I., & Filaire, E., (2017). Marine algae as attractive source to skincare. *Free Radical Research, 51*(6), 555–567. doi: 10.1080/10715762.2017.1355550.

Bhalamurugan, G. L., Valerie, O., & Mark, L., (2018). Valuable bioproducts obtained from microalgal biomass and their commercial applications: A review. *Environmental Engineering Research, 23*(3), 229–241. doi: 10.4491/eer.2017.220.

Bhandari, K., Chaurasia, S. P., & Dalai, A. K., (2015). Lipase-catalyzed esterification of docosahexaenoic acid-rich fatty acids with glycerol. *Chemical Engineering Communications, 202*(7), 920–926. doi: 10.1080/00986445.2014.891505.

Bleakley, S., & Hayes, M., (2017). Algal proteins: Extraction, application, and challenges concerning production. *Foods, 6*(5). doi: 10.3390/foods6050033.

Borowitzka, L. J., & Borowitzka, M. A., (1989). β-carotene (provitamin a) production with algae. In: Vandamme, E. J., (ed.), *Biotechnology of Vitamins, Pigments and Growth Factors* (pp. 15–26). Springer, Dordrecht, Netherlands.

Borowitzka, M. A., (1995). Microalgae as sources of pharmaceuticals and other biologically active compounds. *Journal of Applied Phycology, 7*(1), 3–15. doi: 10.1007/BF00003544.

Borowitzka, M. A., (2013). High-value products from microalgae - their development and commercialization. *Journal of Applied Phycology, 25*(3), 743–756. doi: 10.1007/s10811-013-9983-9.

Brennan, L., & Owende, P., (2010). Biofuels from microalgae: A review of technologies for production, processing, and extractions of biofuels and co-products. *Renewable and Sustainable Energy Reviews, 14*(2), 557–577. doi: https://doi.org/10.1016/j.rser.2009.10.009.

Burja, A. M., Banaigs, B., Abou-Mansour, E., Grant, B. J., & Wright, P. C., (2001). Marine cyanobacteria - a prolific source of natural products. *Tetrahedron, 57*(46), 9347–9377. doi: https://doi.org/10.1016/S0040-4020(01)00931-0.

Bwapwa, J. K., Jaiyeola, A. T., & Chetty, R., (2017). Bioremediation of acid mine drainage using algae strains: A review. *South African Journal of Chemical Engineering, 24*, 62–70. doi: https://doi.org/10.1016/j.sajce.2017.06.005.

Cáceres, T., Megharaj, M., & Naidu, R., (2008). Toxicity and transformation of fenamiphos and its metabolites by two microalgae *Pseudokirchneriella subcapitata* and *Chlorococcum* sp. *Science of the Total Environment, 398*(1), 53–59. doi: https://doi.org/10.1016/j.scitotenv. 2008.03.022.

Cardozo, K. H. M., Guaratini, T., Barros, M. P., Falcão, V. R., Tonon, A. P., Lopes, N. P., Campos, S., et al., (2007). Metabolites from algae with economical impact. *Comparative Biochemistry and Physiology Part C: Toxicology and Pharmacology, 146*(1), 60–78. doi: https://doi.org/10.1016/j.cbpc.2006.05.007.

Carney, L. T., & Lane, T. W., (2014). Parasites in algae mass culture. *Frontiers in Microbiology, 5*, 278. doi: 10.3389/fmicb.2014.00278.

Challouf, R., Trabelsi, L., Ben, D. R., El Abed, O., Yahia, A., Ghozzi, K., Ben, A. J., et al., (2011). Evaluation of cytotoxicity and biological activities in extracellular polysaccharides released by cyanobacterium *Arthrospira platensis*. *Brazilian Archives of Biology and Technology, 54*, 831–838.

Chan, K. C., Mong, M. C., & Yin, M. C., (2009). Antioxidative and anti-inflammatory neuro-protective effects of astaxanthin and canthaxanthin in nerve growth factor differentiated pc12 cells. *Journal of Food Science, 74*(7), H225–H231. doi: 10.1111/j.1750-3841.2009.01274.x.

Chebolu, S., & Daniell, H., (2009). Chloroplast-derived vaccine antigens and biopharma-ceuticals: Expression, folding, assembly and functionality. In: Karasev, A. V., (ed.), *Plant-Produced Microbial Vaccines* (pp. 33–54). Springer, Berlin, Heidelberg.

Cheirsilp, B., Thawechai, T., & Prasertsan, P., (2017). Immobilized oleaginous microalgae for production of lipid and phytoremediation of secondary effluent from palm oil mill in fluidized bed photobioreactor. *Bioresource Technology, 241*, 787–794. doi: https://doi.org/10.1016/j.biortech.2017.06.016.

Chen, Z., Lei, X., Zhang, B., Yang, L., Zhang, H., Zhang, J., Li, Y., et al., (2014). First report of *Pseudobodo* sp., a new pathogen for a potential energy-producing algae: *Chlorella vulgaris* cultures. *PLoS One, 9*(3), e89571. doi: 10.1371/journal.pone.0089571.

Cheng, J., Li, K., Zhu, Y., Yang, W., Zhou, J., & Cen, K., (2017). Transcriptome sequencing and metabolic pathways of astaxanthin accumulated in *Haematococcus pluvialis* mutant under 15% CO_2. *Bioresource Technology, 228*, 99–105. doi: 10.1016/j.biortech.2016.12.084.

Choi, J. M., Lee, E. O., Lee, H. J., Kim, K. H., Ahn, K. S., Shim, B. S., Kim, N. I., et al., (2007). Identification of campesterol from *Chrysanthemum coronarium* L. and its antiangiogenic activities. *Phytotherapy Research, 21*(10), 954–959. doi: 10.1002/ptr.2189.

Chronakis, I. S., & Madsen, M., (2011). Algal proteins. In: Phillips, G. O., & Williams, P. A., (eds.), *Handbook of Food Proteins* (pp. 353–394). Woodhead Publishing, Cambridge, United Kingdom.

Ciferri, O., (1983). *Spirulina*, the edible microorganism. *Microbiological Reviews, 47*(4), 551–578.

Conde, E., Balboa, E. M., Parada, M., & Falqué, E., (2013). Algal proteins, peptides, and amino acids. In: Domínguez, H., (ed.), *Functional Ingredients from Algae for Foods and Nutraceuticals* (pp. 135–180). Woodhead Publishing, Cambridge, United Kingdom.

Cooper, M. B., & Smith, A. G., (2015). Exploring mutualistic interactions between microalgae and bacteria in the omics age. *Current Opinion in Plant Biology, 26*, 147–153. doi: https://doi.org/10.1016/j.pbi.2015.07.003.

Cordero, B. F., Couso, I., León, R., Rodríguez, H., & Vargas, M. Á., (2011). Enhancement of carotenoids biosynthesis in *Chlamydomonas reinhardtii* by nuclear transformation using a phytoene synthase gene isolated from *Chlorella zofingiensis*. *Applied Microbiology and Biotechnology, 91*(2), 341–351. doi: 10.1007/s00253-011-3262-y.

Cravotto, G., Boffa, L., Mantegna, S., Perego, P., Avogadro, M., & Cintas, P., (2008). Improved extraction of vegetable oils under high-intensity ultrasound and/or microwaves. *Ultrasonics Sonochemistry, 15*(5), 898–902. doi: 10.1016/j.ultsonch.2007.10.009.

Cysewski, G. R. A. T. L., (2007). Industrial production of microalgal cell-mass and secondary products-species of high potential: *Haematococcus*. In: Richmond, A., (ed.), *Handbook of Microalgal Culture: Biotechnology and Applied Phycology* (pp. 281–288). Blackwell Publishing Ltd, UK.

Das, P., Aziz, S. S., & Obbard, J. P., (2011). Two phase microalgae growth in the open system for enhanced lipid productivity. *Renewable Energy, 36*(9), 2524–2528. doi: https://doi.org/10.1016/j.renene.2011.02.002.

Das, P., Lei, W., Aziz, S. S., & Obbard, J. P., (2011). Enhanced algae growth in both phototrophic and mixotrophic culture under blue light. *Bioresource Technology, 102*(4), 3883–3887. doi: https://doi.org/10.1016/j.biortech.2010.11.102.

Dauvillée, D., Delhaye, S., Gruyer, S., Slomianny, C., Moretz, S. E., D'Hulst, C., Long, C. A., et al., (2010). Engineering the chloroplast targeted malarial vaccine antigens in *Chlamydomonas* starch granules. *PLoS One, 5*(12), e15424. doi: 10.1371/journal.pone.0015424.

De Jesus, R. M. F., De Morais, A. M. M. B., & De Morais, R. M. S. C., (2014). Bioactivity and applications of polysaccharides from marine microalgae. *Polysaccharides: Bioactivity and Biotechnology*, 1–38. doi: 10.1007/978-3-319-03751-6_47-1.

De Philippis, R., Margheri, M. C., Pelosi, E., & Ventura, S., (1993). Exopolysaccharide production by a unicellular *Cyanobacterium* isolated from a hypersaline habitat. *Journal of Applied Phycology, 5*(4), 387–394. doi: 10.1007/BF02182731.

Del, C. J. A., García-González, M., & Guerrero, M. G., (2007). Outdoor cultivation of microalgae for carotenoid production: Current state and perspectives. *Applied Microbiology and Biotechnology, 74*(6), 1163–1174.

Demirbaş, A., & Demirbas, M., (2010). *Algae Energy: Algae as a New Source of Biodiesel* (Vol. 36, pp. 1–200). Springer-Verlag, London.

Demirbaş, A., & Demirbas, M., (2011). Importance of algae oil as a source of biodiesel. *Energy Conversion and Management, 52*, 163–170. doi: 10.1016/j.enconman.2010.06.055.

Draaisma, R. B., Wijffels, R. H., Slegers, P. M., Brentner, L. B., Roy, A., & Barbosa, M. J., (2013). Food commodities from microalgae. *Current Opinion in Biotechnology, 24*(2), 169–177. doi: https://doi.org/10.1016/j.copbio.2012.09.012.

Dreesen, I. A. J., Hamri, G. C. E., & Fussenegger, M., (2010). Heat-stable oral alga-based vaccine protects mice from *Staphylococcus aureus* infection. *Journal of Biotechnology, 145*(3), 273–280. doi: https://doi.org/10.1016/j.jbiotec.2009.12.006.

Edelmann, M., Aalto, S., Chamlagain, B., Kariluoto, S., & Piironen, V., (2019). Riboflavin, niacin, folate, and vitamin B12 in commercial microalgae powders. *Journal of Food Composition and Analysis, 82*, 103226. doi: https://doi.org/10.1016/j.jfca. 2019.05.009.

Ejike, C. E. C. C., Collins, S. A., Balasuriya, N., Swanson, A. K., Mason, B., & Udenigwe, C. C., (2017). Prospects of microalgae proteins in producing peptide-based functional foods for promoting cardiovascular health. *Trends in Food Science and Technology, 59*, 30–36. doi: https://doi.org/10.1016/j.tifs.2016.10.026.

Falkowski, P., Katz, M., Knoll, A., Quigg, A., Raven, J., Schofield, O., & Taylor, F., (2004). The evolution of modern eukaryotic phytoplankton. *Science, 305*, 354–360. doi: 10.1126/science.1095964.

Feng, S., Feng, W., Zhao, L., Gu, H., Li, Q., Shi, K., Guo, S., & Zhang, N., (2014). Preparation of transgenic *Dunaliella salina* for immunization against white spot syndrome virus in crayfish. *Archives of Virology, 159*(3), 519–525. doi: 10.1007/s00705-013-1856-7.

Fernandes, B. D., Mota, A., Teixeira, J. A., & Vicente, A. A., (2015). Continuous cultivation of photosynthetic microorganisms: Approaches, applications, and future trends. *Biotechnology Advances, 33*, 1228–1245. doi: 10.1016/j.biotechadv.2015.03.004.

Francavilla, M., Trotta, P., & Luque, R., (2010). Phytosterols from *Dunaliella tertiolecta* and *Dunaliella salina*: A potentially novel industrial application. *Bioresource Technology, 101*(11), 4144–4150. doi: https://doi.org/10.1016/j.biortech.2009.12.139.

Galasso, C., Gentile, A., Orefice, I., Ianora, A., Bruno, A., Noonan, D. M., Sansone, C., et al., (2019). Microalgal derivatives as potential nutraceutical and food supplements for human health: A focus on cancer prevention and interception. *Nutrients, 11*(6), 1226.

Galasso, C., Nuzzo, G., Brunet, C., Ianora, A., Sardo, A., Fontana, A., & Sansone, C., (2018). The marine dinoflagellate *Alexandrium minutum* activates a mitophagic pathway in human lung cancer cells. *Marine Drugs, 16*(12), 502. doi: 10.3390/md16120502.

Gammone, M. A., Riccioni, G., & D'Orazio, N., (2015). Carotenoids: Potential allies of cardiovascular health? *Food and Nutrition Research, 59*, 26762. doi: 10.3402/fnr.v59.26762.

Gantar, M., & Svirčev, Z., (2008). Microalgae and cyanobacteria: Food for thought. *Journal of Phycology, 44*(2), 260–268. doi: 10.1111/j.1529-8817.2008.00469.x.

Gantar, M., Simović, D., Djilas, S., Gonzalez, W. W., & Miksovska, J., (2012). Isolation, characterization and antioxidative activity of C-phycocyanin from *Limnothrix* sp. strain 37-2-1. *Journal of Biotechnology, 159*(1), 21–26. doi: https://doi.org/10.1016/j.jbiotec.2012.02.004.

García, J. L., De Vicente, M., & Galán, B., (2017). Microalgae, old sustainable food, and fashion nutraceuticals. *Microbial Biotechnology, 10*(5), 1017–1024. doi: 10.1111/ 1751-7915.12800.

Gargouch, N., Karkouch, I., Elleuch, J., Elkahoui, S., Michaud, P., Abdelkafi, S., Laroche, C., & Fendri, I., (2018). Enhanced B-phycoerythrin production by the red microalga *Porphyridium marinum*: A powerful agent in industrial applications. *International Journal of Biological Macromolecules, 120*, 2106–2114. doi: https://doi.org/10.1016/j.ijbiomac.2018.09.037.

Gellenbeck, K., (2011). Utilization of algal materials for nutraceutical and cosmeceutical applications - what do manufacturers need to know? *Journal of Applied Phycology, 24*, 1–5. doi: 10.1007/s10811-011-9722-z.

Geng, D., Wang, Y., Wang, P., Li, W., & Sun, Y., (2003). Stable expression of hepatitis B surface antigen gene in *Dunaliella salina* (Chlorophyta). *Journal of Applied Phycology, 15*(6), 451–456. doi: 10.1023/B0000004298.89183.

Gerken, H. G., Donohoe, B., & Knoshaug, E. P., (2013). Enzymatic cell wall degradation of *Chlorella vulgaris* and other microalgae for biofuels production. *Planta, 237*(1), 239–253. doi: 10.1007/s00425-012-1765-0.

Ghosh, T., Paliwal, C., Maurya, R., & Mishra, S., (2015). Microalgal rainbow colors for nutraceutical and pharmaceutical applications. *Plant Biology and Biotechnology*, 777–791. doi: 10.1007/978-81-322-2286-6_32.

Goiris, K., Muylaert, K., Fraeye, I., Foubert, I., De Brabanter, J., & De Cooman, L., (2012). Antioxidant potential of microalgae in relation to their phenolic and carotenoid content. *Journal of Applied Phycology, 24*(6), 1477–1486. doi: 10.1007/s10811-012-9804-6.

Goto, M., Kanda, H., Wahyudiono, & Machmudah, S., (2015). Extraction of carotenoids and lipids from algae by supercritical CO_2 and subcritical dimethyl ether. *The Journal of Supercritical Fluids, 96*, 245–251. doi: 10.1016/j.supflu.2014.10.003.

Granum, E., & Myklestad, S. M., (2002). A simple combined method for determination of β-1,3-glucan, and cell wall polysaccharides in diatoms. *Hydrobiologia, 477*(1), 155–161. doi: 10.1023/A:1021077407766.

Grobbelaar, J. U., (2007). Algal nutrition-mineral nutrition. In: Richmond, A., (ed.), *Handbook of Microalgal Culture: Biotechnology and Applied Phycology* (pp. 91–115). Blackwell Publishing Ltd, Israel.

Guarnieri, M. T., & Pienkos, P. T., (2015). Algal omics: Unlocking bioproduct diversity in algae cell factories. *Photosynthetic Research, 123*(3), 255–263. doi: 10.1007/s11120-014-9989-4.

Guzmán, S., Gato, A., Lamela, M., Freire-Garabal, M., & Calleja, J., (2003). Anti-inflammatory and immunomodulatory activities of polysaccharide from *Chlorella stigmatophora* and *Phaeodactylum tricomutum*. *Phytotherapy Research, 17*, 665–670.

Haimeur, A., Ulmann, L., Mimouni, V., Guéno, F., Pineau-Vincent, F., Meskini, N., & Tremblin, G., (2012). The role of *Odontella aurita*, a marine diatom rich in EPA, as a dietary supplement in dyslipidemia, platelet function and oxidative stress in high-fat-fed rats. *Lipids in Health and Disease, 11*(1), 147. doi: 10.1186/1476-511X-11-147.

Halim, R., Gladman, B., Danquah, M. K., & Webley, P. A., (2011). Oil extraction from microalgae for biodiesel production. *Bioresource Technology, 102*(1), 178–185. doi: https://doi.org/10.1016/j.biortech.2010.06.136.

Hamed, I., (2016). The evolution and versatility of microalgal biotechnology: A review. *Comprehensive Reviews in Food Science and Food Safety, 15*(6), 1104–1123. doi: 10.1111/1541-4337.12227.

Hamilton, M. L., Haslam, R. P., Napier, J. A., & Sayanova, O., (2014). Metabolic engineering of *Phaeodactylum tricornutum* for the enhanced accumulation of omega-3 long-chain polyunsaturated fatty acids. *Metabolic Engineering, 22*(100), 3–9. doi: 10.1016/j.ymben.2013.12.003.

Hanagata, N., & Dubinsky, Z., (2002). Secondary carotenoid accumulation in *Scenedesmus komarekii* (Chlorophyceae, Chlorophyta). *Journal of Phycology, 35*, 960–966. doi: 10.1046/j.1529-8817.1999.3550960.x.

Hayashi, K., Hayashi, T., & Kojima, I., (1996). A natural sulfated polysaccharide, calcium spirulan, isolated from *Spirulina platensis*: *In vitro* and *ex vivo* evaluation of anti-herpes simplex virus and anti-human immunodeficiency virus activities. *AIDS Research and Human Retroviruses, 12*(15), 1463–1471. doi: 10.1089/aid.1996.12.1463.

He, D. M., Qian, K. X., Shen, G. F., Zhang, Z. F., Li, Y. N., Su, Z. L., & Shao, H. B., (2007). Recombination and expression of classical swine fever virus (CSFV) structural protein E2 gene in *Chlamydomonas reinhardtii* chloroplasts. *Colloids and Surfaces B: Biointerfaces, 55*(1), 26–30.

He, Q., (2007). Microalgae as platforms for recombinant proteins. In: Richmond, A., (ed.), *Handbook of Microalgal Culture: Biotechnology and Applied Phycology* (pp. 471–484). Blackwell Publishing Ltd, UK.

Henley, W. J., Litaker, R. W., Novoveská, L., Duke, C. S., Quemada, H. D., & Sayre, R. T., (2013). Initial risk assessment of genetically modified (GM) microalgae for commodity-scale biofuel cultivation. *Algal Research, 2*(1), 66–77. doi: https://doi.org/10.1016/j.algal.2012.11.001.

Ho, S. H., Nakanishi, A., Ye, X., Chang, J. S., Chen, C. Y., Hasunuma, T., & Kondo, A., (2015). Dynamic metabolic profiling of the marine microalga *Chlamydomonas* sp. JSC4 and enhancing its oil production by optimizing light intensity. *Biotechnology for Biofuels, 8*, 48. doi: 10.1186/s13068-015-0226-y.

Hoffman, Y., Aflalo, C., Zarka, A., Gutman, J., James, T. Y., & Boussiba, S., (2008). Isolation and characterization of a novel chytrid species (phylum Blastocladiomycota), parasitic on the green alga *Haematococcus*. *Mycological Research, 112*(1), 70–81. doi: https://doi.org/10.1016/j.mycres.2007.09.002.

Hu, L., Huang, B., Zuo, M. M., Guo, R. Y., & Wei, H., (2008). Preparation of the phycoerythrin subunit liposome in a photodynamic experiment on liver cancer cells. *Acta Pharmacologica Sinica, 29*(12), 1539–1546. doi: 10.1111/j.1745-7254.2008.00886.x.

Hudek, K., Davis, L. C., Ibbini, J., & Erickson, L., (2014). Commercial products from algae. In: Bajpai, R., Prokop, A., & Zappi, M., (eds.), *Algal Biorefineries: Cultivation of Cells and Products* (Vol. 1, pp. 275–295). Springer, Netherlands.

Hung, K. M., Chiu, S. T., & Wong, M. H., (1996). Sludge-grown algae for culturing aquatic organisms: Part I. Algal growth in sludge extracts. *Environmental Management, 20*(3), 361–374. doi: 10.1007/BF01203844.

Iwamoto, H., (2007). Industrial production of microalgal cell-mass and secondary products-major industrial species: *Chlorella*. In: Richmond, A., (ed.), *Handbook of Microalgal Culture: Biotechnology and Applied Phycology* (pp. 253–263). Blackwell Publishing Ltd, UK.

Jha, D., Jain, V., Sharma, B., Kant, A., & Garlapati, V. K., (2017). Microalgae-based pharmaceuticals and nutraceuticals: An emerging field with immense market potential. *ChemBioEng Reviews, 4*(4), 257–272. doi: 10.1002/cben.201600023.

Ji, B., Zhang, W., Zhang, N., Wang, J., Lutzu, G. A., & Liu, T., (2014). Biofilm cultivation of the oleaginous microalgae *Pseudochlorococcum* sp. *Bioprocess and Biosystems Engineering, 37*(7), 1369–1375. doi: 10.1007/s00449-013-1109-x.

Jin, E., Feth, B., & Melis, A., (2003). A mutant of the green alga *Dunaliella salina* constitutively accumulates zeaxanthin under all growth conditions. *Biotechnology and Bioengineering, 81*(1), 115–124.

Jo, W. S., Choi, Y. J., Kim, H. J., Nam, B. H., Hong, S. H., Lee, G. A., Lee, S. W., Seo, S. Y., & Jeong, M. H., (2010). Anti-inflammatory effect of microalgal extracts from *Tetraselmis suecica*. *Food Science and Biotechnology, 19*(6), 1519–1528.

Joshi, N., & Rahimbhai, M., (2019). Microalgae and its use in nutraceuticals and food supplements. In: Vítová, M., (ed.), *Microalgae-from Physiology to Application* (pp. 1–11). IntechOpen, UK.

Kadam, S. U., Tiwari, B. K., & O'Donnell, C. P., (2013). Application of novel extraction technologies for bioactives from marine algae. *Journal of Agricultural and Food Chemistry, 61*(20), 4667–4675. doi: 10.1021/jf400819p.

Kathiresan, S., Chandrashekar, A., Ravishankar, G. A., & Sarada, R., (2015). Regulation of astaxanthin and its intermediates through cloning and genetic transformation of β-carotene ketolase in *Haematococcus pluvialis*. *Journal of Biotechnology, 196, 197*, 33–41. doi: 10.1016/j.jbiotec.2015.01.006.

Katiyar, R., & Arora, A., (2020). Health-promoting functional lipids from microalgae pool: A review. *Algal Research, 46*, 101800.

Kim, D. H., Kim, Y. T., Cho, J. J., Bae, J. H., Hur, S. B., Hwang, I., & Choi, T. J., (2002). Stable integration and functional expression of flounder growth hormone gene in transformed micro-alga, *Chlorella ellipsoidea*. *Marine Biotechnology, 4*(1), 63–73. doi: 10.1007/s1012601-0070-x.

Kim, D. Y., Vijayan, D., Praveenkumar, R., Han, J. I., Lee, K., Park, J. Y., Chang, W. S., et al., (2016). Cell-wall disruption and lipid/astaxanthin extraction from microalgae: *Chlorella* and *Haematococcus*. *Bioresource Technology, 199*, 300–310. doi: https://doi.org/10.1016/j.biortech.2015.08.107.

Kim, H. J., Choi, Y. K., Jeon, H. J., Bhatia, S. K., Kim, Y. H., Kim, Y. G., Choi, K. Y., et al., (2015). Growth promotion of *Chlorella vulgaris* by modification of nitrogen source composition with symbiotic bacteria, *Microbacterium* sp. HJ1. *Biomass and Bioenergy, 74*, 213–219. doi: https://doi.org/10.1016/j.biombioe.2015.01.012.

Kim, M., Ahn, J., Jeon, H., & Jin, E., (2017). Development of a *Dunaliella tertiolecta* strain with increased zeaxanthin content using random mutagenesis. *Marine Drugs, 15*(6), 189. doi: 10.3390/md15060189.

Kim, S. K., & Kang, K. H., (2011). Medicinal effects of peptides from marine microalgae. *Advances in Food and Nutrition Research, 64,* 313–323.

Koo, S. Y., Cha, K. H., Song, D. G., Chung, D., & Pan, C. H., (2012). Optimization of pressurized liquid extraction of zeaxanthin from *Chlorella ellipsoidea*. *Journal of Applied Phycology, 24*(4), 725–730.

Krichnavaruk, S., Shotipruk, A., Goto, M., & Pavasant, P., (2008). Supercritical carbon dioxide extraction of astaxanthin from *Haematococcus pluvialis* with vegetable oils as co-solvent. *Bioresource Technology, 99*(13), 5556–5560. doi: https://doi.org/10.1016/j.biortech.2007.10.049.

Le, T. M., Knulst, A., & Röckmann, H., (2014). Anaphylaxis to *Spirulina* confirmed by skin prick test with ingredients of *Spirulina* tablets. *Food and Chemical Toxicology, 74,* 309, 310. doi: 10.1016/j.fct.2014.10.024.

Lee, J. B., Hayashi, K., Hirata, M., Kuroda, E., Suzuki, E., Kubo, Y., & Hayashi, T., (2006). Antiviral sulfated polysaccharide from *Navicula directa*, a diatom collected from deep-sea water in Toyama Bay. *Biological and Pharmaceutical Bulletin, 29,* 2135–2139.

Lee, J. B., Hayashi, T., Hayashi, K., Sankawa, U., Maeda, M., Nemoto, T., & Nakanishi, H., (1998). Further purification and structural analysis of calcium spirulan from *Spirulina platensis*. *Journal of Natural Products, 61*(9), 1101–1104. doi: 10.1021/np980143n.

Lee, S., Lee, Y. S., Jung, S. H., Kang, S. S., & Shin, K. H., (2003). Anti-oxidant activities of fucosterol from the marine algae *Pelvetia siliquosa*. *Arch Pharm Res., 26*(9), 719–722. doi: 10.1007/bf02976680.

Lee, Y. S., Shin, K. H., Kim, B. K., & Lee, S., (2004). Anti-diabetic activities of fucosterol from *Pelvetia siliquosa*. *Arc Pharm Res., 27*(11), 1120–1122. doi: 10.1007/bf02975115.

Légeret, B., Schulz-Raffelt, M., Nguyen, H. M., Auroy, P., Beisson, F., Peltier, G., Blanc, G., & Li-Beisson, Y., (2016). Lipidomic and transcriptomic analyses of *Chlamydomonas reinhardtii* under heat stress unveil a direct route for the conversion of membrane lipids into storage lipids. *Plant Cell and Environment, 39*(4), 834–847. doi: 10.1111/pce.12656.

León, R., Marti, X N M., Vigara, J., Vilchez, C., & Vega, J. M., (2003). Microalgae mediated photoproduction of β-carotene in aqueous-organic two-phase systems. *Biomolecular Engineering, 20*(4), 177–182. doi: https://doi.org/10.1016/S1389-0344(03)00048-0.

Leu, S., & Boussiba, S., (2014). Advances in the production of high-value products by microalgae. *Industrial Biotechnology, 10,* 169–183. doi: 10.1089/ind.2013.0039.

Levasseur, W., Perre, P., & Pozzobon, V., (2020). A review of high value-added molecules production by microalgae in light of the classification. *Biotechnology Advances, 107*545. doi: 10.1016/j.biotechadv.2020.107545.

Li, H. B., Cheng, K. W., Wong, C. C., Fan, K. W., Chen, F., & Jiang, Y., (2007). Evaluation of antioxidant capacity and total phenolic content of different fractions of selected microalgae. *Food Chemistry, 102*(3), 771–776. doi: https://doi.org/10.1016/j.foodchem.2006.06.022.

Liu, T., Liu, W. H., Zhao, J. S., Meng, F. Z., & Wang, H., (2017). Lutein protects against β-amyloid peptide-induced oxidative stress in cerebrovascular endothelial cells through modulation of Nrf-2 and NF-κb. *Cell Biology and Toxicology, 33*(1), 57–67. doi: 10.1007/s10565-016-9360-y.

Lu, C., Rao, K., Hall, D., & Vonshak, A., (2001). Production of eicosapentaenoic acid (EPA) in *Monodus subterraneus* grown in a helical tubular photobioreactor as affected by cell density and light intensity. *Journal of Applied Phycology, 13*(6), 517–522.

Lu, Y., Zhou, W., Wei, L., Li, J., Jia, J., Li, F., Smith, S. M., & Xu, J., (2014). Regulation of the cholesterol biosynthetic pathway and its integration with fatty acid biosynthesis in the oleaginous microalga *Nannochloropsis oceanica*. *Biotechnology for Biofuels, 7*, 81. doi: 10.1186/1754-6834-7-81.

Luo, X., Su, P., & Zhang, W., (2015). Advances in microalgae-derived phytosterols for functional food and pharmaceutical applications. *Marine Drugs, 13*(7), 4231–4254. doi: 10.3390/md13074231.

Ma, Z., Tian, M., Tan, Y., Cui, G., Feng, Y., Cui, Q., & Song, X., (2017). Response mechanism of the docosahexaenoic acid producer *Aurantiochytrium* under cold stress. *Algal Research, 25*, 191–199. doi: https://doi.org/10.1016/j.algal.2017.05.021.

Manirafasha, E., Ndikubwimana, T., Zeng, X., Lu, Y., & Jing, K., (2016). Phycobiliprotein: Potential microalgae derived pharmaceutical and biological reagent. *Biochemical Engineering Journal, 109*, 282–296. doi: https://doi.org/10.1016/j.bej.2016.01.025.

Mansouri, H., & Talebizadeh, R., (2017). Effects of indole-3-butyric acid on growth, pigments, and UV-screening compounds in *Nostoc linckia*. *Phycological Research, 65*(3), 212–216. doi: 10.1111/pre.12177.

Manuell, A., Beligni, M., Elder, J., Siefker, D., Tran, M., Weber, A., McDonald, T., & Mayfield, S., (2007). Robust expression of a bioactive mammalian protein in *Chlamydomonas* chloroplast. *Plant Biotechnology Journal, 5*, 402–412. doi: 10.1111/ j.1467-7652.2007.00249.x.

Marles, R., Barrett, M., Barnes, J., Chavez, M., Gardiner, P., Ko, R., Mahady, G., et al., (2011). United States Pharmacopeia safety evaluation of *Spirulina*. *Critical Reviews in Food Science and Nutrition, 51*, 593–604. doi: 10.1080/10408391003721719.

Martins, A., Vieira, H., Gaspar, H., & Santos, S., (2014). Marketed marine natural products in the pharmaceutical and cosmeceutical industries: Tips for success. *Marine Drugs, 12*(2), 1066–1101. doi: 10.3390/md12021066.

Matsui, M., Muizzuddin, N., Arad, S., & Marenus, K., (2003). Sulfated polysaccharides from red microalgae have anti-inflammatory properties *in vitro* and *in vivo*. *Applied Biochemistry and Biotechnology, 104*, 13–22.

Mayfield, S. P., Franklin, S. E., & Lerner, R. A., (2003). Expression and assembly of a fully active antibody in algae. *Proceedings of the National Academy of Sciences of the United States of America, 100*(2), 438–442. doi: 10.1073/pnas.0237108100.

Meenakshi, B., (2016). Pharmaceutically valuable bioactive compounds of algae. *Asian Journal of Pharmaceutical and Clinical Research, 9*(6). doi: 10.22159/ajpcr.2016.v9i6.14507.

Milledge, J., (2011). Commercial application of microalgae other than as biofuels: A brief review. *Reviews in Environmental Science and Biotechnology, 10*, 31–41.

Mobin, S., & Alam, F., (2017). Some promising microalgal species for commercial applications: A review. *Energy Procedia, 110*, 510–517. doi: https://doi.org/10.1016/j.egypro.2017.03.177.

Molino, A., Iovine, A., Casella, P., Mehariya, S., Chianese, S., Cerbone, A., Rimauro, J., & Musmarra, D., (2018). Microalgae characterization for consolidated and new application in human food, animal feed and nutraceuticals. *International Journal of Environmental Research and Public Health, 15*(11). doi: 10.3390/ijerph15112436.

Molino, A., Mehariya, S., Di Sanzo, G., Larocca, V., Martino, M., Leone, G. P., Marino, T., et al., (2020). Recent developments in supercritical fluid extraction of bioactive compounds from microalgae: Role of key parameters, technological achievements, and challenges. *Journal of CO$_2$ Utilization, 36*, 196–209. doi: https://doi.org/10.1016/j.jcou.2019.11.014.

Moreno-Garrido, I., & Canavate, J. P., (2001). Assessing chemical compounds for controlling predator ciliates in outdoor mass cultures of the green algae *Dunaliella salina*. *Aquacultural Engineering,* 107–114. doi: 10.1016/S0144-8609(00)00067-4.

Mourelle, M., Gómez, C., & Legido, J., (2017). The potential use of marine microalgae and cyanobacteria in cosmetics and thalassotherapy. *Cosmetics, 4*(4), 46. doi: 10.3390/cosmetics4040046.

Mouritsen, O., Dawczynski, C., Duelund, L., Jahreis, G., Vetter, W., & Schröder, M., (2013). On the human consumption of the red seaweed dulse, *Palmaria palmata. Journal of Applied Phycology, 25*, 1777–1791 doi: 10.1007/s10811-013-0014-7.

Nasrabadi, M., & Razavi, S., (2010). Enhancement of canthaxanthin production from *Dietzia natronolimnaea* HS-1 in a fed-batch process using trace elements and statistical methods. *Brazilian Journal of Chemical Engineering, 27*, 517–529. doi: 10.1590/S0104-66322010000400003.

Nauroth, J. M., Liu, Y. C., Van, E. M., Bell, R., Hall, E. B., Chung, G., & Arterburn, L. M., (2010). Docosahexaenoic acid (DHA) and docosapentaenoic acid (DPAN-6) algal oils reduce inflammatory mediators in human peripheral mononuclear cells *in vitro* and paw edema *in vivo. Lipids, 45*(5), 375–384. doi: 10.1007/s11745-010-3406-3.

Navarro, F., Forján, E., Vázquez, M., Montero, L. Z., Bermejo, E., Castaño, M., Toimil, A., et al., (2016). Microalgae as a safe food source for animals: Nutritional characteristics of the acidophilic microalga *Coccomyxa onubensis. Food and Nutrition Research, 60*. doi: 10.3402/fnr.v60.30472.

Nethravathy, M. U., Mehar, J. G., Mudliar, S. N., & Shekh, A. Y., (2019). Recent advances in microalgal bioactives for food, feed, and healthcare products: Commercial potential, market space, and sustainability. *Comprehensive Reviews in Food Science and Food Safety, 18*(6), 1882–1897. doi: 10.1111/1541-4337.12500.

Nicoletti, M., (2016). Microalgae nutraceuticals. *Foods (Basel, Switzerland), 5*(3), 54. doi: 10.3390/foods5030054.

Nurra, C., Torras, C., Clavero, E., Ríos, S., Rey, M., Lorente, E., Farriol, X., & Salvadó, J., (2014). Biorefinery concept in a microalgae pilot plant. Culturing, dynamic filtration and steam explosion fractionation. *Bioresource Technology, 163*, 136–142. doi: https://doi.org/10.1016/j.biortech.2014.04.009.

Odjadjare, E. C., Mutanda, T., & Olaniran, A. O., (2017). Potential biotechnological application of microalgae: A critical review. *Critical Reviews in Biotechnology, 37*(1), 37–52. doi: 10.3109/07388551.2015.1108956.

Odjadjare, E., Mutanda, T., & Olaniran, A., (2015). Potential biotechnological application of microalgae: A critical review. *Critical Reviews in Biotechnology, 37*, 1–16.

Olaizola, M., (2003). Commercial development of microalgal biotechnology: From the test tube to the marketplace. *Biomolecular Engineering, 20*(4), 459–466. doi: https://doi.org/10.1016/S1389-0344(03)00076-5.

Oppen-Bezalel, L., & Shaish, A., (2009). Application of the colorless carotenoids, phytoene, and phytofluene in cosmetics, wellness, nutrition, and therapeutics. In: Ben-Amotz, A., (ed.), *The Alga Dunaliella* (pp. 423–444), CRC Press, Boca Raton.

Paliwal, C., Mitra, M., Bhayani, K., Bharadwaj, S. V. V., Ghosh, T., Dubey, S., & Mishra, S., (2017). Abiotic stresses as tools for metabolites in microalgae. *Bioresource Technology, 244*(2), 1216–1226. doi: 10.1016/j.biortech.2017.05.058.

Pelah, D., Sintov, A., & Cohen, E., (2004). The effect of salt stress on the production of canthaxanthin and astaxanthin by *Chlorella zofingiensis* grown under limited light intensity. *World Journal of Microbiology and Biotechnology, 20*(5), 483–486. doi: 10.1023/B:WIBI.0000040398.93103.21.

Piccaglia, R., Marotti, M., & Grandi, S., (1998). Lutein and lutein ester content in different types of *Tagetes patula* and *T. erecta*. *Industrial Crops and Products, 8*(1), 45–51.

Piironen, V., Lindsay, D. G., Miettinen, T. A., Toivo, J., & Lampi, A. M., (2000). Plant sterols: Biosynthesis, biological function, and their importance to human nutrition. *Journal of the Science of Food and Agriculture, 80*, 939–966.

Plaza, M., Santoyo, S., Jaime, L., García-Blairsy, R. G., Herrero, M., Señoráns, F. J., & Ibáñez, E., (2010). Screening for bioactive compounds from algae. *Journal of Pharmaceutical and Biomedical Analysis, 51*(2), 450–455. doi: https://doi.org/10.1016/j.jpba.2009.03.016.

Pouvreau, J. B., Morançais, M., Taran, F., Rosa, P., Dufossé, L., Guérard, F., Pin, S., Fleurence, J., & Pondaven, P., (2008). Antioxidant and free radical scavenging properties of marennine, a blue-green polyphenolic pigment from the diatom *Haslea ostrearia* (gaillon/bory) Simonsen responsible for the natural greening of cultured oysters. *Journal of Agricultural and Food Chemistry, 56*(15), 6278–6286. doi: 10.1021/jf073187n.

Prakash, D., Verma, S., Bhatia, R., & Tiwary, B., (2011). Risks and precautions of genetically modified organisms. *ISRN Ecology, 2011*, 369573. doi: 10.5402/2011/369573.

Pruvost, J., Le Gouic, B., Lepine, O., Legrand, J., & Le Borgne, F., (2016). Microalgae culture in building-integrated photobioreactors: Biomass production modeling and energetic analysis. *Chemical Engineering Journal, 284*, 850–861. doi: 10.1016/j.cej.2015.08.118.

Pulz, O., & Gross, W., (2004). Valuable products from biotechnology of microalgae. *Applied Microbiology and Biotechnology, 65*(6), 635–648. doi: 10.1007/s00253-004-1647-x.

Rafika, C., Trabelsi, L., Rym, B. D., Abed, O., Yahia, A., Khemissa, G., Ammar, J., et al., (2011). Evaluation of cytotoxicity and biological activities in extracellular polysaccharides released by cyanobacterium *Arthrospira platensis*. *Brazilian Archives of Biology and Technology, 54*, 831–838.

Rammuni, M. N., Ariyadasa, T. U., Nimarshana, P. H. V., & Attalage, R. A., (2019). Comparative assessment on the extraction of carotenoids from microalgal sources: Astaxanthin from *H. pluvialis* and β-carotene from *D. salina*. *Food Chemistry, 277*, 128–134. doi: https://doi.org/10.1016/j.foodchem.2018.10.066.

Raposo, M. F. D. J., De Morais, R. M. S. C., & Bernardo De, M. A. M. M., (2013). Bioactivity and applications of sulphated polysaccharides from marine microalgae. *Marine Drugs, 11*(1), 233–252. doi: 10.3390/md11010233.

Raposo, M. F., De Morais, A. M., & De Morais, R. M., (2014). Influence of sulphate on the composition and antibacterial and antiviral properties of the exopolysaccharide from *Porphyridium cruentum*. *Life Science, 101*(1, 2), 56–63. doi: 10.1016/j.lfs.2014.02.013.

Rasala, B. A., Muto, M., Lee, P. A., Jager, M., Cardoso, R. M. F., Behnke, C. A., Kirk, P., et al., (2010). Production of therapeutic proteins in algae, analysis of expression of seven human proteins in the chloroplast of *Chlamydomonas reinhardtii*. *Plant Biotechnology Journal, 8*(6), 719–733. doi: 10.1111/j.1467-7652.2010.00503.x.

Rasmussen, H. E., Blobaum, K. R., Park, Y. K., Ehlers, S. J., Lu, F., & Lee, J. Y., (2008). Lipid extract of *Nostoc commune* var. *Sphaeroides Kutzing*, a blue-green alga, inhibits the activation of sterol regulatory element-binding proteins in HepG2 cells. *Journal of Nutrition, 138*(3), 476–481. doi: 10.1093/jn/138.3.476.

Rasmussen, H. M., & Johnson, E. J., (2013). Nutrients for the aging eye. *Clinical Interventions in Aging, 8*, 741–748. doi: 10.2147/cia.s45399.

Rath, B., (2012). Commercial and industrial applications of microalgae: A review. *Journal of Algal Biomass Utilization, 3*, 89–100.

Richardson, J. W., Johnson, M. D., Zhang, X., Zemke, P., Chen, W., & Hu, Q., (2014). A financial assessment of two alternative cultivation systems and their contributions to algae biofuel economic viability. *Algal Research, 4,* 96–104. doi: https://doi.org/10.1016/j.algal.2013.12.003.

Roberts, J. E., & Dennison, J., (2015). The photobiology of lutein and zeaxanthin in the eye. *Journal of Ophthalmology, 2015,* 687173. doi: 10.1155/2015/687173.

Roh, M. K., Uddin, M. S., & Chun, B. S., (2008). Extraction of fucoxanthin and polyphenol from *Undaria pinnatifida* using supercritical carbon dioxide with co-solvent. *Biotechnology and Bioprocess Engineering, 13*(6), 724–729. doi: 10.1007/s12257-008-0104-6.

Rojas-Franco, P., Franco-Colín, M., Camargo, M. E. M., Carmona, M. M. E., Ortíz-Butrón, M. D. R. E., Blas-Valdivia, V., & Cano-Europa, E., (2018). Phycobiliproteins and phycocyanin of *Arthrospira maxima* (*Spirulina*) reduce apoptosis promoters and glomerular dysfunction in mercury-related acute kidney injury. *Toxicology Research and Application, 2.* doi: 10.1177/2397847318805070.

Sadovskaya, I., Souissi, A., Souissi, S., Grard, T., Lencel, P., Greene, C. M., Duin, S., et al., (2014). Chemical structure and biological activity of a highly branched (1→3,1→6)-β-D-glucan from *Isochrysis galbana*. *Carbohydrate Polymers, 111,* 139–148.

Sánchez, J. F., Fernández-Sevilla, J. M., Acién, F. G., Cerón, M. C., Pérez-Parra, J., & Molina-Grima, E., (2008). Biomass and lutein productivity of *Scenedesmus almeriensis*: Influence of irradiance, dilution rate and temperature. *Applied Microbiology and Biotechnology, 79*(5), 719–729.

Santoyo, S., Jaime, L., Plaza, M., Herrero, M., Rodriguez-Meizoso, I., Ibáñez, E., & Reglero, G., (2012). Antiviral compounds obtained from microalgae commonly used as carotenoid sources. *Journal of Applied Phycology, 24,* 731–741.

Sarma, S. S. S., Jurado, L., Susana, P., & Nandini, S., (2001). Effect of three food types on the population growth of *Brachionus calyciflorus* and *Brachionus patulus* (Rotifera: Brachionidae). *Revista de Biología Tropical, 49,* 77–84.

Sathasivam, R., Radhakrishnan, R., Hashem, A., & Abd Allah, E. F., (2019). Microalgae metabolites: A rich source for food and medicine. *Saudi Journal of Biological Sciences, 26*(4), 709–722. doi: https://doi.org/10.1016/j.sjbs.2017.11.003.

Schmidt, I., Schewe, H., Gassel, S., Jin, C., Buckingham, J., Hümbelin, M., Sandmann, G., & Schrader, J., (2010). Biotechnological production of astaxanthin with *Phaffia rhodozyma/ Xanthophyllomyces dendrorhous*. *Applied Microbiology and Biotechnology, 89,* 555–571.

Seddon, J. M., Ajani, U. A., Sperduto, R. D., Hiller, R., Blair, N., Burton, T. C., Farber, M. D., et al., (1994). Dietary carotenoids, vitamins A, C, and E, and advanced age-related macular degeneration. *The Journal of the American Medical Association, 272*(18), 1413–1420. doi: 10.1001/jama.1994.03520180037032.

Sharma, N. K., & Rai, A. K., (2008). Allergenicity of airborne cyanobacteria *Phormidium fragile* and *Nostoc muscorum*. *Ecotoxicology and Environmental Safety, 69*(1), 158–162. doi: https://doi.org/10.1016/j.ecoenv.2006.08.006.

Sheu, J. H., Wang, G. H., Sung, P. J., & Duh, C. Y., (1999). New cytotoxic oxygenated fucosterols from the brown alga *Turbinaria conoides*. *Journal of Natural Products, 62*(2), 224–227. doi: 10.1021/np980233s.

Shin, W., Boo, S. M., & Longcore, J., (2011). *Entophlyctis apiculata*, a chytrid parasite of *Chlamydomonas* sp. (Chlorophyceae). *Canadian Journal of Botany, 79,* 1083–1089. doi: 10.1139/b01-086.

Singh, D., Gupta, A., Wilkens, S. L., Mathur, A. S., Tuli, D. K., Barrow, C. J., & Puri, M., (2015). Understanding response surface optimization to the modeling of Astaxanthin extraction from a novel strain *Thraustochytrium* sp. S7. *Algal Research, 11*, 113–120. doi: https://doi.org/10.1016/j.algal.2015.06.005.

Singh, R., Parihar, P., Singh, M., Bajguz, A., Kumar, J., Singh, S., Singh, V. P., & Prasad, S. M., (2017). Uncovering potential applications of cyanobacteria and algal metabolites in biology, agriculture, and medicine: Current status and future prospects. *Frontiers in Microbiology, 8*, 515. doi: 10.3389/fmicb.2017.00515.

Skjånes, K., Rebours, C., & Lindblad, P., (2013). Potential for green microalgae to produce hydrogen, pharmaceuticals, and other high value products in a combined process. *Critical Reviews in Biotechnology, 33*(2), 172–215. doi: 10.3109/07388551. 2012.681625.

Smee, D. F., Bailey, K. W., Wong, M. H., O'Keefe, B. R., Gustafson, K. R., Mishin, V. P., & Gubareva, L. V., (2008). Treatment of influenza A (H1N1) virus infections in mice and ferrets with cyanovirin-N. *Antiviral Research, 80*(3), 266–271. doi: 10.1016/j. antiviral.2008.06.003.

Solovchenko, A. E., Khozin-Goldberg, I., Didi-Cohen, S., Cohen, Z., & Merzlyak, M. N., (2008). Effects of light intensity and nitrogen starvation on growth, total fatty acids, and arachidonic acid in the green microalga *Parietochloris incisa*. *Journal of Applied Phycology, 20*(3), 245–251.

Soria-Guerra, R. E., Ramírez-Alonso, J. I., Ibáñez-Salazar, A., Govea-Alonso, D. O., Paz-Maldonado, L. M. T., Bañuelos-Hernández, B., Korban, S. S., & Rosales-Mendoza, S., (2014). Expression of an HBcAg-based antigen carrying angiotensin II in *Chlamydomonas reinhardtii* as a candidate hypertension vaccine. *Plant Cell, Tissue, and Organ Culture, 116*(2), 133–139. doi: 10.1007/s11240-013-0388-x.

Souza, P. O., Ferreira, L. R., Pires, N. R. X., Filho, P. J. S., Duarte, F. A., Pereira, C. M. P., & Mesko, M. F., (2012). Algae of economic importance that accumulate cadmium and lead: A review. *Revista Brasileira de Farmacognosia, 22*(4), 825–837. doi: 10.1590/S0102-695X2012005000076.

Specht, E. A., & Mayfield, S. P., (2014). Algae-based oral recombinant vaccines. *Frontiers in Microbiology, 5*, 60. doi: 10.3389/fmicb.2014.00060.

Spolaore, P., Joannis-Cassan, C., Duran, E., & Isambert, A., (2006). Commercial applications of microalgae. *Journal of Bioscience and Bioengineering, 101*(2), 87–96. doi: https://doi.org/10.1263/jbb.101.87.

Staats, N., De Winder, B., Stal, L., & Mur, L., (1999). Isolation and characterization of extracellular polysaccharides from the epipelic diatoms *Cylindrotheca closterium* and *Navicula salinarum*. *European Journal of Phycology, 34*(2), 161–169. doi: 10.1080/09670269910001736212.

Steinrücken, P., Erga, S. R., Mjøs, S. A., Kleivdal, H., & Prestegard, S. K., (2017). Bioprospecting North Atlantic microalgae with fast growth and high polyunsaturated fatty acid (PUFA) content for microalgae-based technologies. *Algal Research, 26*, 392–401. doi: https://doi.org/10.1016/j.algal.2017.07.030.

Stephenson, A. L., Kazamia, E., Dennis, J. S., Howe, C. J., Scott, S. A., & Smith, A. G., (2010). Life-cycle assessment of potential algal biodiesel production in the United Kingdom: A comparison of raceways and air-lift tubular bioreactors. *Energy and Fuels, 24*(7), 4062–4077. doi: 10.1021/ef1003123.

Streekstra, H., (2010). Arachidonic acid: Fermentative production by *Mortierella* fungi. In: Cohen, Z., & Ratledge, C., (eds.), *Single Cell Oils* (pp. 97–114). AOCS Press, Champaign.

Sun, M., Qian, K., Su, N., Chang, H., Liu, J., & Shen, G., (2003). Foot-and-mouth disease virus VP1 protein fused with *cholera* toxin B subunit expressed in *Chlamydomonas reinhardtii* chloroplast. *Biotechnology Letters, 25*(13), 1087–1092. doi10.1023/A: 1024140114505.

Surai, A. P., Surai, P. F., Steinberg, W., Wakeman, W. G., Speake, B. K., & Sparks, N. H., (2003). Effect of canthaxanthin content of the maternal diet on the antioxidant system of the developing chick. *British Poultry Science, 44*(4), 612–619. doi: 10.1080/00071660310001616200.

Surzycki, R., Greenham, K., Kitayama, K., Dibal, F., Wagner, R., Rochaix, J. D., Ajam, T., & Surzycki, S., (2009). Factors effecting expression of vaccines in microalgae. *Biologicals, 37*(3), 133–138. doi: https://doi.org/10.1016/j.biologicals.2009.02.005.

Szabo, N. J., Matulka, R. A., Kiss, L., & Licari, P., (2012). Safety evaluation of a high lipid Whole Algalin Flour (WAF) from *Chlorella protothecoides*. *Regulatory Toxicology and Pharmacology, 63*(1), 155–165. doi: https://doi.org/10.1016/j.yrtph. 2012.03.011.

Takahashi, K., Hosokawa, M., Kasajima, H., Hatanaka, K., Kudo, K., Shimoyama, N., & Miyashita, K., (2015). Anticancer effects of fucoxanthin and fucoxanthinol on colorectal cancer cell lines and colorectal cancer tissues. *Oncology Letters, 10*(3), 1463–1467. doi: 10.3892/ol.2015.3380.

Takebe, Y., Saucedo, C. J., Lund, G., Uenishi, R., Hase, S., Tsuchiura, T., Kneteman, N., et al., (2013). Antiviral lectins from red and blue-green algae show potent *in vitro* and *in vivo* activity against hepatitis C virus. *PLoS One, 8*(5), e64449. doi: 10.1371/journal.pone.0064449.

Talukdar, J., (2020). COVID-19: Potential of microalgae derived natural astaxanthin as adjunctive supplement in alleviating cytokine storm. *OSF Preprints*. doi: 10.2139/ssrn.3579738.

Tiberg, E., & Einarsson, R., (1989). Variability of allergenicity in eight strains of the green algal genus *Chlorella*. *International Archives of Allergy and Immunology, 90*(3), 301–306. doi: 10.1159/000235042.

Tran, M., Van, C., Barrera, D. J., Pettersson, P. L., Peinado, C. D., Bui, J., & Mayfield, S. P., (2013). Production of unique immunotoxin cancer therapeutics in algal chloroplasts. *Proceedings of the National Academy of Sciences of the United States of America, 110*(1), E15–E22. doi: 10.1073/pnas.1214638110.

Tran, M., Zhou, B., Pettersson, P. L., Gonzalez, M. J., & Mayfield, S. P., (2009). Synthesis and assembly of a full-length human monoclonal antibody in algal chloroplasts. *Biotechnology and Bioengineering, 104*(4), 663–673. doi: 10.1002/bit.22446.

Udayan, A., Arumugam, M., & Pandey, A., (2017). Nutraceuticals from algae and cyanobacteria. In: Rastogi, R. P., Madamwar, D., & Pandey, A., (eds.), *Algal Green Chemistry* (pp. 65–89). Elsevier Publications, Amsterdam.

Van, B. V. A., Spenkelink, B., Mooibroek, H., Sijtsma, L., Bosch, D., Rietjens, I. M. C. M., & Alink, G. M., (2009). An n-3 PUFA-rich microalgal oil diet protects to a similar extent as a fish oil-rich diet against AOM-induced colonic aberrant crypt foci in F344 rats. *Food and Chemical Toxicology, 47*(2), 316–320. doi: https://doi.org/10.1016/j.fct.2008.11.014.

Van, D. S. M., Noordam, M. Y., & Fels-Klerx, H. J. V. D., (2013). Safety of novel protein sources (insects, microalgae, seaweed, duckweed, and rapeseed) and legislative aspects for their application in food and feed production. *Comprehensive Reviews in Food Science and Food Safety, 12*(6), 662–678. doi: 10.1111/1541-4337.12032.

Vigani, M., Parisi, C., Rodriguez-Cerezo, E., Barbosa, M. J., Sijtsma, L., Ploeg, M., & Enzing, C., (2015). Food and feed products from micro-algae: Market opportunities and challenges for the EU. *Trends in Food Science and Technology, 42*. doi: 10.1016/j.tifs.2014.12.004.

Vo, T. S., & Kim, S. K., (2010). Potential anti-HIV agents from marine resources: An overview. *Marine Drugs, 8*(12), 2871–2892.

Volkman, J., (2003). Sterols in microorganisms. *Applied Microbiology and Biotechnology, 60*(5), 495–506.

Wang, X., & Zhang, X., (2013). Separation, antitumor activities, and encapsulation of polypeptide from *Chlorella pyrenoidosa. Biotechnology Progress, 29.*

Wang, X., Brandsma, M., Tremblay, R., Maxwell, D., Jevnikar, A. M., Huner, N., & Ma, S., (2008). A novel expression platform for the production of diabetes-associated autoantigen human glutamic acid decarboxylase (hGAD65). *BMC Biotechnology, 8*(1), 87. doi: 10.1186/1472-6750-8-87.

Watanabe, F., Takenaka, S., Kittaka-Katsura, H., Ebara, S., & Miyamoto, E., (2002). Characterization and bioavailability of vitamin B12-compounds from edible algae. *Journal of Nutritional Science and Vitaminology, 48*(5), 325–331. doi: 10.3177/jnsv.48.325.

Wijffels, R. H., Barbosa, M. J., & Eppink, M. H. M., (2010). Microalgae for the production of bulk chemicals and biofuels. *Biofuels, Bioproducts and Biorefining, 4*(3), 287–295. doi: 10.1002/bbb.215.

Williams, P. J. I. B., & Laurens, L. M. L., (2010). Microalgae as biodiesel and biomass feedstocks: Review and analysis of the biochemistry, energetics, and economics. *Energy and Environmental Science, 3*(5), 554–590. doi: 10.1039/B924978H.

Wong, M. H., Hung, K. M., & Chiu, S. T., (1996). Sludge-grown algae for culturing aquatic organisms, Part II. Sludge-grown algae as feeds for aquatic organisms. *Environmental Management, 20*(3), 375–384. doi: 10.1007/BF01203845.

Yaakob, Z., Ali, E., Zainal, A., Mohammad, M., & Takriff, M., (2014). An overview: Biomolecules from microalgae for animal feed and aquaculture. *Journal of Biological Research (Thessalonikē, Greece), 21*, 6.

Yao, J., Weng, Y., Dickey, A., & Wang, K. Y., (2015). Plants as factories for human pharmaceuticals: Applications and challenges. *International Journal of Molecular Sciences, 16*(12), 28549–28565. doi: 10.3390/ijms161226122.

Yasukawa, K., Akihisa, T., Kanno, H., Kaminaga, T., Izumida, M., Sakoh, T., Tamura, T., & Takido, M., (1996). Inhibitory effects of sterols isolated from *Chlorella vulgaris* on 12-0-tetradecanoylphorbol-13-acetate-induced inflammation and tumor promotion in mouse skin. *Biological and Pharmaceutical Bulletin, 19*(4), 573–576. doi: 10.1248/bpb.19.573.

Yim, J. H., Kim, S. J., Ahn, S. H., & Lee, H. K., (2007). Characterization of a novel bioflocculant, p-KG03, from a marine dinoflagellate, *Gyrodinium impudicum* KG03. *Bioresource Technology, 98*(2), 361–367. doi: 10.1016/j.biortech.2005.12.021.

Yuan, J. P., Peng, J., Yin, K., & Wang, J. H., (2011). Potential health-promoting effects of astaxanthin: A high-value carotenoid mostly from microalgae. *Molecular Nutrition and Food Research, 55*(1), 150–165. doi: 10.1002/mnfr.201000414.

Yusuf, C., (2013). Raceways-based production of algal crude oil. *Green, 3*(3, 4), 195–216. doi: https://doi.org/10.1515/green-2013-0018.

Zaid, A. A. A., Hammad, D. M., & Sharaf, E. M., (2015). Antioxidant and anticancer activity of *Spirulina platensis* water extracts. *International Journal of Pharmacology, 11*(7), 846–851. doi: 10.3923/ijp.2015.846.851.

Zhao, L., & Sweet, B. V., (2008). Lutein and zeaxanthin for macular degeneration. *American Journal of Health-System Pharmacy, 65*(13), 1232–1238. doi: 10.2146/ahjp080052.

CHAPTER 10

ANTICANCER COMPOUNDS FROM FRESHWATER MICROALGAE

CLARA MARTINS,[1] SAMUEL SILVESTRE,[2,3] and LÍLIA SANTOS[1]

[1]*Coimbra Collection of Algae (ACOI), Department of Life Sciences, University of Coimbra, Coimbra–3000-456, Portugal*

[2]*CICS-UBI-Health Sciences Research Center, University of Beira Interior, Av. Infante D. Henrique, Covilhã–6200-506, Portugal*

[3]*CNC-Center for Neuroscience and Cell Biology, University of Coimbra, Coimbra–3004-517, Portugal*

ABSTRACT

Cancer is a global health problem causing approximately 9.6 million deaths each year. Conventional chemotherapy has known side effects for the patient, and there is a constant search for new and less aggressive approaches. The use of natural compounds with anticancer activity and low toxicity has been regarded as another possibility for the combat of cancer. Microalgae are currently seen as a source of bioactive compounds with many different chemical structures and physiological functions in the cell. Bioactivity studies have focused more on marine microalgae, and there are fewer studies about freshwater microalgae. In this context, it was reported that some microalgal compounds, namely pigments, peptides, phytosterols, and polysaccharides demonstrated anticancer activity. In recent years, nanoparticle technology has been developed for incorporation and successful delivery of bioactive compounds, allowing the treatment of several diseases. Thus, microalgae anticancer compounds-based nanocarriers may be regarded as a promising possibility for the development of safer anticancer therapies. In this chapter, the potential anticancer effect of freshwater microalgae extracts and isolated compounds are developed. In addition, antitumoral nanodelivery systems involving microalgae products are also explored.

10.1   INTRODUCTION

Microalgae are photosynthetic microorganisms, prokaryotes, or eukaryotes, which are able to survive and colonize several environments with extreme conditions (Spolaore et al., 2006; Amaro et al., 2013). These microorganisms survive in almost all biotopes due to their capacity to modulate metabolism. In stress conditions, microalgae produce secondary metabolites that allow them to tolerate these disadvantageous conditions (Olaizola, 2003; Amaro et al., 2013; Martínez Andrade et al., 2018). A variety of compounds produced by both primary and secondary metabolisms of microalgae have been studied due to their unique chemical characteristics, their potential function in nature, and in the use for human benefit (Romano et al., 2017). Microalgal compounds such as pigments, fatty acids, peptides, or polysaccharides (Amaro et al., 2013; Da Silva Vaz et al., 2016) have demonstrated bioactivity effects with applications in the feed, food, nutritional, cosmetic, and pharmaceutical industries (Pulz et al., 2004; Shannon and Abu-Ghannam, 2016). Microalgae are natural sources to discover new interesting compounds or to produce known products at lower costs. In fact, for some microalgae species that can grow on a large scale, it is possible to improve the yield of interesting bioactive compounds by optimization of the culture conditions (Amaro et al., 2013; Borowitzka, 2013; Dewi et al., 2018).

There is an enormous diversity of microalgae, with more than 30,000 species already described. Of these, only a few 1000 strains are kept in collections, a few 100 are under research for biotechnological purposes, and a handful is cultivated in industry (Mobin and Alam, 2017). However, microalgae remain largely an unexplored source for drug discovery.

Cancer is a global health problem causing approximately 9.6 million deaths worldwide, being the second cause of death in the world (World Health Organization, 2018). Cancer includes a group of pathologies related to the uncontrolled proliferation of cells in the body, existing more than 200 known different types of cancers. According to the World Health Organization (WHO), the most common types of cancers in men are lung, prostate, colorectal, stomach, and liver cancers, while breast, colorectal, lung, cervix, and thyroid cancers are the most common among women. It is estimated that by 2030 the incidence of cancer will be reaching to over 21.7 million new cases and 13 million deaths (American Cancer Society, 2018). The majority of anticancer treatments induce DNA damage leading to the death of both healthy and proliferating cancer cells. Therefore, new compounds more effective and with less deleterious side effects are urgently needed. In this context, an agent that can induce apoptosis in cancer without side effects is a good anticancer agent (Azamai et al., 2009).

Due to an enormous capacity to adapt to a wide range of habitats and extreme conditions, microalgae are considered a promising source to discover drug compounds (Amaro et al., 2013; Alves et al., 2018). The anticancer activity of some compounds from microalgae may be due to their capacity to cross the lipophilic membranes and interact with proteins involved in apoptosis (Reyna-Martinez et al., 2018). Further, some microalgae compounds can also lead to DNA-dependent DNA polymerases inhibition or alterations in cyclins expression. In order to evaluate the possible effective application of these biocompounds in cancer treatment, it is important to know if they do not represent a threat to normal cells and if their mechanism of action in cancer cells is via necrosis or apoptosis. In this context, it is known that apoptosis or necrosis can occur independently, sequentially, and/or simultaneously and can affect the tumor growth in one or more stages of carcinogenesis (Dewi et al., 2018).

So far, the search for bioactivity, namely anticancer activity, has occurred mainly in marine microalgae and cyanobacteria, and several studies mention the existence of compounds with anticancer activity in these two groups of microorganisms (Costa et al., 2012; Alves et al., 2018). However, interesting compounds with possible applications in several areas can also be found in freshwater microalgae (Soares et al., 2019). The demand for bioactive compounds from freshwater microalgae has increased with the need for new, more effective, and safer compounds for the treatment of cancer. Several reports of microalgae with anticancer activity have been published in recent years (Table 10.1) and provide essential information that could lead to future research and a possible incorporation of compounds from this source into anticancer medicines. Furthermore, it is already known that the different cultivation conditions (e.g., nutrient availability, temperature, light intensity) and growth phase may influence the production of bioactive compounds, as has been reported in some studies (Table 10.2).

In this chapter, the anticancer effect against several cancer cell lines of freshwater microalgae extracts and isolated compounds are discussed. In addition, the use of microalgae products in nanodelivery systems with antitumoral activity is also developed.

TABLE 10.1 Microalgae Species, Anticancer Compounds, and Affected Cell Lines

Microalgae	Compound or Fraction	Target Cells	Mechanism of Action	References
Botryidiopsidaceae species	Ethanol extract	A375, Hs578T, HCT116, HeLa	Decrease of expression of anti-apoptotic Bcl-2 protein; increase of apoptosis-inducing genes caspase-3 and p53	Suh et al. (2017)
Chlamydomonas reinhardtii	Ethanol-extracted sulfated polysaccharides	MDA-MB-231	Decrease in the number of colonies; structural damage of cellular microtubule network; induction of apoptosis	Kamble et al. (2018)
Chlorophyta strains from three genera (*Desmococcus*, *Chlorella*, and *Scenedesmus*)	Ethanol extracts	MCF-7, CEM, G361	–	Ördög et al. (2004)
Chlorella ellipsoidea	Lutein	HCT116	Antiproliferation via apoptosis induction	Cha et al. (2008)
Chlorella pyrenoidosa	Polysaccharide fractions	A549	–	Sheng et al. (2007)
	Isolated polypeptide CPAP	HepG2	Apoptosis; necrotic death	Wang and Zhang (2013)
Chlorella sorokiniana	Water extracts	A549, CL1-5	Caspase-dependent mitochondrial dysfunction	Lin et al. (2017)
Chlorella sorokiniana, Scenedesmus sp.	Methanol extracts	Murine L5178Y-R lymphoma cell line	Apoptosis via activation of caspases	Reyna-Martínez et al. (2018)

TABLE 10.1 (Continued)

Microalgae	Compound or Fraction	Target Cells	Mechanism of Action	References
Chlorella vulgaris	Ergosterol peroxide	12-O-tetradecanoylphorbol-13-acetate (TPA)-induced inflammation and tumor promotion in mouse skin	—	Yasukawa et al. (1996)
	Aqueous ethanol extracts	H1299, A549, H1437	—	Wang et al. (2010)
	Hot aqueous extract	HepG2	Inducing apoptosis signaling cascades via expression of p53, Bax, and caspase-3 proteins	Yusof et al. (2010)
	Peptide VECYGPNRPQF	AGS	Dose-dependent antiproliferation activities and induction of post-G1 cell cycle arrest in AGS cells	Sheih et al. (2010)
	Water extracts	EACC, HepG2	—	Shanab et al. (2012)
	Peptides from a hydrolyzed protein	MCF-7	—	Sedighi et al. (2016)
Chlorella sp., Scenedesmus sp.	Water extracts	A549, MCF-7, MDA, MB-435, LNCaP	—	Jabeen et al. (2017)
Chlorella pyrenoidosa, Scenedesmus sp., Chlorococcum sp.	Exopolysaccharides	HCT116, HCT8	—	Zhang et al. (2019)
Graesiella sp.	Aqueous extracellular polysaccharides	HepG2, Caco-2	—	Trabelsi et al. (2016)

TABLE 10.1 *(Continued)*

Microalgae	Compound or Fraction	Target Cells	Mechanism of Action	References
Haematococcus pluvialis	Astaxanthin	HCT116, SW480, HT-29, WiDr; LS-174	Apoptosis induction via cyclin D1 expression decrease; p53 increase and some cyclin kinase inhibition	Palozza et al. (2009)
	Astaxanthin	HepG2	Arrest of the cells in G2/M phase of the cell cycle; apoptosis induction	Nagaraj et al. (2012)
	Carotenoid fractions	HepG2, MCF-7, HCT116, A549	–	El-Baz et al. (2018)
Navicula incerta	Stigmasterol	HepG2	DNA fragmentation; Arrest of the cell cycle; Apoptosis	Kim et al. (2014)
15 strains (6 chlorophytes, 8 diatoms and 1 eustigmatophyte)	Methanol/chloroform extracts	KB, HepG2, SK-LU-1, MCF-7	–	Hoa et al. (2011)
Scenedesmus obliquus	Diethyl ether extracts	HepG2, HCT116, MCF-7	–	Marrez et al. (2019)

TABLE 10.2 Active Microalgal Species, Sources, and Culturing Conditions

Microalgae	Source	Culturing Conditions	Harvesting Time	References
Botryidiopsidaceae sp.	King Sejong Station	–	–	Suh et al. (2017)
Chlamydomonas reinhardtii	Chlamydomonas Genetic Center, Duke University, USA	Tris-acetate phosphate medium (pH 7); continuous light intensity of 300 μmol photons m^{-2} s^{-1}	–	Kamble et al. (2018)
Chlorophyta strains from three genera (*Desmococcus, Chlorella,* and *Scenedesmus*)	Mosonmagyaróvár algal culture collection (MACC)	Batch culture; conditions not provided	Early stationary phase of growth	Ördög et al. (2004)
Chlorella sorokiniana, Scenedesmus sp.	San Juan River, Cadereyta	LC liquid medium; continuous artificial light (1,000 lux); 25°C	Green growth	Reyna-Martinez et al. (2018)
Chlorella vulgaris	University of Malaya Algae Culture Collection (UMACC, Malaysia)	Bold's Basal medium; 12 h:12 h light cycle	–	Yusof et al. (2010)
	Microbiology Department, Soils, Water, and Environment Research Institute (SWERI), Agricultural Research Center (ARC)	Bold medium; continuous shaking (150 rpm); light intensity of 40 μE m^{-2} s^{-1}; 25°C	30 days of growth	Shanab et al. (2012)
	Agricultural Biotechnology Research Institute of Iran (ABRII)	3N-BBM/vit medium	Mid-log growth phase	Sedighi et al. (2016)
Graesiella sp.	Ain Echffa, a hot spring, Tunisia	Bold's basal medium (pH 6.8); light intensity of 20 μmol photons m^{-2} s^{-1}; 16 h: 8 h light cycle; 40°C	–	Trabelsi et al. (2016)

TABLE 10.2 *(Continued)*

Microalgae	Source	Culturing Conditions	Harvesting Time	References
Haematococcus pluvialis	Culture Collection of Algae, Center for Advanced Studies in Botany, University of Madras, Chennai	Bold's Basal medium (pH 7.5); 18 h: 6 h light cycle; 23–25°C	–	Nagaraj et al. (2012)
	River Nile	BG11 medium; Continuous aeration; Light intensity of ≈2,500 lux; 22±3°C	After 10 days of growth	El-Baz et al. (2018)

10.2 ANTICANCER EXTRACTS FROM FRESHWATER MICROALGAE

10.2.1 AQUEOUS EXTRACTS FROM CHLORELLA AND SCENEDESMUS GENERA

Various authors have focused on the screening of anticancer activity of microalgae from Chlorophyta phylum, namely *Chlorella* and *Scenedesmus* genera since they have demonstrated interesting bioactivities. Yusof et al. (2010) and Shanab et al. (2012) studied the cytotoxicity of hot aqueous and aqueous extracts of *Chlorella vulgaris*, respectively, on a human hepatoma cancer cell line (HepG2) and in Ehrlich ascites carcinoma cells (EACC). In the study of Yusof et al. (2010), it was verified that after 24 hours of incubation, the extract inhibited the proliferation of HepG2 cells in a concentration-dependent manner, ranging from 0 to 4 mg ml⁻¹, and the determined IC_{50} value was 1.6 mg ml⁻¹. In the work of Shanab et al. (2012) it was observed that *C. vulgaris* extracts at a concentration of 100 µg ml⁻¹ led to more than 30% inhibition and 60% inhibition of EACC and HepG2 cells proliferation, respectively, after 72 hours of treatment. In addition, Yusof et al. (2010) evaluated the action mechanisms, showing that their *C. vulgaris* extract induced a high apoptotic rate (70%), an increased expression of pro-apoptotic proteins p53, Bax, and caspase-3, and decreased the expression of the anti-apoptotic protein Bcl-2 in HepG2 cells, which subsequently led to increased DNA damage and apoptosis. The cytotoxic effects of *Chlorella sorokiniana* on two human non-small cell lung cancer cells, A549 and CL1-5 human lung adenocarcinoma cells, and its effects on tumor growth in a subcutaneous xenograft tumor model was evaluated by Lin et al. (2017). Interestingly, after 24 hours of exposure to an aqueous extract of *C. sorokiniana* in concentrations between 15.63 and 1,000 ng ml⁻¹, a concentration-dependent reduction in cell viability of the two cancer cell lines was observed. The authors demonstrated that the mechanism through which this *C. sorokiniana* extract induced the apoptosis of human non-small cell lung cancer cells was a caspase-dependent mitochondrial dysfunction. The antitumor activity *in vivo* was also analyzed. For this, 10 days after inoculation of tumoral cells in mice, when the tumor volume was 50 mm³, *C. sorokiniana* extract dissolved in distilled H_2O was administered in a daily dose of 50 mg kg⁻¹ body weight. The results proved that this *C. sorokiniana* extract at 50 mg kg⁻¹ reduced the tumor volume. Thus, the authors concluded that the *in vivo* growth of human non-small cell lung cancer cells was markedly inhibited by oral administration of *C. sorokiniana*. Also, the cytotoxicity of aqueous extracts of *Chlorella* sp. and *Scenedesmus* sp. on four human cancer cell lines, specifically A549, human breast adenocarcinoma (MCF-7), human melanoma (MDA MB-435) and human prostate cancer cells derived from a metastatic site in the lymph node (LNCaP), was evaluated by Jabeen et al. (2017). The effects of these extracts were compared with the observed in the human BPH-1 cell line (non-tumoral human prostate epithelial cells). Extracts were tested in concentrations of 5 and 10 mg ml^{-1} and those treated with a cellulase before extraction showed better toxicity against LNCaP cells (viability of 18% and 10% after treatment with *Chlorella* sp. and *Scenedesmus* sp. extracts, respectively). The efficacy of these extracts against the remaining three cancer cell lines was lower. The authors also treated the extracts with a lysozyme followed by dialysis. *Chlorella* sp. extracts showed the highest efficacy against A549 and LNCaP cell lines (viability of 4 and 6%, respectively). However, its effect on BPH-1 cell line was also significant (viability of 8%). The extract of *Scenedesmus* sp. demonstrated good toxicity against the two tumor cell lines (viability of 45% against A549 and 29% against LNCaP), without toxicity to non-tumoral human cells (viability of 88%). This work also evidenced that the activity of these microalgae may be selective to cancer cells versus non-tumoral cells, and higher effects were observed in certain cancer types.

10.2.2   ORGANIC EXTRACTS

10.2.2.1   ETHANOL AND AQUEOUS ETHANOL EXTRACTS

The anticancer activity of 10 Chlorophyta strains from genera *Desmococcus*, *Chlorella*, and *Scenedesmus* against MCF-7, CEM (human lymphoblastoid leukemia) and G361 (human malignant melanoma) cancer cell lines was studied by Ördög et al. (2004). For this, six concentrations of ethanol extracts were tested on cell lines for 3 days. These algal extracts inhibited the growth of all tumor cell lines with IC$_{50}$ determined values ranging between 0.06 and 1.65 mg well^{-1}. Also, the anticancer activity of aqueous ethanol extracts from *C. vulgaris* in three human lung cancer cell lines, H1299, A549, and H1437 was investigated by Wang et al. (2010). These authors verified that after 24 hours of treatment, the proliferation of the three cancer cell lines was inhibited by the extract of *C. vulgaris* in a dose-dependent manner, in concentrations ranging from 20 to 200 µg ml^{-1}. Furthermore, a higher concentration of extract (200 µg ml^{-1}) caused a significant inhibitory effect on migration of all lung cancer cell lines tested. Also, the ethanol extract from Antarctic freshwater microalga Botrydiopsidaceae species led to a selective cytotoxicity to different cancer cell lines [a human cervical cancer cell line (HeLa), a human colon cancer cell line (HCT116), a human breast cancer cell line (Hs578T) and a human melanoma cell line (A375)] (Suh et al., 2017). In this work, the four cancer cell lines were treated with an ethanol extract of Botrydiopsidaceae species in different concentrations between 0.8 and 50 µg ml^{-1} for 24 hours and the cell proliferation was measured for 72 hours. Interestingly, the cell proliferation of treated HeLa and Hs578T cells was concentration-dependently inhibited at $\geq$ 6.25 µg ml^{-1}. In addition, it was observed that HeLa cells were more sensitive to these ethanol extracts from Botrydiopsidaceae species than the other cancer cell lines. This result indicated that this extract exhibited selective cytotoxicity to different cancer cell lines. Furthermore, the cellular viability of all tested cancer cell lines was significantly decreased in a time-dependent manner after the treatment with an extract concentration of 12.5 µg ml^{-1}. Microscopy observations of the treated cancer cells showed changes in cell morphology evidencing cell death. The ethanol extract of Botrydiopsidaceae species induced cellular apoptosis through the modulation of apoptotic genes (p53, Bcl-2 and caspase-3 gene) whereas decreased the anti-apoptotic Bcl-1 protein in a concentration-dependent manner. These data suggest that the ethanol extract of Botrydiopsidaceae species enhances apoptosis in cancer cells by activating caspase-3 via the apoptotic pathway.

10.2.2.2   METHANOL/CHLOROFORM EXTRACTS

The anticancer activity of methanol/chloroform extracts of 15 microalgal strains from 11 different genera, including 6 chlorophytes, 8 diatoms and 1 eustigmatophyte was assessed against human epidermic carcinoma (KB), HepG2, human lung carcinoma (SK-LU-1) and MCF-7 cell lines by Hoa et al. (2011). The authors tested extracts of methanol/chloroform (1:1, v/v) in different concentrations (128, 32, 8, 2, and 0.5 µg ml^{-1}). After 3 days of treatment, *Ankistrodesmus gracilis* and *Amphiprora alata* showed the strongest inhibition of KB cells with calculated IC$_{50}$ values of 26.50 and 29.82 µg ml^{-1}, respectively. *A. gracilis* also exhibited the strongest activity against HepG2 cells (IC$_{50}$ = 9.64 µg ml^{-1}). Taking into account these results, *A. gracilis* may have potential biocompounds of interest that would be worth of further characterization.

10.2.2.3 METHANOL EXTRACTS

The antitumor activity of methanol extracts of *C. sorokiniana* and *Scenedesmus* sp. was tested against the murine L5178Y-R lymphoma cell line (Reyna-Martinez et al., 2018). Different concentrations of extracts (ranging from 7.8 to 500 µg ml^{-1}) were evaluated *in vitro*. After 48 hours of treatment, the extracts showed a concentration-dependent activity against this murine tumor cell line. *C. sorokiniana* and *Scenedesmus* sp. methanol extracts with concentration of 500 µg ml^{-1} caused 61% and 75% tumor cell cytotoxicity, respectively. Furthermore, the cytotoxicity induced with both extracts was mediated by apoptosis since there was an activation of caspases that caused DNA fragmentation and collapse of the cell.

10.2.2.4 DIETHYL ETHER EXTRACTS

The anticancer activity of *Scenedesmus obliquus* diethyl ether extracts was studied by Marrez et al. (2019). These authors tested the anticancer activity of the *Scenedesmus obliquus* extract on HepG2, HCT116 and MCF-7 cells lines in concentrations between 0.78 and 200 µg ml^{-1}. After 48 hours of incubation, the highest anticancer effect was verified in HCT116 and HepG2 cell lines with IC$_{50}$ values of 24.6 and 42.8 µg ml^{-1}, respectively, while in MCF-7 cells, the determined IC$_{50}$ was 93.8 µg ml^{-1}.

10.3 ACTIVE ANTICANCER COMPOUNDS FROM FRESHWATER MICROALGAE

Microalgae compounds from different chemical classes, including pigments, peptides, phytosterols, and polysaccharides have been shown to display promising antitumor potential and consequently have proven to be effective against several types of cancer (Amaro et al., 2013; Alves et al., 2018). These microalgae compounds modulate several cellular mechanisms such as the involved in cellular cytotoxicity, downregulate the invasion capacity of tumor cells, and enhance cancer cells apoptosis (El-Hack et al., 2019).

10.3.1 CAROTENOIDS

Several freshwater microalgae accumulate large amounts of carotenoids such as astaxanthin, β-carotene, canthaxanthin, zeaxanthin, violaxanthin, and lutein. These carotenoids are important as nutraceuticals in human health care. They can act in cancer prevention and can be applied in its treatment since they protect normal cells from genetic damage as well as have antiproliferative, cytotoxic, and pro-apoptotic effects on cancer cells (Gong and Bassi, 2016).

Astaxanthin extracted from microalgae, namely *Haematococcus pluvialis* has gained attention due to its bioactivity properties. This carotenoid has been shown to be a potent anticancer agent in experimental animal models, displaying an activity much superior than the observed with other carotenoids (Amaro et al., 2013). Palozza et al. (2009) studied the growth-inhibitory effects of an astaxanthin-rich *H. pluvialis* extract in colon cancer cells lines, including, HCT-116, HT-29, LS-174, WiDr, and SW-480 cells, and its mechanism of action. After 24 h of treatment, different concentrations of the extract (5–25 μg ml^{-1}) had relevant cell growth-inhibitory effects and the concentration of 25 μg ml^{-1} led to the highest inhibition of proliferation of all types of colon cancer cells lines. The possible mechanisms responsible for the reduction of cell number were examined. Concentrations of the extract starting from 15 μg ml^{-1} induced the arrest of cell cycle progression at the G0/G1 phase and promoted pro-apoptotic effects in cancer cells. This arrest involved a down-regulation of cyclin D1, which has been implicated in the control of the G0/G1 phase of the cell cycle (Palozza et al., 2009). It is known that this protein is an oncogene and can be over-expressed in several cancer cell lines. Therefore, this study demonstrated that this *H. pluvialis* extract rich in astaxanthin is a potent growth inhibitor of several colon cancer cell lines. It is worth noting that Spiller and Dewell (2003) assessed the safety of an astaxanthin-rich *H. pluvialis* extract and they reported that few milligrams of astaxanthin from a *H. pluvialis* algal extract per day could be safely consumed by healthy adults. The effect of astaxanthin from *H. pluvialis* on proliferation and induction of apoptosis of HepG2 cells was also investigated by Nagaraj et al. (2012). A 25 μg ml^{-1} concentration of astaxanthin led to a 50% of cellular growth inhibition after 48 h of treatment. Further it has been found that HepG2 cells treated with extracts concentrations of 15 and 25 μg ml^{-1} after 24 hours of treatment presented an extensive DNA fragmentation. According to the authors, astaxanthin interfered with the normal reorganization of the microtubule network and inhibits the formation of a normal spindle at metaphase that is required for mitosis and cell proliferation. These effects induced the arrest of the cells in the G2/M phase of the cell cycle and promoted the apoptotic cell death. A promising cytotoxicity of carotenoid-rich fractions from *H. pluvialis* was also found in the work of El-Baz et al. (2018). These authors analyzed the cytotoxic activity of carotenoid-rich fractions from *H. pluvialis* on four cancer cell lines: HepG2, MCF-7, HCT116, and A549. The carotenoid fractions consisting of astaxanthin, zeaxanthin, lutein, canthaxanthin, and β-carotene in free form and bounded to fatty acids demonstrated a potent cytotoxicity activity on HCT116 with 100% inhibition of cell viability at a concentration of 0.1 mg ml^{-1}, similarly to the previously reported by Palozza et al. (2009). Also, it was verified a relevant cytotoxicity on MCF-7 cells, with 91.2% inhibition, and a moderate activity on HepG2 with 63.4% inhibition of cell growth.

The yellowish carotenoid lutein has also been shown to be pharmacologically interesting. In this context, Cha et al. (2008) studied the antiproliferative effect of extracts from *Chlorella vulgaris* on HCT-116 cells. The extracts were analyzed by high performance liquid chromatography (HPLC) and the major carotenoid identified was lutein. After the treatment with various extract concentrations, the HCT-116 cells growth was inhibited with an extract concentration of 40 μg ml^{-1}. Further, the mechanism of action was explored, and the authors verified that the *C. vulgaris* extract induced apoptosis in colon cancer cells. This work is in agreement with has been referred by some authors who reported that a dietary lutein can inhibit the growth of mouse mammary tumors by apoptosis induction (Chew et al., 2003).

10.3.2 PEPTIDES

A few studies reported the anticancer activity of microalgae peptides. However, this research on microalgae as a source of bioactive peptides has triggered a great interest due to their therapeutic potential in the treatment of various diseases (Wang and Zhang, 2013). Sheih et al. (2010) studied the effect of a peptide fraction isolated from a pepsin hydrolysate of *Chlorella vulgaris* against human gastric cancer cell lines (AGS). Interestingly, a strong dose-dependent antiproliferation by induction of the cell cycle arrest was observed with this peptide fraction. The anticancer effects of peptides from a hydrolyzed *C. vulgaris* protein on MCF-7 cells were examined by Sedighi et al. (2016). After 72 hours of treatment, the 3–5 KDa peptide fractions at a concentration of 50 μg μl^{-1} led to a 50% reduction in cell viability. The highest inhibitory effect on MCF-7 cells (cell growth reduction of over 60%) was observed at a concentration of 160 μg μl^{-1}. Wang and Zhang (2013) analyzed the anticancer activity of an isolated polypeptide (CPAP) from *C. pyrenoidosa* on HepG2 cells. These cells were treated with different concentrations of CPAP (0.1, 0.2, 0.3, 0.4, and 0.5 mg ml^{-1}) for 72 hours.

Interestingly, CPAP displayed dose-dependent growth inhibition effects on HepG2 cells and the highest inhibition rate (60.9%) was observed at 0.5 mg ml^{-1}. In addition, the determined IC$_{50}$ was 0.43 mg ml^{-1}. The morphological changes caused by CPAP on HepG2 cells were also analyzed by microscopy. At 0.3–0.5 mg ml^{-1}, the cells exhibited characteristics of apoptosis such as cell membrane shrinkage, condensation, and fragmentation of nuclear chromatin as well as the formation of black apoptotic bodies. These authors concluded that CPAP could induce apoptosis and necrotic death of HepG2 cells.

10.3.3 PHYTOSTEROLS

Several studies have described the bioactivity of phytosterols against tumors. Phytosterols derived from microalgae can be divided into four main groups, 4-desmethyl-Δ^5-sterols, 4-desmethyl-Δ^7-sterols, 4-methyl sterols and dihydroxylated sterols. Considering the great diversity of structures that these steroids may present among species, microalgae may be a promising natural source of phytosterols with potential bioactivity namely anticancer (Luo et al., 2015).

Nine sterols from *Chlorella vulgaris* were isolated, and their inhibitory effects against 12-*O*-tetradecanoylphorbol-13-acetate (TPA)-induced inflammation and tumor promotion in mouse skin were assessed by Yasukawa et al. (1996). Of these nine steroids isolated and identified in the hexane fraction obtained from the methanol extract, six sterols had relevant anti-inflammatory effects. In addition, two μmol of ergosterol peroxide topically applied strongly inhibited the TPA-induced tumor promotion (77% reduction in the average number of tumors per mouse at week 20). The authors compared their results with those of the study of Yasukawa et al. (1991) and verified that the inhibitory effect of ergosterol peroxide was almost comparable with that of ergosterol, and was more effective than sitosterol and stigmasterol. This was the first study to report that ergosterol peroxide showed inhibition of TPA-induced tumor promotion in carcinogenesis of mouse skin.

A significant toxicity of isolated stigmasterol from *Navicula incerta* against HepG2 cells was demonstrated in the study of Kim et al. (2014). After 24 hours of incubation, stigmasterol originated a significant toxicity on HepG2 cells in a dose-dependent manner, with 40%, 43% and 54% toxicity at concentrations of 5, 10, and 20 μM, respectively, which indicated a dose-dependent trend. The authors also verified some morphological changes caused by stigmasterol, including condensation of cytoplasm and nucleus, bleb of cell membrane and DNA fragmentation. This DNA damage led to changes in the life cycle cell and its consequent termination. It was also observed that stigmasterol induced the expression of pro-apoptotic genes while inhibiting the anti-apoptotic factors causing the cell death.

10.3.4   POLYSACCHARIDES

Microalgal polysaccharides of some genera have been described to have anticancer activity (Matos et al., 2017). Polysaccharides from *Chlorella pyrenoidosa* exhibited high antitumor activity against A549 cells (Sheng et al., 2007). In fact, two polysaccharide fractions composed of rhamnose, mannose, glucose, galactose, and an unknown monosaccharide were obtained, named CPPS la and CPPS IIa. After 48 hours of exposure, different concentrations of the polysaccharide fractions (200, 600, 800, and 1000 µg ml^{-1}) presented high antitumor activity against A549 cells in a dose-dependent manner. The fraction CPPS Ia at a concentration of 1000 µg ml^{-1} showed an inhibition rate of 68.7%, significantly higher than that of CPPS IIa (49.5%). Therefore, this study demonstrated that both fractions from *C. pyrenoidosa* can be used for developing natural safer antitumor drugs.

Also, Zhang et al. (2019) investigated the antitumor effects of partially purified exopolysaccharides (EPSs) from *C. pyrenoidosa* and from two other chlorophytes genera, *Scenedesmus* sp. and *Chlorococcum* sp. (EPS-CP, EPS-SS, and EPS-CS, respectively). These microalgal EPSs were tested against the human colon cancer cell lines HCT116 and HCT8. After treatment with EPS-CP, EPS-SS, and EPS-CS for 24 hours, the antitumor effects on the two cell lines at concentration of 6 mg ml^{-1} were significant since they showed inhibitory effects on cell viability of 17.2, 19.2, and 18.7% on HCT116 cells and 35.9, 38.6, and 22.9% on HCT8 cells, respectively.

Aqueous extracellular polysaccharides (AEPS) from the thermophilic microalgae *Graesiella* sp. also exhibited higher antiproliferative effects against HepG2 and human colon cancer cells (Caco-2) (Trabelsi et al., 2016). The two cancer cell lines were treated with AEPS concentrations between 0.01 and 2.5 mg ml^{-1} during 72 hours. The determined IC$_{50}$ was 0.3 mg ml^{-1} and 1.06 mg ml^{-1} in Caco-2 and HepG2 cells, respectively. Further, a 2.5 mg ml^{-1} concentration of AEPS caused 91% and 70.4% of growth inhibition of Caco-2 and HepG2 cells, respectively. These results demonstrated that Caco-2 cells were more sensitive to these saccharides than

HepG2. The authors also inferred that the antiproliferative activity of AEPS may be related with the sulfate groups since they have immunostimulating properties (Gardeva et al., 2009). Since the biological activity of sulfated polysaccharides are dependent on their molecular weight, sulfate content and position, type of sugar residues, type of linkage and molecular geometry, more studies on chemical structure and physicochemical characteristics including rheological properties are necessary (Kamble et al., 2018).

The inhibition of MDA-MB-231 cell growth by ethanol-extracted sulfated polysaccharides from *Chlamydomonas reinhardtii* (Cr-SPs) was observed by Kamble et al. (2018). After 72 hours of treatment with different concentrations (0–500 μg ml^{-1}), the determined IC$_{50}$ was 172 μg ml^{-1} and a concentration of 500 μg ml^{-1} led to 95% of inhibition of cell viability. The authors also analyzed the ability of Cr-SPs to interfere with the clonogenic propagation, to perturb the morphology of microtubules and to induce apoptosis. Interestingly, it was verified a decrease in the number of cell colonies, with concentrations of 100, 200, and 400 μg ml^{-1} reducing the number of colonies in 29, 46, and 69%, respectively, compared to control. This indicates that Cr-SPs prevented the colony-forming ability of MDA-MB-231 cells in a concentration-dependent manner. The normal network morphology of cellular microtubule was also loss, since after 48 h of treatment with 150 μg ml^{-1} of Cr-SPs the microtubule network appeared disorganized and a higher concentration (300 μg ml^{-1}) disrupted the structural integrity of the microtubules. In addition, a concentration of 400 μg ml^{-1} of Cr-SPs induced the apoptosis of MDA-MB-231 cells.

10.4 NANODELIVERY SYSTEMS FOR CONTROLLED RELEASE OF ANTICANCER COMPOUNDS

Nanoparticles containing products derived from natural sources such as microalgae have received much interest in recent years. The purpose of these nanoformulations is to provide a fast and efficient delivery due to their favorable properties for facilitating delivery in a variety of cancer situations (Zhang et al., 2017). Nanoparticles can be produced from several materials, including inorganic (e.g., metals, silica, carbon) and organic (e.g., polymers, and lipids) structures. Their sizes (from a few nanometers to 1 mm), their shapes (smooth or sharp), their plasticity (stiff or transformable) can be manipulated according to the intended purpose (Bajpai et al., 2018). A further advantage of the use of nanoparticles composed of natural anticancer components is their versatility and ability to overcome sequential biological barriers, which have proven to be a major obstacle to the efficacy of nanoparticle-based drug delivery (Bajpai et al., 2018).

However, there are some limitations to the application of bioactive compounds, namely easy degradation of peptides in the stomach by enzymes and gastric acid. A possible solution for this is the encapsulation of bioactive compounds so that they can be protected against degradation in the stomach and be released in the intestinal tract (Wang and Zhang, 2013). The micro- and nanoencapsulation of a polypeptide from *C. pyrenoidosa* (CPAP, *C. pyrenoidosa* antitumor polypeptide) were investigated by Wang and Zhang (2013). These authors explored two methods, complex coacervation involving edible alginate, calcium chloride ($CaCl_2$), chitosan, and ionotropic gelation which involves edible chitosan and sodium tripolyphosphate. The encapsulation efficiency was determined and the results demonstrated that the efficiency of microencapsulation was 2.5-fold higher than that of nanoencapsulation (74.5% vs. 30.1%). The release properties of microencapsulated and nanoencapsulated CPAP were examined in SGI (simulated gastric juice) and SIJ (simulated intestinal juice). The *in vitro* tests revealed that CPAP was well preserved in microencapsulation and nanoencapsulation against gastric enzymatic degradation. Furthermore, the content of the encapsulated CPAP for intestinal absorption was higher than that of non-encapsulated CPAP. However, the results of antitumor activity showed that encapsulated CPAP lost some activity in comparison with the non-encapsulated CPAP. Thus, this study indicates that the encapsulated CPAP improves the bioavailability in the intestinal tract, and it can be developed for treatment of various types of cancer. However, more studies on the encapsulation method will be necessary to avoid any loss of activity of the bioactive compound.

Amongst the metallic nanoparticles, silver nanoparticles are used in biomedical applications since they have shown to have several bioactivities (e.g., antimicrobial, anticancer, anti-inflammatory, antiviral) (Khalid et al., 2017). Although there are chemical methods to prepare silver nanoparticles, some of them usually involve extreme conditions, thus limiting the medical applications (Ebrahiminezhad et al., 2016). The use of biological compounds in the production of nanoparticles has gained interest since there are advantages, namely low cost and milder conditions for their preparation. Furthermore, these biological methods have low or null toxicity for the environment and are economically more viable. In this context, bioactive compounds from microalgae, namely carbohydrates and proteins, can be useful in the reduction and capping of silver nanoparticles, avoiding the need of toxic chemicals and extreme reaction conditions (Ebrahiminezhad et al., 2016; Khalid et al., 2017). Therefore, some algae extracts have been studied aiming a greener preparation of silver nanoparticles which were subsequently therapeutically tested for bioactivities (Khalid et al., 2017).

In fact, Khalid et al. (2017) used ethanolic extracts of three different freshwater microalgae strains - *Dictyosphaerium* sp. strain HM1, *Dictyosphaerium* sp. HM2 and *Pectinodesmus* sp. strain HM3 - to produce silver nanoparticles and analyzed their anticancer activity on HepG2 and MCF-7 cell lines. For this, an aqueous solution of silver nitrate was added to ethanolic extracts of microalgae followed by incubation at 37°C. After purification of synthesized silver nanoparticles, the cancer cell lines were exposed to five concentrations (10–50 µg ml^{-1}) during 24 hours. A dose-dependent trend was observed for all three types of microalgae-mediated silver nanoparticles since its activity was increased with increasing concentration. The best results were observed with *Dictyosphaerium* sp. HM2 nanoparticles, with IC$_{50}$ values of 0.29 and 0.16 µg ml^{-1} in HepG2 and MCF-7 cells, respectively. Furthermore, the authors also verified a reduction of the number of cells in the G$_0$ to G$_1$ phase and an increase in the G$_2$ to M phase after treatment with microalgae silver nanoparticles, indicating a cell cycle arrest at latter stages.

Microalgae-based nanoformulations are a sustainable alternative to other delivery systems and they should be urgently commercialized, taking into account the number of diseases related to cancer that arises annually. However, some characteristics need to be improved to develop stable and effective formulations of microalgae drugs with adequate usages in nanomedicine-based therapies (Bajpai et al., 2018).

10.5 CONCLUSIONS AND FUTURE PERSPECTIVES

Freshwater microalgae have been identified as a promising group of microorganisms containing novel biochemically active natural products against a large variety of cancer cell lines. In fact, and as a relevant example, *Chlorophyta* microalgae have showed very promising anticancer activity against some cancer cell lines. In addition, extracts of *Chlorella*, *Scenedesmus*, and *Ankistrodesmus* genera exhibited significant cytotoxic activities at concentrations between 1 and 500 µg ml^{-1}. Furthermore, extracts of diatom *Amphiprora* displayed interesting antiproliferative effects against specific cancer cell lines, with IC$_{50}$ values of approximately 10 µg ml^{-1}.

Freshwater microalgal compounds such as carotenoids, peptides, phytosterols, and polysaccharides demonstrated promising anticancer activity with potential for the pharmaceutical industry market. For example, carotenoids including astaxanthin from *Haematococcus pluvialis* and lutein from *Chlorella ellipsoidea* showed a strong activity against several cancer cell lines. In addition, carotenoids are also known to have high antioxidant, anti-inflammatory, anti-obesity, anti-diabetic, and cardioprotective effects. Some carotenoids are already on the market, namely astaxanthin, being used as dietary supplements. Peptides from *Chlorella* also showed considerable biotechnology interest. For example, peptide fractions of *C. vulgaris* and *C. pyrenoidosa* with concentration of 50–500 μg ml^{-1} reduced the viability of cancer cells above 50%.

The identification and isolation of phytosterols from freshwater microalgae have also been of high interest. In this context, phytosterols isolated from *C. vulgaris* have demonstrated high cellular inhibition. In addition, stigmasterol isolated from *N. incerta* exhibited a significant cellular toxicity, leading to DNA fragmentation, arrest of the cell cycle and apoptosis. This phytosterol is receiving attention not only due to its anticancer activity but also due to the ability to reduce blood cholesterol concentrations and to prevent cardiovascular disorders.

Polysaccharides isolated from *Chlorella* also demonstrated promising bioactivity. In fact, in concentrations of 1000 μg ml^{-1} led to a cell growth inhibition rate of approximately 68%. In addition, polysaccharides from other *Chlorophyta* freshwater microalgae have also shown promising bioactivity. *Scenedesmus* and *Graesiella* polysaccharides demonstrated inhibitory effects on cell viability about 40% and 90%, respectively. In this point, it is known that microalgal polysaccharides can modulate the immune system, and are interesting not only as natural therapeutic agents but also as cosmetic additives.

This chapter evidences a promising anticancer potential of several freshwater microalgae genera which may have a potential role in the development of future new drugs. However, more studies aiming to clarify the specific targets and the mechanisms of antiproliferative action are needed. Together with the promising anticancer activity of microalgae, these may also be a sustainable alternative to develop new nanoformulated delivery systems. However, the stability of microalgae formulations of drugs and their use in nanomedicine-based therapies should be further explored in the near future.

KEYWORDS

- **action mechanisms**
- **cancer**
- **carotenoids**
- **freshwater microalgae extracts**
- **microalgae**
- **microalgae-based nanoformulations**
- **nanocarriers**
- **new anticancer compounds**
- **peptides**
- **phytosterols**
- **polysaccharides**
- **safer anticancer therapies**
- **target cells**

REFERENCES

Alves, C., Silva, J., Pinteus, S., Gaspar, H., Alpoim, M. C., Botana, L. M., & Pedrosa, R., (2018). From marine origin to therapeutics: The antitumor potential of marine algae-derived compounds. *Frontiers in Pharmacology, 9*, 777. doi: https://doi.org/10.3389/fphar. 2018.00777.

Amaro, H. M., Barros, R., Guedes, A. C., Sousa-Pinto, I., & Malcata, F. X., (2013). Microalgal compounds modulate carcinogenesis in the gastrointestinal tract. *Trends in Biotechnology, 31*, 92–98. doi: https://doi.org/10.1016/j.tibtech.2012.11.004.

American Cancer Society, (2018). *Cancer Basics*. Available online: https://www.cancer.org/ (accessed on 2 July 2021).

Azamai, E. S. M., Sulaiman, S., Habib, S. H. M., Looi, M. L., Das, S., Hamid, N. A. A., Ngah, W. Z. W., & Yusof, Y. A. M., (2009). *Chlorella vulgaris* triggers apoptosis in hepatocarcinogenesis-induced rats. *Journal of Zhejiang University Science B, 10*(1), 14–21. doi: https:// doi.org/10.1631/jzus.B0820168.

Bajpai, V., Shukla, S., Kang, S. M., Hwang, S., Song, X., Huh, Y., & Han, Y. K., (2018). Developments of cyanobacteria for nano-marine drugs: Relevance of nanoformulations in cancer therapies. *Marine Drugs, 16*(6), 179. doi: https://doi.org/10.3390/md16060179.

Borowitzka, M. A., (2013). High-value products from microalgae-their development and commercialization. *Journal of Applied Phycology, 5*(3), 743–756. doi: https://doi.org/10.1007/ s10811-013-9983-9.

Cha, K. H., Koo, S. Y., & Lee, D. U., (2008). Antiproliferative effects of carotenoids extracted from *Chlorella ellipsoidea* and *Chlorella vulgaris* on human colon cancer cells. *Journal of Agricultural and Food Chemistry, 56*(22), 10521–10526. doi: https://doi.org/10.1021/jf802111x.

Chew, B. P., Brown, C. M., Park, J. S., & Mixter, P. F., (2003). Dietary lutein inhibits mouse mammary tumor growth by regulating angiogenesis and apoptosis. *Anticancer Research, 23*(4), 3333–3339.

Costa, M., Costa-Rodrigues, J., Fernandes, M. H., Barros, P., Vasconcelos, V., & Martins, R., (2012). Marine cyanobacteria compounds with anticancer properties: A review on the implication of apoptosis. *Marine Drugs, 10*(10), 2181–2207. doi: https://doi.org/10.3390/md10102181.

Da Silva, V. B., Moreira, J. B., De Morais, M. G., & Costa, J. A. V., (2016). Microalgae as a new source of bioactive compounds in food supplements. *Current Opinion in Food Science, 7*, 73–77. doi: https://doi.org/10.1016/j.cofs.2015.12.006.

Dewi, I. C., Falaise, C., Hellio, C., Bourgougnon, N., & Mouget, J. L., (2018). Anticancer, antiviral, antibacterial, and antifungal properties in microalgae. In: Levine, I. A., & Fleurence, J., (eds.), *Microalgae in Health and Disease Prevention* (pp. 235–261). Academic Press. doi: https://doi.org/10.1016/B978-0-12-811405-6.00012-8.

Ebrahiminezhad, A., Bagheri, M., Taghizadeh, S. M., Berenjian, A., & Ghasemi, Y., (2016). Biomimetic synthesis of silver nanoparticles using microalgal secretory carbohydrates as a novel anticancer and antimicrobial. *Advances in Natural Sciences: Nanoscience and Nanotechnology, 7*(1), 015018. doi: https://doi.org/10.1088/2043-6262/7/1/015018.

El-Baz, F. K., Hussein, R. A., Mahmoud, K., & Abdo, S. M., (2018). Cytotoxic activity of carotenoid-rich fractions from *Haematococcus pluvialis* and *Dunaliella salina* microalgae and the identification of the phytoconstituents using LC-DAD/ESI-MS. *Phytotherapy Research, 32*(2), 298–304. doi: https://doi.org/10.1002/ptr.5976.

El-Hack, M. E. A., Abdelnour, S., Alagawany, M., Abdo, M., Sakr, M. A., Khafaga, A. F., Mahgoub, S. A., et al., (2019). Microalgae in modern cancer therapy: Current knowledge. *Biomedicine and Pharmacotherapy, 111*, 42–50. doi: https://doi.org/10.1016/j.biopha.2018.12.069.

Gardeva, E., Toshkova, R., Minkova, K., & Gigova, L., (2009). Cancer protective action of polysaccharide derived from red microalga *Porphyridium cruentum* - a biological background. *Biotechnology and Biotechnological Equipment, 23*(S1), 783–787. doi: https://doi.org/10.1080/13102818.2009.10818540.

Gong, M., & Bassi, A., (2016). Carotenoids from microalgae: A review of recent developments. *Biotechnology Advances, 34*(8), 1396–1412. doi: https://doi.org/10.1016/j.biotechadv.2016.10.005.

Hoa, L. T. P., Quang, D. N., Ha, N. T. H., & Tri, N. H., (2011). Isolating and screening mangrove microalgae for anticancer activity. *Research Journal of Phytochemistry, 5*, 156–162. doi: https://doi.org/10.3923/rjphyto.2011.156.162.

Jabeen, A., Reeder, B., Hisaindee, S., Ashraf, S., Al Darmaki, N., Battah, S., & Al-Zuhair, S., (2017). Effect of enzymatic pretreatment of microalgae extracts on their antitumor activity. *Biomedical Journal, 40*(6), 339–346. doi: https://doi.org/10.1016/j.bj.2017.10.003.

Kamble, P., Cheriyamundath, S., Lopus, M., & Sirisha, V. L., (2018). Chemical characteristics, antioxidant, and anticancer potential of sulfated polysaccharides from *Chlamydomonas reinhardtii*. *Journal of Applied Phycology, 30*(3), 1641–1653. doi: https://doi.org/10.1007/s10811-018-1397-2.

Khalid, M., Khalid, N., Ahmed, I., Hanif, R., Ismail, M., & Janjua, H. A., (2017). Comparative studies of three novel freshwater microalgae strains for synthesis of silver nanoparticles:

Insights of characterization, antibacterial, cytotoxicity and antiviral activities. *Journal of Applied Phycology, 29*(4), 1851–1863. doi: https://doi.org/10.1007/s10811-017-1071-0.

Kim, Y. S., Li, X. F., Kang, K. H., Ryu, B., & Kim, S. K., (2014). Stigmasterol isolated from marine microalgae *Navicula incerta* induces apoptosis in human hepatoma HepG2 cells. *BMB Reports, 47*(8), 433. doi: https://doi.org/10.5483/BMBRep.2014.47.8.153.

Lin, P. Y., Tsai, C. T., Chuang, W. L., Chao, Y. H., Pan, I. H., Chen, Y. K., Lin, C. C., & Wang, B. Y., (2017). *Chlorella sorokiniana* induces mitochondrial-mediated apoptosis in human non-small cell lung cancer cells and inhibits xenograft tumor growth *in vivo*. *BMC Complementary and Alternative Medicine, 17*(1), 88. doi: https://doi.org/10.1186/s12906-017-1611-9.

Luo, X., Su, P., & Zhang, W., (2015). Advances in microalgae-derived phytosterols for functional food and pharmaceutical applications. *Marine Drugs, 13*(7), 4231–4254. doi: https://doi.org/10.3390/md13074231.

Marrez, D. A., Naguib, M. M., Sultan, Y. Y., & Higazy, A. M., (2019). Antimicrobial and anticancer activities of *Scenedesmus obliquus* metabolites. *Heliyon, 5*(3), e01404. doi: https://doi.org/10.1016/j.heliyon.2019.e01404.

Martínez, A. K., Lauritano, C., Romano, G., & Ianora, A., (2018). Marine microalgae with anti-cancer properties. *Marine Drugs, 16*(5), 165. doi: https://doi.org/10.3390/md16050165.

Matos, J., Cardoso, C., Bandarra, N. M., & Afonso, C., (2017). Microalgae as healthy ingredients for functional food: A review. *Food and Function, 8*(8), 2672–2685. doi: https://doi.org/10.1039/C7FO00409E.

Mobin, S., & Alam, F., (2017). Some promising microalgal species for commercial applications: A review. *Energy Procedia, 110*, 510–517. doi: https://doi.org/10.1016/j.egypro.2017.03.177.

Nagaraj, S., Rajaram, M. G., Arulmurugan, P., Baskaraboopathy, A., Karuppasamy, K., Jayappriyan, K. R., Sundararaj, R., & Rengasamy, R., (2012). Antiproliferative potential of astaxanthin-rich alga *Haematococcus pluvialis* Flotow on human hepatic cancer (HepG2) cell line. *Biomedicine and Preventive Nutrition, 2*(3), 149–153. doi: https://doi.org/10.1016/j.bionut.2012.03.009.

Olaizola, M., (2003). Commercial development of microalgal biotechnology: From the test tube to the marketplace. *Biomolecular Engineering, 20*(4–6), 459–466. doi: https://doi.org/10.1016/S1389-0344(03)00076-5.

Ördög, V., Stirk, W. A., Lenobel, R., Bancířová, M., Strnad, M., Van, S. J., Szigeti, J., & Németh, L., (2004). Screening microalgae for some potentially useful agricultural and pharmaceutical secondary metabolites. *Journal of Applied Phycology, 16*(4), 309–314. doi: https://doi.org/10.1023/B:JAPH.0000047789.34883.aa.

Palozza, P., Torelli, C., Boninsegna, A., Simone, R., Catalano, A., Mele, M. C., & Picci, N., (2009). Growth-inhibitory effects of the astaxanthin-rich alga *Haematococcus pluvialis* in human colon cancer cells. *Cancer Letters, 283*(1), 108–117. doi: https://doi.org/10.1016/j.canlet.2009.03.031.

Pulz, O., & Gross, W., (2004). Valuable products from biotechnology of microalgae. *Applied Microbiology and Biotechnology, 65*(6), 635–648. doi: https://doi.org/10.1007/s00253-004-1647-x.

Reyna-Martinez, R., Gomez-Flores, R., López-Chuken, U., Quintanilla-Licea, R., Caballero-Hernandez, D., Rodríguez-Padilla, C., Beltrán-Rocha, J. C., & Tamez-Guerra, P., (2018). Antitumor activity of *Chlorella sorokiniana* and *Scenedesmus* sp. microalgae native of Nuevo León State, México. *Peer J., 6*, e4358. doi: https://doi.org/10.7717/peerj.4358.

Romano, G., Costantini, M., Sansone, C., Lauritano, C., Ruocco, N., & Ianora, A., (2017). Marine microorganisms as a promising and sustainable source of bioactive molecules. *Marine Environmental Research, 128*, 58–69. doi: https://doi.org/10.1016/j. marenvres.2016.05.002.

Sedighi, M., Jalili, H., Ranaei-Siadat, S. O., & Amrane, A., (2016). Potential health effects of enzymatic protein hydrolysates from *Chlorella vulgaris*. *Applied Food Biotechnology, 3*(3), 160–169.

Shanab, S. M., Mostafa, S. S., Shalaby, E. A., & Mahmoud, G. I., (2012). Aqueous extracts of microalgae exhibit antioxidant and anticancer activities. *Asian Pacific Journal of Tropical Biomedicine, 2*(8), 608–615. doi: https://doi.org/10.1016/S2221-1691(12)60106-3.

Shannon, E., & Abu-Ghannam, N., (2016). Antibacterial derivatives of marine algae: An overview of pharmacological mechanisms and applications. *Marine Drugs, 14*(4), 81. doi: https://doi.org/10.3390/md14040081.

Sheih, I. C., Fang, T. J., Wu, T. K., & Lin, P. H., (2010). Anticancer and antioxidant activities of the peptide fraction from algae protein waste. *Journal of Agricultural and Food Chemistry, 58*(2), 1202–1207. doi: https://doi.org/10.1021/jf903089m.

Sheng, J., Yu, F., Xin, Z., Zhao, L., Zhu, X., & Hu, Q., (2007). Preparation, identification, and their antitumor activities *in vitro* of polysaccharides from *Chlorella pyrenoidosa*. *Food Chemistry, 105*(2), 533–539. doi: https://doi.org/10.1016/j.foodchem.2007.04.018.

Soares, A. T., Da Costa, D. C., Vieira, A. A. H., & Antoniosi, F. N. R., (2019). Analysis of major carotenoids and fatty acid composition of freshwater microalgae. *Heliyon, 5*(4), e01529. doi: https://doi.org/10.1016/j.heliyon.2019.e01529.

Spiller, G. A., & Dewell, A., (2003). Safety of an astaxanthin-rich *Haematococcus pluvialis* algal extract: A randomized clinical trial. *Journal of Medicinal Food, 6*(1), 51–56. doi: https://doi.org/10.1089/109662003765184741.

Spolaore, P., Joannis-Cassan, C., & Isambert, A., (2006). Commercial applications of microalgae. *Journal of Bioscience and Bioengineering, 101*, 87–96. doi: https://doi.org/10.1263/jbb.101.87.

Suh, S. S., Kim, S. M., Kim, J. E., Hong, J. M., Lee, S. G., Youn, U. J., Han, S. J., Kim, I. C. & Kim, S., (2017). Anticancer activities of ethanol extract from the Antarctic freshwater microalga, *Botryidiopsidaceae* sp. *BMC Complementary and Alternative Medicine, 17*(1), 509. doi: https://doi.org/10.1186/s12906-017-1991-x.

Trabelsi, L., Chaieb, O., Mnari, A., Abid-Essafi, S., & Aleya, L., (2016). Partial characterization and antioxidant and antiproliferative activities of the aqueous extracellular polysaccharides from the thermophilic microalgae *Graesiella* sp. *BMC Complementary and Alternative Medicine, 16*(1), 210. doi: https://doi.org/10.1186/s12906-016-1198-6.

Wang, H. M., Pan, J. L., Chen, C. Y., Chiu, C. C., Yang, M. H., Chang, H. W., & Chang, J. S., (2010). Identification of anti-lung cancer extract from *Chlorella vulgaris* CC by antioxidant property using supercritical carbon dioxide extraction. *Process Biochemistry, 45*(12), 1865–1872. doi: https://doi.org/10.1016/j.procbio.2010.05.023.

Wang, X., & Zhang, X., (2013). Separation, antitumor activities, and encapsulation of polypeptide from *Chlorella pyrenoidosa*. *Biotechnology Progress, 29*(3), 681–687. doi: https://doi.org/10.1002/btpr.1725.

World Health Organization, (2018). *Cancer: Key facts.* Available online: http://www.who.int/en/news-room/fact-sheets/detail/cancer (accessed on 2 July 2021).

Yasukawa, K., Akihisa, T., Kanno, H., Kaminaga, T., Izumida, M., Sakoh, T., Tamura, T., & Takido, M., (1996). Inhibitory effects of sterols isolated from *Chlorella vulgaris* on 12-O-tetradecanoylphorbol-13-acetate-induced inflammation and tumor promotion in mouse skin. *Biological and Pharmaceutical Bulletin, 19*(4), 573–576. doi: https://doi.org/10.1248/bpb.19.573.

Yasukawa, K., Takido, M., Matsumoto, T., Takeuchi, M., & Nakagawa, S., (1991). Sterol and triterpene derivatives from plants inhibit the effects of a tumor promoter, and sitosterol and betulinic acid inhibit tumor formation in mouse skin two-stage carcinogenesis. *Oncology, 48*(1), 72–76. doi: https://doi.org/10.1159/000226898.

Yusof, Y. A. M., Saad, S. M., Makpol, S., Shamaan, N. A., & Ngah, W. Z. W., (2010). Hot water extract of *Chlorella vulgaris* induced DNA damage and apoptosis. *Clinics, 65*(12), 1371–1377. doi: https://doi.org/10.1590/S1807-59322010001200023.

Zhang, J., Liu, L., Ren, Y., & Chen, F., (2019). Characterization of exopolysaccharides produced by microalgae with antitumor activity on human colon cancer cells. *International Journal of Biological Macromolecules, 128*, 761–767. doi: https://doi.org/10.1016/j.ijbiomac.2019.02.009.

Zhang, Y. S., Zhang, Y. N., & Zhang, W., (2017). Cancer-on-a-chip systems at the frontier of nanomedicine. *Drug Discovery Today, 22*(9), 1392–1399. doi: https://doi.org/10.1016/j.drudis.2017.03.011.

CHAPTER 11

BIOACTIVE MICROALGAL PEPTIDES FROM *SPIRULINA*

MICHELE GREQUE DE MORAIS,[1] BRUNA DA SILVA VAZ,[1]
ETIELE GREQUE DE MORAIS,[2] CRISTIANE REINALDO LISBOA,[3]
LUCIELEN OLIVEIRA SANTOS,[4] and JORGE ALBERTO VIEIRA COSTA[3]

*[1]Laboratory of Microbiology and Biochemistry,
College of Chemistry and Food Engineering,
Federal University of Rio Grande, Rio Grande, Brazil*

*[2]Center for Marine Sciences, Faculty of Sciences and Technology,
University of Algarve, Algarve, Portugal*

*[3]Laboratory of Biochemical Engineering,
College of Chemistry and Food Engineering,
Federal University of Rio Grande, Rio Grande, Brazil*

*[4]Laboratory of Biotechnology, College of Chemistry and Food
Engineering, Federal University of Rio Grande, Rio Grande, Brazil*

ABSTRACT

Various research centers have focused on a search for natural biopeptide products. One of the objectives of the cultivating microorganisms, such as microalgae, is to obtain compounds with bioactive capacities. Microalgae have been the focus of biotechnology research due to their nutritional, economic, and ecological importance. These microorganisms have been used for many years as a source of food and bioactive compounds. *Spirulina* is one of the most studied strains, and it is used for different purposes worldwide. This microalga, it has been produced industrially for human consumption for several years, is GRAS (Generally Recognized as Safe) certified and is rich in several micronutrients. *Spirulina* biomass is mainly composed of proteins

(50–70%) and is a source of biopeptides with antioxidant, anti-inflammatory, antihypertensive, and antimicrobial activities. In this context, this chapter aims to discuss the production and extraction of biopeptides from *Spirulina* as well as the absorption characteristics of these bioactive compounds and their applications.

11.1 INTRODUCTION

The increased number of stress-related illnesses, poor diet habits, sedentary lifestyles, and processed-product diets has led researchers to increasingly look for compounds that can provide health benefits. Proteins are macromolecules composed of amino acids that are linked by peptide bonds and have important functions in biological processes (Murray et al., 2017). Proteins and peptides may be obtained from different sources, such as meat, milk, eggs, wheat, and microorganisms. In intact proteins, peptides are inactive, but after chemical or enzymatic treatments, they can be released and produce physiological effects.

Peptides have many biological activities depending on their amino acid sequence (Ibañez et al., 2012; Kang and Kim, 2013). Moreover, peptides are suitable sources of protein for human food because they are absorbed more efficiently through the gastrointestinal tract (GIT) in comparison to intact proteins or free amino acids (Lisboa et al., 2014). The high protein content in several species of microalgae has made this microorganism an important source of this nutrient, whether as a nutritional supplement or as a protein source to obtain peptides and amino acids (Koyande et al., 2019).

Microalgae have a high productivity and can double their biomass in an average of 2 to 5 days, achieving high yields with no application of pesticides, herbicides, or fungicides (Vaz et al., 2016). Moreover, they do not need arable land for cultivation and require less water than higher plants (Henrikson, 1994). Furthermore, microalgae do not require a vascular system to transport nutrients because they absorb nutrients directly from their surrounding medium (Geada et al., 2018).

Because of these attributes, microalgae have been used for many years as a source of food and bioactive compounds. *Spirulina* is one of the most studied microalgae strains, and it is used for different purposes all over the world. In this context, this review aims to discuss the production and extraction of biopeptides from *Spirulina* as well as the absorption characteristics of these bioactive compounds and their applications.

11.2 SPIRULINA MICROALGAE: BIOMASS APPLICATION AND BIOACTIVE COMPOUNDS

Microalgae are photosynthetic microorganisms that can be categorized in a variety of classes but are generally defined by their pigments, cellular structure (eukaryotic or prokaryotic) and life cycle. The prokaryotic group contains cyanobacteria, which have structural characteristics similar to bacteria, such as the presence of peptidoglycan in the cell wall. These microorganisms are called microalgae due to the presence of photosynthesis-related compounds (chlorophylls) (Masojídek et al., 2013).

The morphological, physiological, and genetic differences of each microalgal species have the capacity to produce various biocompounds. Therefore, these microorganisms are the focus of study in many research centers, mainly due to the products of their metabolism, the majority of which have bioactivities (Morais et al., 2015). *Spirulina* (Figure 11.1) is a filamentous cyanobacterium with a spiral structure, bluish-green color, and can be grown at a large scale, especially for use as food. This microalgal species is mainly known for its high production of proteins, which can range from 50 to 70% of the dry cell weight. The first reports of *Spirulina* cultivation were by the Aztec people. However, this microalga only began to be used at an industrial scale in the 20th century (Vonshak, 1997).

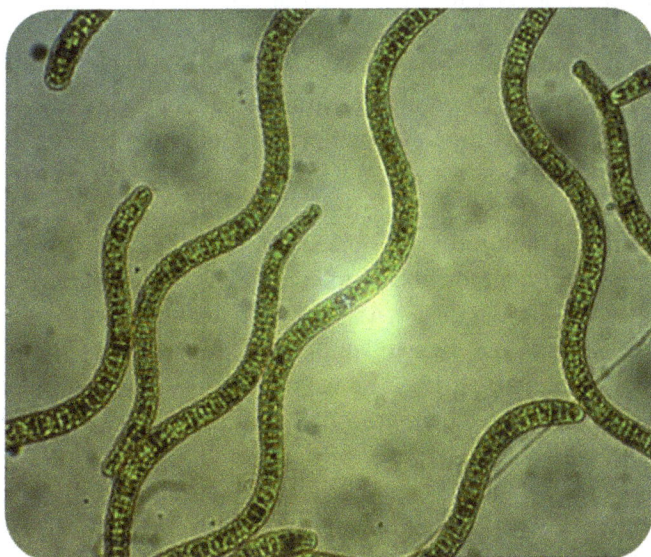

FIGURE 11.1 *Spirulina* sp. LEB 18 strain.

The microalgal proteins derived from *Spirulina* all contain essential amino acids (EAAs) and have high digestibility (85%). Furthermore, this microorganism is a source of biopeptides. Other bioactive compounds in the microalgal biomass include essential fatty acids (EFAs) (omega-6 and omega-3), carotenoids (astaxanthin), and vitamins and minerals (Gutiérrez-Salmeán et al., 2015; Yücetepe and Özçelik, 2016). Toxicological and nutritional evaluations showed that the microalgal biomass is beneficial as a food supplement or as a replacement for conventional protein sources. The use of *Spirulina* biomass is regulated by the United States Department of Agriculture, which awarded it the GRAS (Generally Recognized as Safe) certification, thus permitting its application in human consumption since it is produced under good manufacturing practices (GMP) (Karkos et al., 2011).

11.3 BIOPEPTIDES FROM MICROALGAE

Peptides are active substances composed of more than two amino acids bound to one another through amide bonds (Saadi et al., 2015). These compounds can be obtained from raw materials that contain a high protein concentration. Recently, much attention has been paid to using microalgae proteins as a source of bioactive peptides. These compounds hold a high biological potential due to their broad spectrum of bioactivities (Fan et al., 2014).

Bioactive peptides are usually 3–20 amino acid residues in length and have activities that are based on the amino acid sequence and composition. Peptides are inactive within an intact protein sequence but can be released to stimulate their effects (Kang and Kim, 2013). These peptides can exhibit bioactivities when released through digestion, hydrolysis, or by technological processes, such as high pressure (Ibañez et al., 2012).

The amino acid composition of microalgae is of great interest, not only because it possesses all of the EAAs for human health, but also because these amino acids have a great bioavailability in comparison with other protein sources (Koyande et al., 2019). The major fraction of *Spirulina* protein has been found in the form of supramolecular arrangements known as phycobilisomes (PBS), which are composed of pigmented (phycobiliproteins: PBPs) and nonpigmented (binding peptides) (Martinez-Palma et al., 2015). Thus, proteins obtained from the microalga *Spirulina* can be hydrolyzed in bioactive peptides (Lisboa et al., 2014).

11.4 CULTURE CONDITIONS FOR PROTEIN PRODUCTION

The culture conditions used for growing microalgae influence the metabolism of these microorganisms and can direct the synthesis of specific compounds of interest. The control of pH is essential for effective absorption of nutrients from the culture medium because it directly affects the availability of various chemical elements (Morais et al., 2015). Sornchai and Iamtham (2013) reported that the maximum protein content of microalgae *Spirulina maxima* was obtained by cultivation at pH 9.0.

Temperature is one of the main factors that regulate the morphology and cellular physiology, as well as the production of by-products, of microalgal biomass. High temperature can accelerate the microalgae metabolism and reduced temperatures can inhibit cell growth. The optimum temperature for most microalgae species is 30–37°C (Pandey et al., 2014). Pandey and Tiwari (2010) observed that the optimum temperature for protein biosynthesis by *Spirulina* sp. was 30°C.

The light intensity is one of the major factors that influence chlorophyll biosynthesis, and changes in biochemical content can alter the growth of microalgae in conditions of limited or high illumination (Ajayan et al., 2012). The maximum specific growth rate is increased when high intensity light is used, while when low intensity light is used, a high protein biomass is obtained (Danesi et al., 2011; Cruz-Martínez et al., 2015). The culture medium composition can improve protein biosynthesis in *Spirulina* (Ajayan et al., 2012). The protein content in *S. platensis* biomass can increase when this cyanobacterium is cultivated under non-limiting nitrogen conditions (Sassano et al., 2010).

The reactor configuration is another factor that can affect cell growth and protein composition. *Spirulina* cultivation may be carried out in open or closed bioreactors. The closed tubular photobioreactors (PBRs) allow for the better distribution of light and can consequently allow for greater photosynthetic efficiency to be achieved compared with open bioreactors (Bezerra et al., 2012). The utilization of tubular PBRs for cultivating *S. platensis* is recommended for ammonium compounds because nitrogen loss by off-gassing and water evaporation can be avoided (Cruz-Martínez et al., 2015).

11.5 BIOPEPTIDE METABOLISM IN MICROALGAE

The synthesis of biopeptides by cyanobacteria/microalgae begins through nitrogen catabolism. Nitrogen (N$_2$) of atmospheric origin can be used for

N_2 fixation, or it can be obtained from components in the culture medium. Regardless of the source, to be catabolized, it is necessary that it first be reduced to ammonium (NH_4^+). This molecule is incorporated into biomolecules within glutamate and glutamine. Amino acids are derived from intermediates of glycolysis, the citric acid cycle, or the pentose phosphate pathway, and N_2 enters into these pathways in the form of glutamate or glutamine. The synthesis of polypeptides using these amino acids occurs in four stages, which is one less than protein synthesis (Figure 11.2).

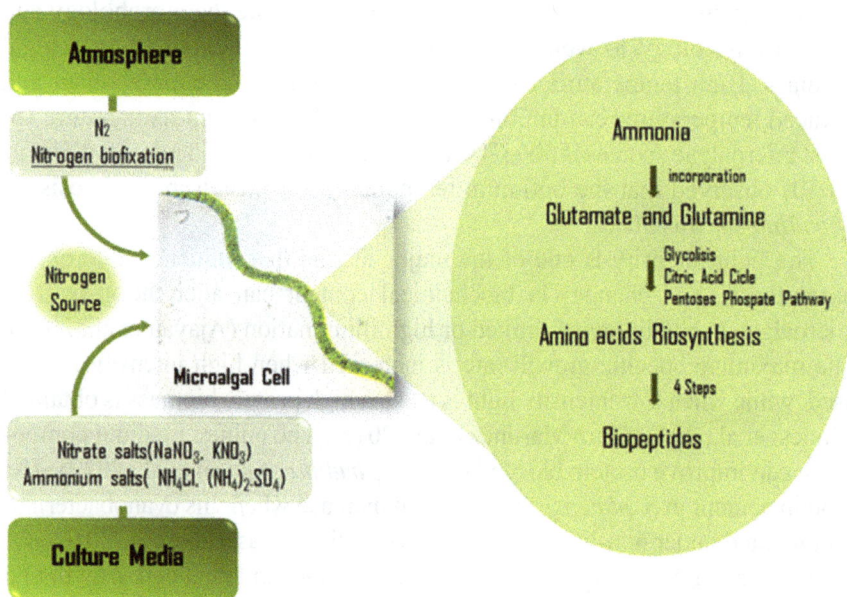

FIGURE 11.2   Synthesis of peptides by microalgae from nitrogen catabolism.

11.5.1 BIOFIXATION, REDUCTION OF NITROGEN AND INCORPORATION OF AMMONIUM INTO BIOMOLECULES

N_2 fixation is accomplished by a protein complex (nitrogenase), whose crucial components are reductase dinitrogenase and dinitrogenase. N_2 fixation occurs by a highly reduced dinitrogenase and requires eight electrons, six for the reduction of N_2 and two for the production of a hydrogen molecule (H_2). The dinitrogenase is reduced by electron transfer with the reductase dinitrogenase. A reduced reductase molecule binds to dinitrogenase and transfers a single electron. Then, the oxidized reductase dissociates from dinitrogenase and the cycle repeats. The overall equation of reducing ammonium during nitrogen fixation can be written as follows: $N_2 + 10\,H^+ + 8e^- + 16\,ATP \rightarrow 2\,NH_4^+ + 16\,ADP + 16\,P_i + H_2$ (Nelson and Cox, 2011).

The nitrate salts of the culture medium are reduced to ammonium by microalgae through a two-stage metabolic enzyme complex. In the first step, the cytosol nitrate is reduced to nitrite by the action of nitrate reductase with a worn molecule of nicotinamide adenine dinucleotide (NADH or NADPH) according to the following equation: $NO_3^- + 2H^+ + 2e^- \rightarrow NO_2^- + H_2O$. The second reaction occurs in the chloroplast of chlorophyll tissues or in plastids of chlorophyll-containing tissues, where nitrate is reduced to ammonium by the action of nitrite reductase. In photosynthetic cells, this reduction can be considered a genuinely photosynthetic process; it consumes reducing power directly from photochemical electron flow through ferredoxin. The nitrite reduction reaction can be observed in the equation: $NO_2^- + 8H^+ + 6e^- \rightarrow NH_4^+ + 2H_2O$ (Taiz and Zeiger, 2003).

The N_2 that has been reduced to NH_4^+ is incorporated into amino acids and other nitrogen biomolecules. Two amino acids, glutamine, and glutamate, provide a critical entry point. The incorporation of ammonium involves the action of glutamine synthetase, and glutamate, which combines with ammonium to generate glutamine, is produced at the expense of an ATP (adenosine triphosphate) (Nelson and Cox, 2011).

11.5.2   AMINO ACID BIOSYNTHESIS

Amino acids are intermediate derivatives of glycolysis, the citric acid cycle, and the pentose phosphate pathway. Nitrogen enters into these pathways through glutamate or glutamine. *Spirulina* can produce 20 EAAs, the amounts of which have been reported by the Food and Agricultural Organization (FAO) for use in feeding 2- to 5-year-old children (Table 11.1). Glutamine, proline, and arginine originate from the glutamate generated from α-ketoglutarate. Proline is derived from the cyclization of glutamate. Arginine is synthesized from glutamate in the form which the amino group is temporarily acetylated from acetyl-CoA. Then, through transamination, the use of ornithine and deacetylation, this acetylated glutamate form is processed to generate arginine (Nelson and Cox, 2011).

TABLE 11.1   Amino Acid Biomass and Content in *Spirulina* sp. LEB 18*

Amino acids	*Spirulina* sp. LEB 18 (%, w/w)	FAO (%, w/w)
Glutamate	10.7	–
Aspartate	9.2	–
Leucine**	8.0	6.6
Alanine	6.5	–
Phenylalanine	5.8	–
Glycine	5.2	–
Arginine	4.9	–
Threonine**	4.9	3.4
Valine**	4.6	3.5
Isoleucine**	4.4	2.8
Serine	4.3	–
Proline	4.0	–
Tyrosine	3.2	–
Lysine**	3.0	5.8
Histidine**	2.7	1.9
Tryptophan**	2.5	1.1
Methionine	1.6	–
Cysteine	0.5	–

As reported by the FAO for Use in Feeding 2- to 5-Year-Old Children.

**Essential Amino Acids (Morais et al., 2009).

Serine, glycine, and cysteine are synthesized from 3-phosphoglycerate. During synthesis, the serine hydroxyl group of the 3-phosphoglycerate is oxidized, generating 3-phosphohydroxypyruvate, which is transaminated to 3-phosphoserine and hydrolyzed. Serine is the precursor of glycine, which is formed by the removal of a carbon and water molecule through the action of serine hydroxymethyltransferase, which requires tetrahydrofolate and pyridoxal phosphate. Cysteine is produced from serine and sulfide, the latter of which is produced by reducing sulfate obtained from the environment (Nelson and Cox, 2011).

Pyruvate and oxaloacetate are precursors of nine amino acids, with alanine and aspartate being products of their direct transamination. Asparagine is generated by aspartate amidation using glutamine as an NH_4^+ donor. Aspartate is also used as a precursor to generate methionine, lysine, and threonine through complex paths that vary significantly between different organisms. Isoleucine and valine are generated from threonine by the insertion of a pyruvate and a reduction reaction. Leucine is generated from a valine precursor. Among the EAAs, aromatics such as phenylalanine, tyrosine, and tryptophan are produced by a route in which chorismate represents a key branching point. The phosphoribosyl pyrophosphate precursor used is tryptophan or histidine, and the path of histidine biosynthesis is already interconnected with that of purines. Tyrosine may also be produced by the hydroxylation of phenylalanine (Campbell and Farrell, 2006).

11.5.3 BIOSYNTHESIS OF BIOPEPTIDES

The biosynthesis of proteins is performed by ribosomes that consist of both proteins and ribosomal ribonucleic acid (rRNA). Prokaryotic microorganisms have 70S ribosomes consisting of a large subunit (50S) and a small subunit (30S). Eukaryotic ribosomes are significantly larger (80S) and contain more protein. The growth of the polypeptide chain in the ribosome starts with amino-terminal amino acid and proceeds through the successive addition of new residues at the carboxyl extremity. In the first step, amino acids are activated in the cytosol by specific aminoacyl-tRNA (transporter ribonucleic acid) synthetases. These enzymes catalyze the formation of aminoacyl-tRNA, with the simultaneous cleavage of ATP, generating adenosine monophosphate and inorganic phosphate. Synthesis fidelity depends on the accuracy of this reaction, and some of these enzymes perform additional steps at separate active sites (Nelson and Cox, 2011).

The second stage, called initiation, involves the formation of a complex between the 30S subunit of the ribosome, the messenger ribonucleic acid (mRNA), GTP (guanosine triphosphate) N-formyl-methionyl-tRNA (fMet-tRNA^{fmet}), three initiation factors and the 50S subunit; GTP is hydrolyzed to guanosine diphosphate and P$_i$ (phosphate). During the third stage, called stretching, the GTP and stretching factors are necessary for connecting the new aminoacyl-tRNA that enters the active site of the ribosome. In the first peptide transfer reaction, a methionine residue is transferred to the amino group in the new aminoacyl-tRNA. Ribosome movement along the mRNA translocates the aminoacyl-tRNA from the A to the P site, a process that requires GTP hydrolysis. The deacylated tRNA then dissociates from the site and the ribosome. In the 4th step, after many cycles of elongation, the synthesis of the biopeptide is completed with the aid of releasing factors. At least four equivalents of high energy phosphate (ATP or GTP) are required to form each peptide bond, an energy investment that is required to ensure translation fidelity (Nelson and Cox, 2011).

11.6   BIOPEPTIDE CATABOLISM

The potential effect of peptides depends on their ability to reach their target organs (Choonara et al., 2014). The conditions of the GIT, such as the presence of gastrointestinal digestive enzymes and an acidic pH (approximately 1.5 to 2) in the stomach, may affect the structure and function of peptides (Ao and Li, 2013). Peptides that resist the digestion process and arrive intact in the gut may have a local function or may be able to cross the epithelium to enter the bloodstream and exert a systemic effect (Segura-Campos et al., 2011).

The possible mechanisms by which peptides are absorbed from the GIT include the transcellular and paracellular pathways. The paracellular route is based on diffusion through elongated structures connecting neighboring cells. This mode of transport is highly dependent on water-soluble peptides but is independent of energy requirements (Saadi et al., 2015). The transcellular route involves the transport of molecules across the cell membrane, which can occur by passive diffusion, endocytosis or through a specific transporter. The transcellular passive transport of molecules across the intestinal membrane occurs through simple or facilitated diffusion along a concentration gradient (Choonara et al., 2014).

The molecules can be transported via endocytosis based on the binding affinity of the molecules to cognate cell receptors and cell penetration is maintained for vascularization (Saadi et al., 2015). Dipeptides and tripeptides can be absorbed intact through the intestinal membrane via the specific transporter peptide (PepT1) using the transmembrane electrochemical proton gradient as a broad specificity transport force. PepT1 facilitates the exit of short peptides resistant to hydrolysis from enterocytes into the bloodstream (Segura-Campos et al., 2011).

Absorption through each mechanism is determined by the physico-chemical properties of the transported radical, such as its molecular weight, structural orientation, and hydrophilic-lipophilic balance, which are related to the pKa and pH solubility profile of the biomolecule (Pawar et al., 2014). The absorption capacity of the peptides within the intestine increases as the chain length decreases and the hydrophobicity increases (Segura-Campos et al., 2011).

11.7   METHODS OF BIOPEPTIDE OBTAINMENT: CHEMICAL AND ENZYMATIC HYDROLYSIS

The production of bioactive peptides by chemical hydrolysis is based on the cleavage of peptide bonds with acids and alkalis (Figure 11.3). In industrial processes, this technique has advantages compared with enzymatic hydrolysis, due to lower associated costs and ease of execution. However, it has some limitations for use in generating food ingredients. Chemical hydrolysis is difficult to control because acids, strong alkalis, and solvents are used at high temperatures and pH values (Kristinsson and Rasco, 2000).

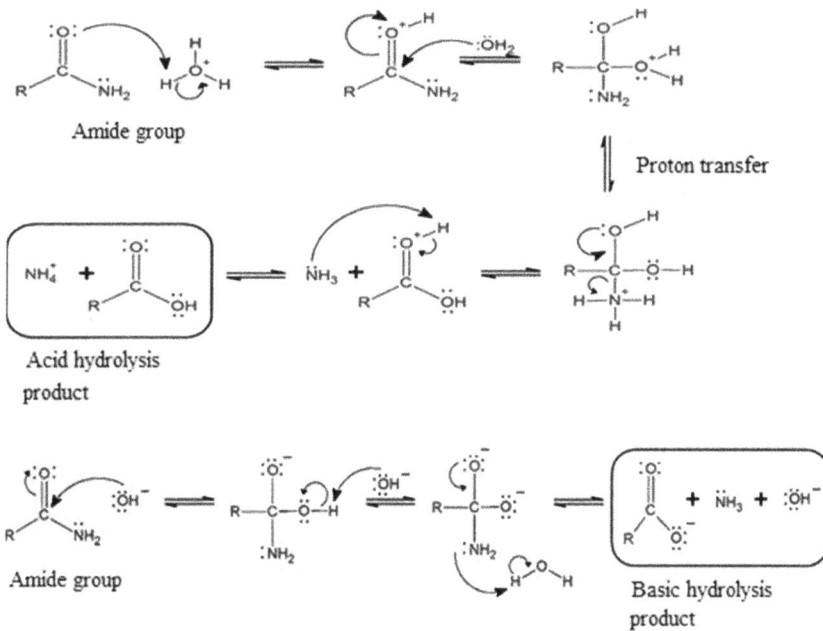

FIGURE 11.3   Representation of the chemical hydrolysis of proteins (acid and alkaline hydrolysis).

Enzymatic hydrolysis has advantages compared to the chemical modification of proteins because residual organic solvents or toxic chemicals are not used in this process. In addition, enzymes have a greater specificity relative to the substrate, reducing the probability of undesirable reactions that result in the formation of toxic products. The enzymatic hydrolysis of protein results in a reduction in molecular weight and an increase in the number of

ionizable groups and exposed hydrophobic groups, which were protected in
the original protein structure (Ovissipour et al., 2012).

During enzymatic hydrolysis, protease enzymes are used that break down
the peptide bonds of protein chains. A representative reaction catalyzed by
these enzymes is shown in Figure 11.4. In this reaction, for each peptide
bond cleaved by the protein, 1 mol of carboxyl groups (COOH) and 1 mol of
amino groups (NH_2) are released. The functional properties of the peptides
obtained by this process depend on the degree of hydrolysis (Damodaran et
al., 2007; Klein et al., 2018).

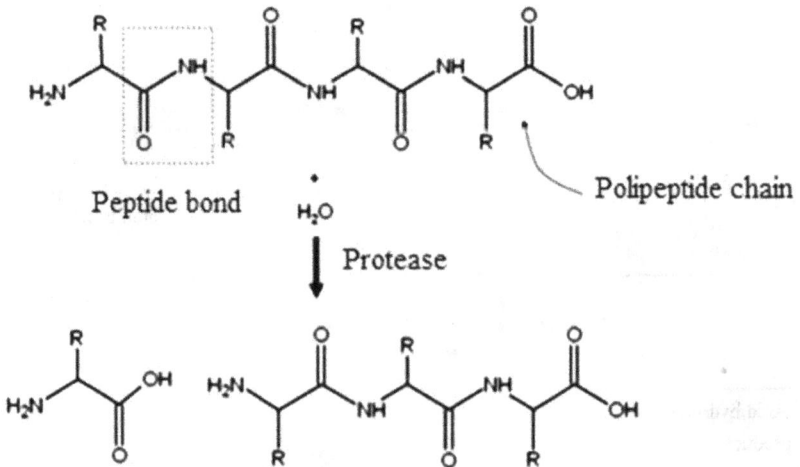

FIGURE 11.4   Representation of protein hydrolysis catalyzed by a protease.

Peptides with a purity equal to or above 95% are required for clinical
use, food consumption, biological assays, and structural studies. Thus, it has
become very common to combine chromatographic techniques with low,
medium, and high resolution to obtain high-quality peptides (Saadi et al.,
2015). Reverse-phase, ion-exchange, size exclusion-ion and affinity liquid
chromatography are the most commonly used techniques in the analysis and
purification of the crude products obtained from chemical and enzymatic
synthesis (Cavaliere et al., 2018).

11.8   APPLICATION OF PEPTIDES DERIVED FROM MICROALGAE

Biopeptides obtained from natural sources, such as microalgae, can be
used as nutritional supplements and nutricosmetics and for pharmaceutical

production since they have physiological properties. Depending on the amino acid sequence, these peptides may be involved in various biological functions, and some of them have multifunctional activities, such as antioxidant (Lisboa et al., 2016), antimicrobial (Guzmán et al., 2019), antihypertensive (Barkia et al., 2019) and anti-cholesterol activities (Li et al., 2019). The activity of these peptides will not only depend on their structure, composition, and the sequence of their amino acids but also on other factors, such as hydrophobicity, charge, or binding properties with microelements (Barkia et al., 2019).

The use of synthetic antioxidants has been evaluated due to their carcinogenic effects on the body when added to foods, nutricosmetics, and pharmaceutical compounds. Thus, increased attention has been drawn to industries that obtain antioxidants from natural and economically viable sources, such as microalgae (Guedes et al., 2013). Several assays have been used to evaluate the antioxidant capacity of peptides because there is no standard method. Some methods determine the ability of antioxidants to scavenge free radicals generated in the medium; others evaluate the effectiveness of antioxidants to inhibit lipid peroxidation by quantification of the reaction products or by measuring the inhibition of oxidation (Lisboa et al., 2016).

Peptides obtained from natural sources exhibiting antihypertensive activity are receiving attention as an alternative to the use of synthetic drugs in the treatment of hypertension (Samarakoon et al., 2013). The angiotensin-I converting enzyme (ACE) (dipeptidyl carboxypeptidase) was originally isolated from horse blood, and whose main function is to cleave the C-terminal dipeptide from angiotensin I (AI) to produce the octa-peptide vasoconstrictor angiotensin II (AII) (Cozzolino, 2016). This enzyme is related to the renin-angiotensin system (RAS) and regulates arterial blood pressure. In humans, the RAS system plays a key role in promoting the regulation of blood pressure and cardiovascular diseases, such as heart failure and hypertension (Saadi et al., 2015). Therefore, the inhibition of this enzyme can generate an antihypertensive effect. Biopeptide inhibitors of ACE act specifically on the angiotensin-converting enzyme, which catalyzes the conversion of AI to AII and controls the degradation of bradykinin and other vasoactive substances (Sheiha et al., 2009).

Smaller companies and research institutes are attempting to identify and develop alternative anti-infective peptides. Among these agents, the most promising in the current post-antibiotic world are antimicrobial peptides that are biologically active fragments composed of less than 50 amino acid residues with a molecular mass of less than 10 kDa. Most amino acid residues are hydrophobic in nature. The potential of these peptides is enhanced, due to the ability of these compounds to rapidly destroy large numbers of drug-resistant microorganisms, such as bacteria, viruses, and fungi. Several methods have been reported to identify the biological role of bioactive peptides from protein hydrolysates. Included among these methods is the agar diffusion assay, also known as the inhibition zone assay. This test is a physical method in which the microorganism is exposed to a biologically active substance on a solid medium, and the size of the inhibition zone of the microorganism and the concentration of the active substance used are related to its biological activity (Ngo et al., 2012; Saadi et al., 2015).

11.9   CONCLUSION

Bioactive peptides obtained from *Spirulina* provide a great biotechnological and biochemical advancement due to their potential biological activities and consequent health benefits. This development is characterized by the high protein content present in the biomass of this microalgae, which is a great source of biopeptides with bioactive properties, such as antioxidant, anti-inflammatory, antihypertensive, and antimicrobial activities, among others. Furthermore, during the cultivation of microalgae, there is the possibility of altering the extracellular media to direct the metabolism of these microorganisms to produce bioactive compounds of interest. The absorption of these biopeptides may have local target functions, or they may enter the bloodstream and provide a systemic effect. These bioactive compounds may contribute to the proper functioning of the organs as well as may help combat diseases and can be added to the formulation of foods, cosmetics, and medicines.

KEYWORDS

- amino acid biosynthesis
- bioactive compounds
- bioactive properties
- biopeptides
- chemical hydrolysis
- enzymatic hydrolysis

- **essential amino acids**
- **microalgae**
- **peptide culture conditions**
- **peptides absorption**
- **proteins**
- ***Spirulina* sp. LEB 18**

REFERENCES

Ajayan, K. V., Selvaraju, M., & Thirugnanamoorthy, K., (2012). Enrichment of chlorophyll and phycobiliproteins in *Spirulina platensis* by the use of reflector light and nitrogen sources: An *in-vitro* study. *Biomass and Bioenergy, 47*, 436–441.

Ao, J., & Li, B., (2013). Stability and antioxidative activities of casein peptide fractions during simulated gastrointestinal digestion *in vitro*: Charge properties of peptides affect digestive stability. *Food Research International, 52*(1), 334–341.

Barkia, I., Saari, N., & Manning, S. R., (2019). Microalgae for high-value products towards human health and nutrition. *Marine Drugs, 17*(5), 304. doi: https://doi.org/ 10.3390/md17050304.

Bezerra, R. P., Matsudo, M. C., Sato, S., Perego, P., Converti, A., & Carvalho, J. C. M., (2012). Effects of photobioreactor configuration, nitrogen source and light intensity on the fed-batch cultivation of *Arthrospira* (*Spirulina*) *platensis*. Bioenergetic aspects. *Biomass and Bioenergy, 37*, 309–317.

Campbell, M. K., & Farrell, S. O., (2006). *Bioquímica* (Vol. 3, p. 360). Thomson Learning, São Paulo.

Cavaliere, C., Capriotti, A. L., La Barbera, G., Montone, C. M., Piovesana, S., & Laganà, A., (2018). Liquid chromatographic strategies for separation of bioactive compounds in food matrices. *Molecules, 23*(12), 3091. doi: https://doi.org/10.3390/molecules23123091.

Choonara, B. F., Choonara, Y. E., Kumar, P., Bijukumar, D., Du Toit, L. C., & Pillay, V., (2014). A review of advanced oral drug delivery technologies facilitating the protection and absorption of protein and peptide molecules. *Biotechnology Advances, 32*(7), 1269–1282.

Cozzolino, S. M. F., (2016). *Biodisponibilidade de nutrientes* (5th edn., p. 1482). Editora Manole, São Paulo.

Cruz-Martínez, L. C., Jesus, C. K. C., Matsudo, M. C., Danesi, E. D. G., Sato, S., & Carvalho, J. C. M., (2015). Growth and *composition of Arthrospira* (*Spirulina*) *platensis* in a tubular photobioreactor using ammonium nitrate as the nitrogen source in a fed-batch process. *Brazilian Journal of Chemical Engineering, 32*(2), 347–356.

Damodaran, S., Parkin, K. L., & Fennema, O. R., (2007). *Fennema's Food Chemistry* (4th edn., p. 1160). CRC Press, New York.

Danesi, E. D. G., Rangel-Yagui, C. O., Sato, S., & Carvalho, J. C. M., (2011). Growth and content of *Spirulina platensis* biomass chlorophyll cultivated at different values of light intensity and temperature using different nitrogen sources. *Brazilian Journal of Microbiology, 42*(1), 362–373.

Fan, X., Bai, L., Zhu, L., Yang, L., & Zhang, X., (2014). Marine algae-derived bioactive peptides for human nutrition and health. *Journal of Agricultural and Food Chemistry, 62*(38), 9211–9222.

Geada, P., Rodrigues, R., Loureiro, L., Pereira, R., Fernandes, B., Teixeira, J. A., Vasconcelos, V., & Vicente, A. A., (2018). Electrotechnologies applied to microalgal biotechnology: Applications, techniques, and future trends. *Renewable and Sustainable Energy Reviews, 94*, 656–668.

Guedes, A. C., Gião, M. S., Seabra, R., Ferreira, A. C. S., Tamagnini, P., Moradas-Ferreira, P., & Malcata, F. X., (2013). Evaluation of the antioxidant activity of cell extracts from microalgae. *Marine Drugs, 11*(4), 1256–1270.

Gutiérrez-Salmeán, G., Fabila-Castillo, L., & Chamorro-Cevallos, G., (2015). Nutritional and toxicological aspects of *Spirulina (Arthrospira)*. *Nutrición Hospitalaria, 32*(1), 34–40.

Guzmán, F., Wong, G., Román, T., Cárdenas, C., Alvárez, C., Schmitt, P., Albericio, F., & Rojas, V., (2019). Identification of antimicrobial peptides from the microalgae *Tetraselmis suecica* (Kylin) Butcher and bactericidal activity improvement. *Marine Drugs, 17*(8), 453. doi: https://doi.org/ 10.3390/md17080453.

Henrikson, R., (1994). *Microalga Spirulina· Superalimento Del Future* (p. 224). Ediciones S. A. Urano, Barcelona.

Ibañez, E., Herrero, M., Mendiola, J. A., & Castro-Puyana, M., (2012). Extraction and characterization of bioactive compounds with health benefits from marine resources: Macro and microalgae, cyanobacteria, and invertebrates. In: Hayes, M., (ed.), *Marine Bioactive Compounds: Sources, Characterization and Applications* (pp. 55–98). Springer, New York.

Kang, K. H., & Kim, S. K., (2013). Beneficial effect of peptides from microalgae on anticancer. *Current Protein and Peptide Science, 14*(3), 212–217.

Karkos, P. D., Leong, S. C., Karkos, C. D., Sivaji, N., & Assimakopoulos, D. A., (2011). *Spirulina* in clinical practice: Evidence-based human applications. *Evidence-Based Complementary and Alternative Medicine, 2011*, 531053. doi: https://doi.org/ 10.1093/ecam/nen058.

Klein, T., Eckhard, U., Dufour, A., Solis, N., & Overall, C. M., (2018). Proteolytic cleavage-mechanisms, function, and "omic" approaches for a near-ubiquitous posttranslational modification. *Chemical Reviews, 118*, 1137–1168.

Koyande, A. K., Chew, K. W., Rambabu, K., Tao, Y., Chu, D. T., & Show, P. L., (2019). Microalgae: A potential alternative to health supplementation for humans. *Food Science and Human Wellness, 8*, 16–24.

Kristinsson, H. G., & Rasco, B. A., (2000). Kinetics of the hydrolysis of Atlantic salmon (*Salmo salar*) muscle proteins by alkaline proteases and a visceral serine protease mixture. *Process Biochemistry, 36*(1, 2), 131–139.

Li, Y., Lammi, C., Boschin, G., Arnoldi, A., & Aiello, G., (2019). Recent advances in microalgae peptides: Cardiovascular health benefits and analysis. *Journal of Agricultural and Food Chemistry, 67*(43), 11825–11838.

Lisboa, C. R., Pereira, A. M., & Costa, J. A. V., (2016). Biopeptides with antioxidant activity extracted from the biomass of *Spirulina* sp. LEB 18. *African Journal of Microbiology Research, 10*(3), 79–86.

Lisboa, C. R., Pereira, A. M., Ferreira, S. P., & Costa, J. A. V., (2014). Utilization of *Spirulina* sp. and *Chlorella pyrenoidosa* biomass for the production of enzymatic protein hydrolysates. *Journal of Engineering Research and Applications, 4*(5), 29–38.

Martinez-Palma, N., Martinez-Ayala, A., & Davila-Ortiz, G., (2015). Determination of antioxidant and chelating activity of protein hydrolysates from *Spirulina (Arthrospira maxima*) obtained by simulated gastrointestinal digestion. *Revista Mexicana de Ingeniería Química, 14*(1), 25–34.

Masojídek, J., Torzillo, G., & Koblížek, M., (2013). Photosynthesis in microalgae. In: Richmond, A., & Hu, C., (eds.), *Handbook of Microalgal Culture: Applied Phycology and Biotechnology* (p. 17). Wiley Blackwell, New Jersey.

Morais, M. G., Radmann, E. M., Andrade, M. R., Teixeira, G. G., Brusch, L. R. F., & Costa, J. A.V., (2009). Pilot scale semicontinuous production of *Spirulina* biomass in southern Brazil. *Aquaculture, 294*, 60–64.

Morais, M. G., Vaz, B. S., Morais, E. G., & Costa, J. A. V., (2015). Biologically active metabolites synthesized by microalgae. *Biomed Research International, 2015*, 835761. doi: https://doi.org/ 10.1155/2015/835761.

Murray, J. E., Laurieri, R., & Delgoda, R., (2017). Proteins. In: Badal, S., & Delgoda, R., (eds.), *Pharmacognosy, Fundamentals, Applications and Strategies* (pp. 477–494), Academic Press.

Nelson, D. L., & Cox, M. M., (2011). *Princípios de Bioquímica de Lehninger* (5th edn., p. 1304). Artmed, Porto Alegre.

Ngo, D. H., Vo, T. S., Ngo, D. N., Wijesekara, I., & Kim, S. K., (2012). Biological activities and potential health benefits of bioactive peptides derived from marine organisms. *International Journal of Biological Macromolecules, 51*(4), 378–383.

Ovissipour, M., Kenari, A., Motamedzadegan, A. A., & Nazari, R. M., (2012). Optimization of enzymatic hydrolysis of visceral waste proteins of yellowfin tuna (*Thunnus albacares*). *Food and Bioprocess Technology, 5*(2), 696–705.

Pandey, A., Lee, D. J., Chisti, Y., & Soccol, C., (2014). *Biofuels From Algae* (1st edn., p. 348). Elsevier.

Pandey, J. P., & Tiwari, A., (2010). Optimization of biomass production of *Spirulina maxima*. *Journal of Algal Biomass Utilization, 1*(2), 20–32.

Pawar, V. K., Meher, J. G., Singh, Y., Chaurasia, M., Reddy, S., & Chourasia, M. K., (2014). Targeting of gastrointestinal tract for amended delivery of protein/peptide therapeutics: Strategies and industrial perspectives. *Journal of Controlled Release, 196*(28), 168–183.

Saadi, S., Saari, N., Anwar, F., Hamid, A. A., & Ghazali, H. M., (2015). Recent advances in food biopeptides: Production, biological functionalities, and therapeutic applications. *Biotechnology Advances, 33*(1), 80–116.

Samarakoon, K. W., O-Nam, K., Ko, J. Y., Lee, J. H., Kang, M. C., Kim, D., Lee, J. B., et al., (2013). Purification and identification of novel angiotensin-I converting enzyme (ACE) inhibitory peptides from cultured marine microalgae (*Nannochloropsis oculata*) protein hydrolysate. *Journal of Applied Phycology, 25*, 1595–1606.

Sassano, C. E. N., Gioielli, L. A., Ferreira, L. S., Rodrigues, M. S., Sato, S., Converti, A., & Carvalho, J. C. M., (2010). Evaluation of the composition of continuously-cultivated *Arthrospira* (*Spirulina*) *platensis* using ammonium chloride as nitrogen source. *Biomass and Bioenergy, 34*(12), 1732–1738.

Segura-Campos, M., Chel-Guerrero, L., Betancur-Ancona, D., et al., (2011). Bioavailability of bioactive peptides. *Food Reviews International, 27*(3), 213–226.

Sheiha, C., Fanga, T. J., & Wub, T., (2009). Isolation and characterization of a novel angiotensin I-converting enzyme (ACE) inhibitory peptide from the algae protein waste. *Food Chemistry, 115*(1), 279–284.

Sornchai, P., & Iamtham, S., (2013). Effects of different initial pH of modified Zarrok medium on large-scale *Spirulina maxima* culture. *Journal of Medical and Bioengineering, 2*(4), 266–269.

Taiz, L., & Zeiger, E., (2003). *Fisiologia Vegetal* (p. 719). Artmed, Porto Alegre.

Vaz, B. S., Costa, J. A. V., & Morais, M. G., (2016). CO_2 biofixation by the cyanobacterium *Spirulina* sp. LEB 18 and the green alga *Chlorella fusca* LEB 111 grown using gas effluents and solid residues of thermoelectric origin. *Applied Biochemistry and Biotechnology, 178*(2), 418–429.

Vonshak, A., (1997). *Spirulina platensis (Arthrospira) Physiology, Cell-Biology and Biotechnology* (p. 233). Taylor and Francis, New York.

Yücetepe, A., & Özçelik, B., (2016). Bioactive peptides isolated from microalgae *Spirulina platensis* and their biofunctional activities. *Akademik Gıda, 14*(4), 412–417.

CHAPTER 12

BIOTECHNOLOGICAL POTENTIALS OF MACROALGAE AS A SOURCE OF PHYCOBILIPROTEINS

RUPALI KAUR,[1] MD. AKHLAQUR RAHMAN,[2] and SHANTHY SUNDARAM[1]

[1]Center of Biotechnology, Nehru Science Center, University of Allahabad, Prayagraj, Uttar Pradesh–211002, India

[2]Department of Biotechnology, S.S. Khanna Girls' Degree College, Prayagraj, Uttar Pradesh–211003, India

ABSTRACT

Phycobiliproteins (PBPs) are a unique class of water-soluble proteins linked with chromophore, which is mainly responsible for the light capture in cyanobacteria. Four main class types are of phycocyanin (PC), phycoerythrin (PE), phycoerythrocyanin, and allophycocyanin (APC). These can also be categorized depending on their structure and light quality capture. PBPs from cyanobacteria have been known to have high value products and bioactive compounds. These biliproteins have anti-inflammatory, antioxidant, anticancer activities. So, for these biliproteins, cyanobacteria have been exploited in industrial applications. The aim of this chapter is to give an overview of PBPs not only in respect of their chemistry but also in terms of their biotechnological applications and the advancements, opportunities in the production of such compounds.

12.1 INTRODUCTION

Cyanobacteria are important photosynthetic oxygen-producing prokaryotes that came into existence during the Precambrian era and are considered as the oldest known fossils with >3.5 billion years old (Fischer, 2008; Chaffey et al., 2014). They are one of the largest and most important groups of bacteria on earth. These primeval groups of Gram-negative prokaryotes have a cosmopolitan distribution ranging from hot springs to the Arctic and Antarctic regions (Stanier and Cohen-Bazire, 1977). Such phototrophs are found in harsh environmental conditions as in deserts, mountains, and hot-springs with their ability of high adaptation potentials (Pagel et al., 2013). Being the primary producers of both aquatic and terrestrial communities, they also have the ability to produce chlorophyll and a number of valuable compounds such as carotenoids (used as antioxidants) as accessory pigments to capture incident light energy. Solar energy at wavelengths of 400–700 nm (photosynthetically active radiation) is absorbed by photosynthetic organisms and transformed into chemical energy that can be used directly by living organisms. Distinct from higher plants and green algae, major light-harvesting antennae in cyanobacteria and red algae are a large multi-subunit protein complex called phycobilisome (PBS), which was discovered in the 1960s (Gantt and Conti, 1965, 1966). PBS which captures solar energy and transfer it into photosystems (PSs) with particularly high efficiency can be categorized into three morphological types: hemiellipsoidal, hemi-discoidal, and bundle shaped (Gantt and Lipschultz et al., 1972; Guglielmi et al., 1981; Elmorjani et al., 1986; Glauser et al., 1992) (Figure 12.1). All three types of PBSs are made up of water-soluble phycobiliproteins (PBPs) and hydrophobic linker peptides (Liu et al., 2016). These blue-green algae have also been studied for their morphology, their mechanism of photosynthesis, nitrogen fixation, and also for certain characteristics of their structure (Yadav et al., 2011). Many different activities are shown by these PBPs like anticancerous, neuroprotective, anti-inflammatory, hepatoprotective, and hypocholesterolemic activities. Recent research studies have found that cell extracts of red pigment from the macroalgae *Gracilaria vermiculophylla* and the blue pigment from the *Arthrospira platensis* showed uniform distribution on the cotton and wool fabrics, with results representing the sustainability and the quality of naturally dyed textiles (Ferrándiz et al., 2016; Moldovan et al., 2017). Using acetone/ammonium sulfate precipitation, gel filtration and ion-exchange chromatography techniques can lead to enhancement for the purification of PBPs (Kannaujiya and Sinha, 2016). Recent research reports have shown that PC has health likely and a broad range of potential pharmaceutical applications. Due to the improvements in purification, there is an increase in the application of individual PC, PE, and APC in the field of biomedical and biotechnological sciences. High-quality value-added products from PBP are mainly dependent on industrialization and commercialization. These can aid as potential fluorescent labels or markers proteins in immunological laboratory. Many private companies have conventional research and development wing primarily focusing on commercial exploitation of these PBPs (Thajuddin and Subramanian et al., 2005). This chapter outlines widespread and prosperous applications of PBPs in the fields of biotechnology, pharmacology, and food applications.

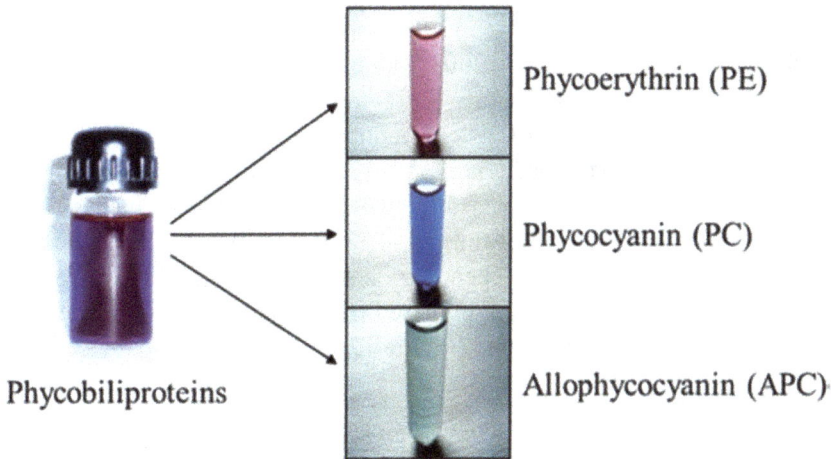

FIGURE 12.1   Color compositions of phycobiliproteins.

12.2   PHYCOBILIPROTEIN (PBP): DEFINITION, TYPES, STRUCTURE, AND STABILITY

Cyanobacteria photosynthetic apparatus comprises of three principal light-harvesting systems: two main PSs (also found in other photosynthetic organisms) and a PBS (Masojıdek et al., 2013). Light-harvesting antenna complexes (PBPs) are the main factor in cyanobacteria and red algae that display different colors and play an important role in the capturing of light energy from the sun. Near about 50% of energy captured is moved to PS for fixation of carbon during photosynthesis (Tavanandi et al., 2018). Water-soluble, colored, autofluorescent pigment proteins carry covalently attached linear tetrapyrrole pigments linked structurally to biliverdin (BV). These are isolated from organisms either as trimers or hexamers. These biliproteins are composed of proteins and are covalently bound with cysteine amino acid chromophores called phycobilins, belonging to open-chain tetrapyrrole (Eriksen, 2008; de Jesus Raposo et al., 2013; Watanabe and Ikeuchi, 2013; Mulders et al., 2014). These proteins constitute up to 60% of total soluble proteins (TSPs) of the cell (Grossman et al., 1993). These biliproteins are named for their unique color. There are three main classes of PBPs which are phycocyanin (PC, blue pigment), allophycocyanin (APC, bluish-green pigment) and phycoerythrin (PE, red pigment) which have maximum absorbance (Amax) at 620 nm, 650 nm, and 565 nm respectively. Also, on the basis of the species, PEs are subdivided into three groups: B-phycoerythrin (B-PE; λ_{max} = 546–565 nm), C-phycoerythrin (C-PE; (λ_{max} = 565 nm), and R-phycoerythrin (R-PE; (λ_{max} = 545–565 nm) (Bermejo et al., 2003; Sun et al., 2003). B-PE is purified and isolated from red alga *P. cruentum*, R-PE found in red macroalgae however CU-PE is found in marine forms of cyanobacteria (Six et al., 2007). C-PE is found in most of cyanobacteria (Table 12.1). All PBS complexes that are associated with PSII are composed of two structures: a core that is always made up of APC and a suite of rods arising out the core, which may contain PC, PC, and PEC and/or PE depending on species and growth environment (Ho et al., 2017, 2019). The color of these PBPs is mainly due to the prosthetic group, which is covalently bound (Figure 12.2).

TABLE 12.1 Spectral Attributes of Phycobiliproteins

Types of Phycobiliproteins	Absorption Maxima (nm)	Absorption Color (Visible Spectra)	Fluorescence Emission Maxima (nm)	Fluorescence Color (UV Radiation)
C-Phycoerythrin (C-PE)	540–565	Pink, red	568–575	Yellow-orange
Phycoerythrocyanin (PEC)	570–590	Purple, light red	600–610	Orange
R-Phycoerythrin (R-PE)	562–565	Red	570–575	Yellow-orange
C-Phycocyanin (C-PC)	610–620	Violet, purple	630–650	Strong red
R-Phycocyanin (R-PC)	617–620	Purple, violet	635–638	Strong red
Allophycocyanin (APC)	650–655	Violet, gray	660–665	Frail red

Blue-colored phycocyanobilin (PCB), purple-colored phycobiliviolin (PVB), red-colored phycoerythrobilin (PEB), and yellow-colored phycourobilin (PUB) are the open-chain tetrapyrrole rings named A, B, C, and D phycobilin rings due to which these biliproteins have characteristic color. These biliproteins are organized in supramolecular complexes known as PBS. These can be found arranged in rows and attached to photosystem II (PS II) on the external surface of thylakoid membranes (Glazer, 1985).

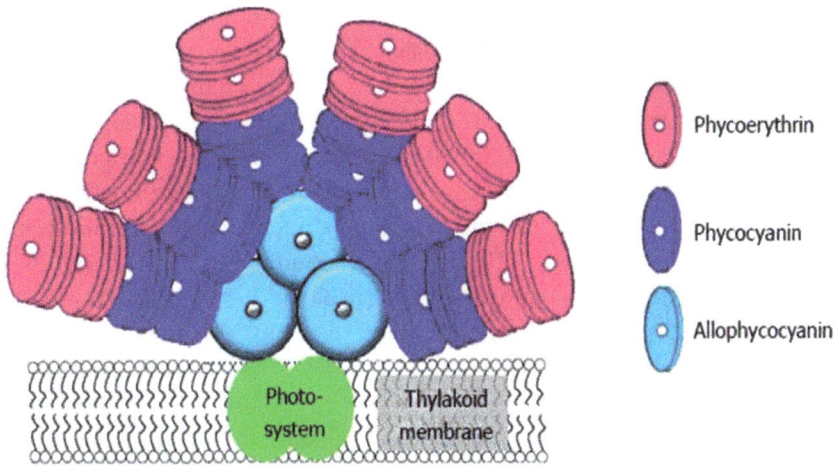

FIGURE 12.2 Structure of phycobiliproteins subunits phycocyanin (PC), phycoerythrin (PE), and allophycocyanin (APC).

Each PBP is made by two different polypeptides denoted α and β, which are similar in sequence and originated from ancient gene duplication events. Each α or β-subunit has at least one phycobilin. PBPs are usually represented as trimers $(\alpha\beta)_3$ or hexamers $(\alpha\beta)_6$, linked by specific proteins. The structure of PBPs is based on the type of phycobilin in the molecule. The formation of PBP needs posttranslational modifications with the addition of at least one of the different phycobilins by lyases and thereby following the methylation of an asparagine residue on β-subunits (Dehne, 1986; Farrar and Lewis, 1987; Snellgrove et al., 1987), adenosylmethionine-dependent methyltransferase, and leading to final proper structure of PBP (Schluchter et al., 2010). Most commonly occurring PBP in cyanobacteria is PC. This blue PC is made up of two α and β subunits with a hexameric conformation $(\alpha\beta)_6$ at pH 5.0–6.0 and a trimeric conformation $(\alpha\beta)_3$ at pH 7.0. The only chromophore present in this biliprotein is phycocyanobilin (Glazer, 1994; Dumay and Morançais, 2016). In the case of PE, it is mostly found in red algae (R-PE) and also in cyanobacteria (C-PE). It has a hexameric structure $(\alpha\beta)_6$ with four PEB bound to the α subunit and two to the β subunits (Glazer, 1994; Dumay and Morançais, 2016). In some cyanobacteria, phycoerythrocyanin may be present as an additional form of PE. Phycoerythrocyanin is present in a trimeric $(\alpha\beta)_3$ or hexameric $(\alpha\beta)_6$ form and has phycoviolobilin as chromophore (Glazer, 1994; Dumay and Morançais, 2016). Whereas, APC is composed of a trimer $(\alpha\beta)_3$ and is used as a channel in the energy transfer through the PSs and absorb the energy from the previous PBPs and then transfer the energy to the photosystem I (PSI) (Lundell and Glazer, 1981; Glazer, 1994; Dumay and Morançais, 2016). The evolutionary pattern of core chains of PBP is from the same ancestor (Thomas and Passaquet, 1999). For proper effectiveness of PC, PE, and APC both α and β-chains are important components. A diverse group of amino acids are also found attached with linear tetrapyrrole chromophore at cysteine positions (Kannaujiya et al., 2014). The molecular weights of α subunits are 12–20 kDa and β subunits is 15–22 kDa in PBPs (Sinha et al., 1995; Kannaujiya et al., 2014). It has been also reported that phycobilins maybe also one of the main components responsible for the bioactivity of PBPs (Hirata et al., 2000). Having these properties, the pigment-protein complexes have special potential biotechnological applications in nutraceuticals and pharmaceuticals, cosmetics, food industry, biomedical research, and clinical laboratories.

12.3   CULTIVATION OF CYANOBACTERIA

With the increase in world population and its demand for food has led to the development of more natural proteins (Foley et al., 2011). Keeping this thing in mind, cyanobacteria can be considered as good source of proteins. So, increase in biomass of cyanobacteria can indirectly be a rich source of carbon for feedstock. Many countries including the USA, Japan, Germany are cultivating cyanobacteria in bulk (Borowitzka, 1999). The production of microalgal biomass is very economical (Vandamme et al., 2013). For the production of biomass light energy, macronutrients (carbon, hydrogen, nitrogen, oxygen, etc.), and micronutrients, and different amounts of CO_2 concentration are needed. Some microalgal species need organic carbon which is more advantageous than autotrophic culture (Perez-Garcia et al., 2011). Seeing the increasing demand for microalgal biomass, its production can be done through photobioreactor (Morweiser et al., 2010). Some researchers are trying to reduce the cost of biomass production by downstream production of biomass (Chen et al., 2010; Greenwell et al., 2010). Production of microalgae using standard culture media can lead to its high cost. The cost for production can be minimized by changing media culture concentrations which can be achieved through dilution (Delrue et al., 2017) or it can be done through alternative components, including wastewater (Van Den Hende et al., 2016b; Arashiro et al., 2019; Kumar et al., 2020). Cyanobacteria like *Spirulina* sp. and *Nostoc* sp. have been produced on a commercial scale in several countries from the point of view of health and food (Dillon et al., 1995). For growth of microalgae, three systems are heterotrophically, mixotrophically or photoautotrophically. In recent years many new technologies have defined an improved version of closed photobioreactor (PBR) to get high value products for commercial-scale (Figure 12.3). Considering different advantages like clean culture, proper source of light, maintenance of biomass and temperature control makes closed PBR a good option (Chrismadha and Borowitzka, 1994). One of the best methods for production of microalgae is done by photoautotrophic means, which is an outdoor method in open sunlight (Jiménez et al., 2003). *Spirulina* is very easy to grow in an open environment as it highly tolerable even at alkaline pH. Most recently closed PBR are being designed of high value food, pharma, and cosmetic products (Xie et al., 2015). Some of microalgae are also dependent on light sources that are heterotrophic growth and mode shows higher growth rate when compared to light-dependent condition (Kuddus et al., 2013). For better efficiency by adjusting the reactor size/volume heterotrophic culture of microalgae can be scaled up at a large scale. *Galdieria sulphuraria* 074G is a prospective organism for better productivity of PC in heterotrophic cultures (Schmidt et al., 2005; Graverholt and Eriksen, 2007; Kuddus et al., 2013). But from point of view of growth, hetertophic growth of *Galdieria sulphuraria* was less in comparison to *Arthrospira platensis*; yet PC production was found to somewhat high (0.86 g L^{-1} day^{-1}) (Graverholt and Eriksen, 2007). The importance of nutrient optimization is one strategy to enhance biomass and PBPs production by cyanobacteria (Kuffner et al., 2001; Li et al., 2019; Arashiro et al., 2020). The nutrients present in the feedstock are utilized by

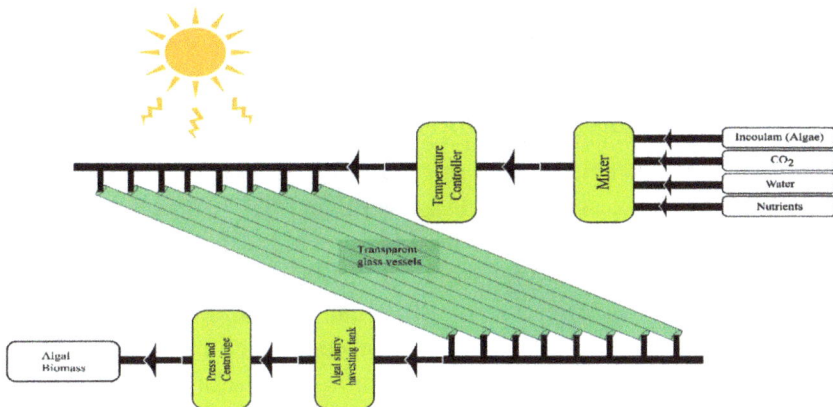

FIGURE 12.3 The graphical illustration of the arrangement and operation of mostly used photo-bioreactor using sunlight for the large-scale production of algal biomass.

microalgae to produce value products such as nutraceuticals, food additives, biopolymers, pigments, growth promoters, energy (biogas) or biofertilizers (Renuka et al., 2018; Arashiro et al., 2019; Hemalatha et al., 2019; Karan et al., 2019; Khan et al., 2019). Furthermore, microalgal photosynthesis adds to CO_2 sequestration and their cultivation does not compete for land with other agricultural activities (Singh et al., 2017; Costa et al., 2018).

For mixotrophic culture, the optimal growth rate of cyanobacteria was improved by the addition of glucose syrup (Marquez et al., 1993) when linked to photo and heterotrophic conditions. So, the growth rate of cyanobacteria in mixotrophic cultivation is faster in closed bioreactor when compared to phototrophic reactors (Chojnacka and Noworyta et al., 2004).

12.4 PURIFICATION OF PBP

Coming to commercial scale for any product it needs to be purified to its purest and finest form. Similarly, these PBP need to be purified for its wide applications. But the extraction process of these pigments is very difficult as these have small cell size and extremely thick mucilaginous cell wall which is difficult to break (Stewart and Farmer, 1984; Wyman, 1992). Different physical and chemical methods have been employed for the extraction of PBPs depending on the type of cyanobacteria (Kannaujiya et al., 2017a). Methods like ultrasonication, liquid nitrogen, freeze-thaw, nitrogen cavitation, microwave-assisted extraction (MAE) and aqueous two-phase system (APTS) are being employed for proper extraction of PBP from red algae and cyanobacteria. Various chemicals are used for proper disruption of thick cell wall and for inhibition of protease degrading enzyme that denatures PBPs like 25 mM PB (phosphate buffer) with 10 mg/ml lysozyme (Kilpatrick, 1985), 1% cellulose with 0.1% pectolyase or lysozyme with acetate buffer, 250 mM Trizma with 10 mM ethylene diamine tetraacetic acid (EDTA) in PB (Viskari and Colyer, 2003), 1% rivanol in PB (Minkova et al., 2003), 1 M PB and NP-40 detergent (Zhao et al., 2014), EDTA, microwave method (Juin et al., 2015) and combination with phenylmethanesulfonyl fluoride (PMSF) with sucrose (Kannaujiya and Sinha, 2015). Purification of PBPs can be achieved by techniques like gel filtration, ion exchange, hydrophobic interaction, hydroxyapatite, and affinity chromatography (Kannaujiya et al., 2017a). Many alternate techniques for quick purification of PBPs is done by help of PEG 4000 with PB (Zhao et al., 2014). In recent times, affinity chromatography has been employed for the purification of PBPs components at small-scale level (Munier et al., 2015). For PE purification, the combination of gel filtration (Sephacryl S-100 HR) and HIC (HiTrap Phenyl FF) chromatography has attained the highest purity index which is 11.53 (Kannaujiya and Sinha, 2016a). Some researchers have also reported that cyanobacteria and red algae can be freeze-dried before mixing with water or extraction buffer, thereby showing enhancement in the productivity of crude extract (Kissoudi et al., 2018).

12.5   APPLICATIONS OF PHYCOBILIPROTEINS (PBPs)

For many years researchers are trying to use microalgae in conventional agriculture for the manufacture of value-added biomass for human benefit. These biliproteins are in great demand these days due to their natural origin. These PBP have been developed on a commercial scale for the manufacture of high value products for different sectors like food, pigments, pharmaceutics, biomedical, nutraceuticals, and in various biotechnology industries (Kannaujiya et al., 2017b) (Figure 12.4).

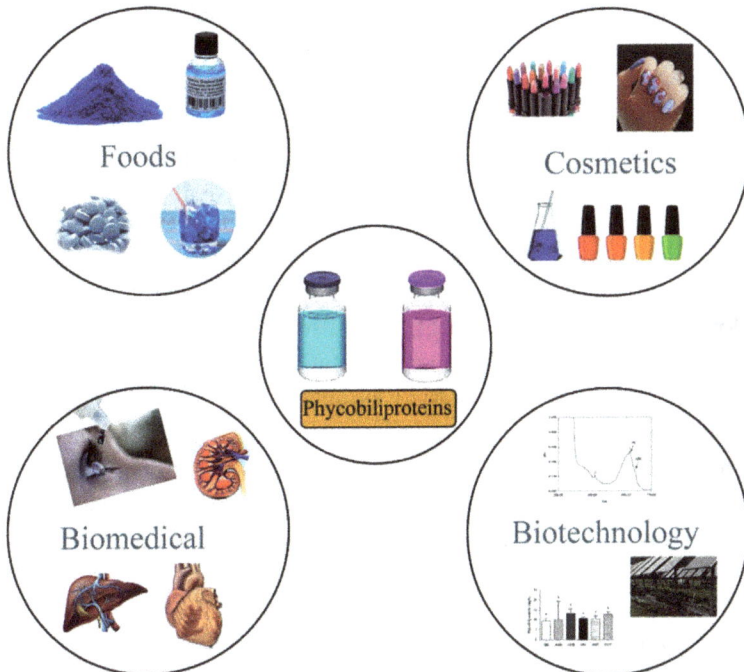

FIGURE 12.4   Role of phycobiliproteins in various fields.

12.5.1 FOOD INDUSTRY

The growing demand for more natural food products is pushing the food and beverage industry towards the use of natural preservatives which are non-toxic in nature. Mostly these coloring agents are used to improve the overall attractiveness of the food for the customers. Nowadays, a number of natural and synthetic additives are used to color foods. These biliproteins are generally regarded as safe (GRAS). Seaweeds are very important foods, containing great amounts of polysaccharides, minerals, proteins, lipids, polyphenols, and vitamins with necessary properties (Arasaki and Arasaki, 1983; Kumar et al., 2008). In the food industry, the utilization of PBP nowadays is increasing due to their many known properties. These water-soluble biliproteins are protein bound forms having enhanced solubility and stability at pH 4 to 7. One of the biliproteins that is PE used widely for red coloration instead of anthocyanin. This red color chromoprotein has yellow fluorescence and is being exploited for a new range of food additives like cakes, jellies, jam, lollipops, and alcoholic beverages. The red PE is mainly produced by microalgae (*Phorphyridium* species) with a yield of 200 mg/L and up to 15% of the dry weight (Duffos et al., 2005). C-PC from *Spirulina platensis* is marketed as a food and cosmetics colorant in Japan (Prasanna et al., 2007). Recent research on yogurt contained PC is showing beneficial satisfaction to their health (Mohammadi-Gouraji et al., 2018). Skim milk containing cyanobacterial PBPs is a new innovation for the food industry. PBP stability improvements by changes at its primary structure and the incorporation of freeze-dried PBPs into pouches should be considered as alternatives for their future commercialization (Galetović et al., 2020). Light, heat, pH, color, stability, and temperature are the important factors for the use as natural colorant in food items. Different research related to the color stability (Jespersen et al., 2005; Mishra et al., 2008; Chaiklahan et al., 2012) and rheological properties (Batista et al., 2006) of PC are been done. In the industrial sector, food supplements and nutraceuticals from *Spirulina* sp. are being employed in medicine. DIC International (formerly Dainippon Ink and Chemicals) is one of the biggest companies involved in food product production from *Spirulina* sp.; its estimated gross production is over 300 tons per annum and is increasing over time (Lee, 1997; DIC, 2009). Linablue, a product of PC, manufactured commercially by Dainippon Ink and Chemicals, Tokyo, Japan, is being marketed as food colorant and cosmetics (Dainippon Ink and Chemicals, 1985). Commercially, PBPs are manufactured from *Rhodella* sp., *Porphyridium* sp. and *Spirulina* sp. for preparation of food additives and dyes (Singh et al., 2005; Spolaore et al.,

2006). However, blue food coloring (PC) is not considered to be commercial food grade in European countries (Prasanna et al., 2007; Eriksen, 2008). Recently the FDA and EFSA have given approval for the use of a *Spirulina* (*Arthrospira*) *platensis* extract containing high levels of PC as natural blue food colorant for coloring candy and chewing gum (Code of Federal Regulation, 2016). The world's largest *Spirulina* farms in California, USA are being owned by Earthrise® Nutritionals covering about 444,000 m² area. Marketing of *Spirulina*-based tablets by Earthrise Nutritionals is done in more than 20 countries. Production of biscuits, cereals, pasta, desserts, and many more *Spirulina*-based food products are being manufactured by Cyanotech Corporation situated in Island of Hawaii, which are being a big part of European countries. Currently, PBPs are used as dye for developing colored food items such as confectionery and dairy products (Sekar and Chandramohan, 2008). In confectionary jelly gums and candies are produced by use of PC is increasing nowadays. Jelly gum centers become discolored when exposed to intense solar radiation for 24 h. For the reduction of degradation of colored products, light and pH play an important role. The hot spring cyanobacterium *Nostoc* sp. strain HKAR-2 isolated from hot spring produces PC which shows more stability to heat and sensitivity to pH and light (Kannaujiya and Sinha, 2016). The thermoacidophilic red microalga *Cyanidioschyzon merolae* might provide an alternative source of PC (Ciniglia et al., 2004). PE requires the use of long-term preservatives to keep it safe and unspoiled, and it has high sensitivity to light and the oxygen activity of food products (Mishra et al., 2010). Detailed knowledge of natural colorant is very important as a point of food so that large scale production of it can be carried.

The use of dried biomass of *Spirulina* for PC production is gaining more attention. The addition of blue color to soft drinks like Pepsi, Bacardi Breezer can retain the color for one month at optimum temperature. This naturally obtained PC is nowadays used in dry confectionary to make sugar candies, flower cake decoration, sugar lollipops as its color are stable for a year. Factors like pH, temperature, light makes these PBP very sensitive as they are being degraded by alterations in the above factors. Some of the preservatives used in the food industry play a significant role in enrichment of the stability of food products under storage conditions (Brul and Coote, 1999). Preservatives such as sucrose, sodium chloride, and citric acid have been reported for long-term heat stability of PC (Chaiklahan et al., 2012). Even the use of crude PC isolated from *Arthrospira platensis* is full of antioxidants and considered as a healthy food (Estrada et al., 2001; Bermejo et al., 2008). This non-purified form of PC contains different kinds of metabolites and its intake gives a positive response on lowering cholesterol, having antiviral, anticancer, and anti-inflammatory effects (Jensen et al., 2001; Singh et al., 2005). Common food preservatives, such as citric and benzoic acids, have also been shown to improve the thermostability of PC, which may be linked to a reduction in pH of the solution to levels that favor stability (e.g., pH 5.5) (Mishra et al., 2008; Mogany et al., 2019) which is due to the fact that benzoic acid acts as an antioxidant and its antimicrobial activity inhibits growth of microbes (Dong and Wang, 2006; Lino and Pena, 2010).

12.5.2   USE OF PHYCOBILIPROTEIN (PBP) IN INDUSTRY

In the last few years, the greener methodology for the generation of antibiotics, pharmaceutical products, and fluorescent tags has gained a lot of attention for industrial production (Sekar and Chandramohan, 2008). PC in terms of antioxidant, anticancer, anti-aging, anti-inflammatory, hepatoprotective, neuroprotective, and pharmacological properties has gained a great advantage over other compounds. These PBPs have been found to have a strong antioxidant effect through elimination of ROS by use of antioxidant enzymes (Wu et al., 2016). Property of scavenging hydroxyl, alkoxyl, peroxide, superoxide, and nitrogen-containing radicals and a role in lipid peroxidation inhibition is very clearly seen from *Aphanizomenon flos-aquae* PC which has strong antioxidative properties (Bhat and Madyastha, 2000; Romay et al., 2003) and has the ability to shield against several types of oxidative damage (Benedetti et al., 2004). Generation of peroxyl radicals due to hemolysis of red blood cells is also being inhibited by PC. Lipid peroxidation in plasma samples exposed to synthetic chemicals such as cupric chloride is also being reduced by PC. Many pathological disorders which generated due to antioxidative mechanism can be prevented by this blue colored pigment. During the process of inflammation, excessive release of histamine, prostaglandin (PGE2), myeloperoxide (MPO) and leukotriene B4 (LTB4) are inhibited by PC indirectly stopping edema formation (Richa et al., 2011). Also, PC isolated from *Spirulina platensis* shows important inhibitory effects on the progression and enlargement of cancer in terms of dosage and time duration of human leukemia K562 cell lines (Liu et al., 2000). Reduction in cholesterol levels by use of PC indirectly helps to inhibit chances of heart disease (Nagaoka et al., 2005). Enzymes such as alanine aminotransferase (ALT), malondialdehyde (MDA) and aspartate aminotransferase (AST) are also maintained at a reduced level in blood by usage of PC (González et al., 2003). Thyroid hormone (T3) is kept in range by the help of PC through the enhancement of serum nitrite (Remirez et al., 2002). R-PE isolated from red algae is a good option as an antioxidant for use in photodynamic therapy and cancer prevention. Third biliprotein APC also has an important role in the inhibition of enterovirus 71 in host cells (Shih et al., 2003). Anti-oxidation mechanism of PC shows much similarity to ascorbic acid and tocopherol. It has been shown that both subunits of PBPs, the apoprotein (α and β subunits) and phycobilins are involved in antioxidant effect by the mechanism of stabilizing and detoxification of ROS (Pleonsil et al., 2013). Light also has an influence on the antioxidant activity of PBPs, by the production of free radicals under light conditions.

12.5.3   USES IN COSMETIC INDUSTRIES

Commercially the demand for PBPs in beauty care products is increasing due to its herbal and safe uses (Richa et al., 2011). Products to be used on day to day on our skin should be safe enough not to cause any allergic reactions. Nowadays, usage of PBPs, particularly PC and PE in Japanese based cosmetics is done (Prasanna et al., 2007). After getting approved through food and drug administration (FDA), these herbal cosmetic products are being marketed in Japan. A natural commercial dye that is Lina Blue is being developed by DIC Corporation, which is being used for natural cosmetics products (Kannaujiya et al., 2017b). Seeing the effectiveness of these natural products, companies are trying to make more products like soaps, creams, lotions, eyeliner, lipsticks for healthy and glowing skin (Figure 12.5). Even some of cyanobacterial extracts contain proteins and peptides that are frequently used in hair care products such as lotions, shampoos, solutions for permanent hair wave, and hair coloring products. *Chlorogloeopsis* spp. and *Spirulina* extracts in hair care products have given positive results such as gloss, smooth combing, hair restoration and moisturizing (Ariede et al., 2017). The blue-green algae hair rescue conditioning mask from Aubrey Organics Company, Ltd. (USA) has been productively applied in hair strengthening and end breakage and splitting prevention (Joshi et al., 2018).

12.5.4   USES OF PBP IN FLUORESCENT INDUSTRY

Seeing the autofluorescence property of PBPs, their usage has acquired more commercialization and their interest in scientific research industries. The individual subunits of PBPs that is PE, APC, and PC also have the property to fluoresce after disintegration. The vital property of these PBP is due to their presence of chromophores, high stokes shift, rate of quenching, high quantum yield, fluorescence emission rate, and medium solubility (Sekar and Chandramohan, 2008). These properties make these PBP act as strong fluorophore for various industries. This property has also been exploited in immunological research laboratories as a marker in antibody receptor and donor. Usage of PBP in such as flow cytofluorimetry, fluorescence microscopy, single-cell analysis, fluorescence-activated cell sorting (FACS) and immunoassays are frequently used as the principal fluorescent agents (Sun et al., 2003; Eriksen, 2008; Sekar and Chandramohan, 2008). Enhancement in the molar coefficients is due to the association of $\alpha6\beta6$ hexamer of every subunit of PBP (Thoren et al., 2006). During denaturation of PBP, the extinction coefficients are also reduced and fluorescence property also declines (Fukui et al., 2004; Kupka and Scheer, 2008). PE ($\alpha6\beta6$ hexamers) is widely used in fluorescent agents in industry due to its intense yellow fluorescence (Glazer, 1994). In comparison to other subunits, the quantum yield of PE was found to be in the range of 82–98%, while that of APC and PC was greatly reduced to about 68% and 50% (Oi et al., 1982).

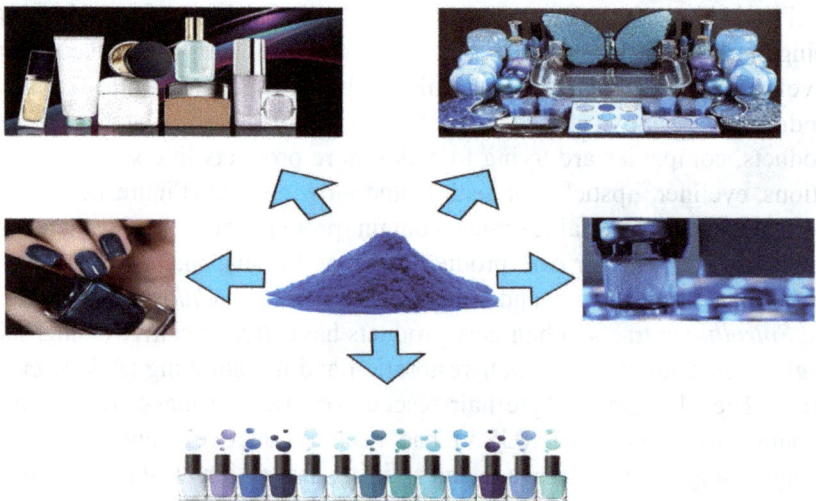

FIGURE 12.5   Potential application of phycobiliproteins in cosmetic industries.

The significant nature of PE fluorescence is due to the far shifting of emissions at 580 nm and excitation at 488 nm. For various deadly diseases like HIV (human immunodeficiency virus), cancer fluorescein-labeled antibodies in conjugation with PE is being used in molecular biology. This PE is being considered as the second-best colored fluorescent products with the emission wavelength at 580 nm and excitation at 488 nm. Association of donor PE and acceptor cyanine dyes Cy5/Cy7 (indodicarbocyanine/indotricarbocyanine) gives multicolor fluorescence analysis (Waggoner, 2006). Strong bonding of PBP with protein A, immunoglobulins, avidin, and biotin is responsible for dual-color single-cell analysis by FACS (Oi et al., 1982). For immune detection of immunoglobulins in cells a fluorochrome (PBXL-3L) has been designed (Telford et al., 2001a). Studies have found that cryptomonad-derived PBPs have low molecular weights and are being used in flow cytometry for extra- and intracellular labeling by fluorescent probes (Telford et al., 2001b). The blue color of PC has been utilized for ecological monitoring (Sode et al., 1991; Simis et al., 2005) and detection of harmful cyanobacteria (Izydorczyk et al., 2005) in the natural environment. Companies such as Boehringer Ingelheim in Germany and Sigma and Molecular Probes in the USA have industrialized PBP-related probe products. APC is commonly used as a fluorescent probe to detect apoptosis (Tang et al., 2017; Li et al., 2018). However, because of the stabilizing effect of its γ subunits, PE is the ideal fluorescent probe, and is more frequently used than other PBPs (Leney et al., 2018). Detection of IgG antibodies to Hendra virus in serum is important to help monitor outbreaks of the virus. The commonly used enzyme-linked immunosorbent assays and fluorescence-based Luminex assays consist of three main steps and take few hours to complete the process. R-PE was used as a fluorescent label which binds directly to IgG protein and due to its large specific surface area of the magnetic nanoparticles, it was likely to reduce each step in the detection process to 20 min (Gao et al., 2015). Recombinant PBPs can be made by using genetic engineering techniques and large-scale fermentation at a lower cost with improved fluorescence characteristics. Wu et al. (2017) co-expressed streptomycin and a fusion protein (SLA) from the APC α subunit of *Thermosynechococcus elongatus* BP-1, along with PEB synthase (Ho1 and PebS) or PCB synthetases (Ho1 and PcyA) in *E. coli*, and two recombinant PBPs (SLA-PEB and SLA-PCB) capable of binding biotin were attained. The detection limits of these fusion proteins in tumor marker alpha-fetoprotein assays were 0.11 and 0.35 ng/mL, respectively. The extraction and purification of PBPs is challenging, and there is no advanced process for industrial production, causing in expensive products and making it problematic to develop and apply PBP-based fluorescent probes. But the characteristic fluorescence peak of some PBPs containing PCB is at 660 nm, which lies within visual imaging window for living tissue at 650–1100 nm

(Shcherbo et al., 2007). Hence, PBPs might be used as a development of near-infrared fluorescent probes. Higher organisms cannot biosynthesize PCB of PBPs, greatly confines the application of the fluorescent proteins in mammalian cells, but mammalian cells are rich in heme, which can synthesize BV under the catalysis of heme enzymes. The development of BV-based PBPs as near-infrared fluorescent probes has become a trend.

12.5.5 ROLE OF PBP IN MEDICAL SCIENCES

Recent research and development about the biliproteins structure have led to further applications of PBP in therapeutic sciences. PC has played very essential role in heart, anti-inflammatory, kidney, lungs, cataract, neurological, and liver diseases (Figure 12.6).

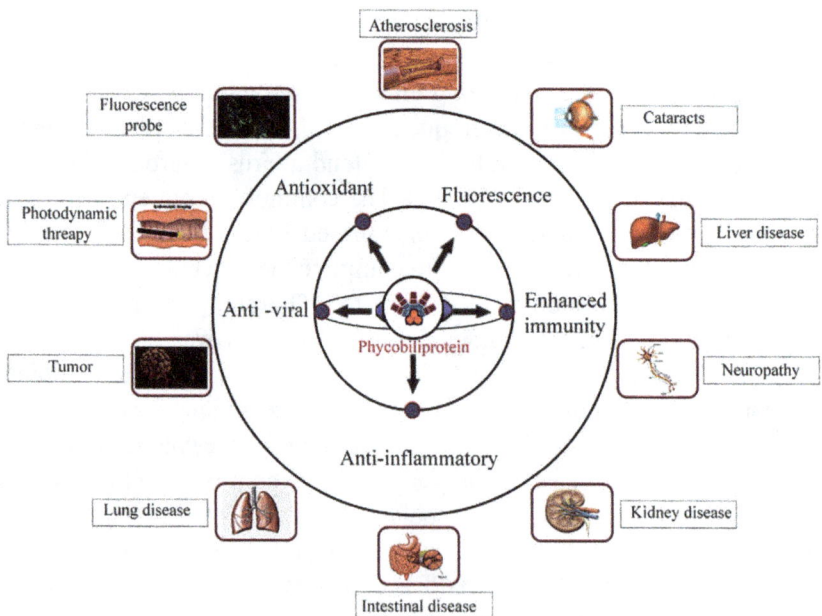

FIGURE 12.6   Role of phycobiliproteins in therapeutic sciences.

12.5.5.1 AS AN ANTICANCER AGENT

PC is widely used as a blue colorant for numerous food and cosmetics due to its high nutraceutical value, protein nature, high-stroke shift, and strong fluorescence in the ultraviolet (UV) spectrum. PC has a strong anticancer effect in different types of cancer cells including lung cancer (Li et al., 2015), breast cancer (Li et al., 2010), bone marrow cancer (Gardeva et al., 2014) and colon cancer (Cai et al., 1995). Distinct anticancer activity is shown by PC purified from *Spirulina platensis*. Being used as a drug carrier PC is disintegrated into low molecular weight peptide by enzymatic hydrolysis and column chromatography and then subjected to tumor inhibitory response on HeLa and 293T tumor cells which shows an improved response as compared to non-disintegrated PC (Wang et al., 2012). Growth response of lung cancer cell line SPC-A-1 is also influenced by the components of alpha and beta subunits of this blue pigment PC (Sun et al., 2010; Zhang et al., 2010). C-phycocyanin (C-PC) is also used as a natural radiosensitizer for enhancement of colon cancer by radiation therapy (RT) efficacy with the inhibition of COX-2 (Kefayat et al., 2019).

12.5.5.2   AS AN ANTI-INFLAMMATORY AGENTS

Anti-inflammatory responses of PC have been well discussed (Liu et al., 2016) and it was first informed by Remirez et al. (2002a). Treatment of osteoarthritis by reducing various inflammatory cytokines, interleukin-6 (IL-6), TNF-, NO, MMP-3 and sulfated glycosaminoglycans is also done by PC (Martinez et al., 2015). PC is capable to impede cyclooxygenase-2 activity and promotes LTB4 formation in mouse ear inflammation test (Romay et al., 1999; Reddy et al., 2000). CCl_4 induced damage is also being relieved by PC in liver cells (Ou et al., 2010).

Being a neuroprotective agent PC also shows protective effects on tributyltin chloride-induced neurotoxicity along with N-acetylcysteine (Mitra et al., 2015). Reduction in free radicals and reduction in high density cholesterol is also reported as a protective effect (Riss et al., 2007). Inhibition of atherosclerosis with the help of PC is done by promoting expression of CD 59 and also reducing the amount of blood fat, muscle cell proliferation, and apoptosis of endothelial cells (Li et al., 2013). Further, PC is able to stop cellular damage by increasing oxalic acid-mediated oxidative stress in canine kidney cells (Farooq et al., 2014). C-PC decreases carrageenan-evoked thermal hyperalgesia. Shih et al. (2009) found that PC can stop excessive levels of NO and PGE2 by downregulation of iNOS and COX-2 expression, which causes the reduction of formation of TNF- and infiltration of neutrophils to inflammation sites. Remirez et al. (2002b) had revealed inhibitory effects on allergic inflammatory response caused by inhibition of histamine release from mast cells. Usage of PC is as antiplatelet therapy in arterial thrombosis is also reported (Hsiao et al., 2005) (Table 12.2).

12.5.5.3   AS AN ANTIOXIDANT AGENT

It has been seen that both subunits of PBPs, the apoprotein (α and β subunits) and phycobilins are involved in antioxidant activity by ROS detoxification (Pleonsil et al., 2013). In recent times, PC is exploited in a variety of biotechnology and pharmacological products. This PC is also approved as a powerful antioxidant, neuroprotective, and hepatoprotective role in various diseases (Romay et al., 2003). It has capability to scavenge hydroxyl, alkoxyl, peroxyl, peroxynitrite ($ONOO-$), and hypochlorous acid (HOCl) and plays an important role as antioxidant molecules (Bermejo et al., 2008). Hydrolysis of PC with trypsin there is partial loss of antioxidant activity of apoprotein part (Zhou et al., 2005). Various types of disorders, including inflammation, atherosclerosis, cancer, reperfusion injury is caused by generation of free radicals which are very dangerous (Kehrer, 1993; Lynch, 1998; Gupta et al., 2011). PC is important neuroprotective agent. They show shielding effects on tributyltin chloride-induced neurotoxicity together with N-acetylcysteine (neuroprotective drug) (Mitra et al., 2015). PC is also known to inhibit Ca^{2+}/phosphate-induced mitochondrial damage in rat brain by inhibition of membrane potential, ROS levels and proapoptotic Cyt c (Marín-Prida et al., 2012). PC is also used to treat neurodegenerative diseases, including ischemic stroke, Alzheimer's disease, and Parkinson's disease (Rimbau et al., 1999; Marín-Prida et al., 2012). PC can considerably lessen liver toxicity and also protects the liver enzymes like P450, aminopyrine-N-demethylase, and glucose-6-phosphatase (Liu et al., 2016). So, this blue pigment also plays a very important role as a hepatoprotective agent. Very recent research has shown that PC also has effects on Kupffer cells functioning indirectly inhibiting phagocytosis, which is initiated in response to oxidative stress-induced production of tumor necrosis factor (TNF) alpha and nitric oxide by hyperthyroid state (Remirez et al., 2002b). PC shows inhibition of cisplatin-induced renal toxicity and oxidative stresses, which leads to reduction of oxidative stress (Fernandez-Rojas et al., 2014). PC is able to inhibit cellular damage by increasing oxalic acid-mediated oxidative stress in canine kidney cells (Farooq et al., 2014). In renal mesangial cells, diabetic nephropathy is prevented by inhibition of NADPH-dependent superoxide production with the help of PC (Zheng et al., 2013). Also, an important role

TABLE 12.2 Pharmacological and Biological Properties of PBPs

Protein or Gene Source	PBP	Pharmacological and Biological Properties	References
Arthospira maxima	C-PC	Antioxidant	Romay et al. (1998b)
Arthospira maxima	PC	Antioxidant, anti-inflammatory ability	Romay et al. (1998a)
AfaMax	PC	Antioxidant, anti-inflammatory ability	Castangia et al. (2016)
	C-PC	Antioxidant, reduce osteoarthritis	Young et al. (2016)
	C-PC	Antioxidant, antihyperlipidemic	Sheu et al. (2013)
Spirulina maxima (dried marine algae)	C-PC	Antioxidant, anti-inflammation activities	Choi and Lee (2018)
Portieria homemannii	R-PE	Antioxidant, anticancer	Senthilkumar et al. (2013a, b)
Synechococcus sp.	PC	Antioxidant, radical scavenging activity	Sonani et al. (2017)
Spirulina platensis	C-PC	Anti-inflammatory ability	Chen et al. (2014)
Limnothrix sp. strain 37-2-1	PC	Antioxidant	Gantar et al. (2012a, b)
Lyngbya sp. A09DM	PE	Antioxidant	Sonani et al. (2014)
Spirulina	PC	Antioxidant	Romay and Gonzalez (2000)
Spirulina platensis	Apo-c-PC β subunit and C-PC	Antioxidant	Pleonsil et al. (2013)
Spyrulina species C	PC	Antioxidant	Lissi et al. (2000)
Spirulina fussiformis	C-PC	Antioxidant	Madhyastha et al. (2009)
Spirulina platensis	Selenium containing PC	Antioxidant	Huang et al. (2007)
Aphanizomenon flos-aquae	PC	Antioxidant	Benedetti et al. (2004)
Spirulina maxima	PC	Antioxidant	Nakagawa et al. (2016)
Spirulina platensis	Se-APC	Antioxidant, inhibition of cancer cells	Fan et al. (2012)

TABLE 12.2 *(Continued)*

Protein or Gene Source	PBP	Pharmacological and Biological Properties	References
Anacystis nidulans UTEX 625	Recombinant APC	Antioxidant	Ge et al. (2006)
Spirulina platensis	Recombinant α PC	Antioxidant	Guan et al. (2009)
Porphyra haitanensis	R-PC	Anti-allergy, anti-inflammation activities	Liu et al. (2015)
Spirulina platensis	C-PC	Decrease the blood glucose level	Setyaningsih et al. (2015); Ou et al. (2016)
Spirulina platensis	C-PC	Anti-atherosclerosis	Riss et al. (2007); Strasky et al. (2013)
Spirulina maxima	C-PC	Reduce liver damage	Ou et al. (2010)
Spirulina platensis	C-PC	Reduce liver damage	Pak et al. (2012); Hussein et al. (2015); Xia et al. (2016)
Arthlpira maxima SAG 25780	C-PC	Reduce liver damage	Nagaraj et al. (2012)
Spirulina platensis	C-PC	Cataract treatment	Kothadia and Shabaraya (2011); Kumari et al. (2013)
Spirulina (Arthrospira) species	C-PC	Reduce nerve damage	Rimbau et al. (1999)
Spirulina platensis	C-PC	Reduce nerve damage	Bermejo-Bescos et al. (2008); Min et al. (2015); Mitra et al. (2015)
	PC	Reduce neurodegenerative diseases associated with proteotoxicity	Macedo et al. (2017)
Spirulina platensis	PC and PCB	Protect kidneys	Zheng et al. (2013)
Spirulina platensis	C-PC	Protect kidneys	Farooq et al. (2014); Fernandez-Rojas et al. (2014)
Spirulina platensis	C-PC	Reduces pulmonary fibrosis	Sun et al. (2011); Sun et al. (2012); Li et al. (2017a, b)

2

TABLE 12.2 *(Continued)*

Protein or Gene Source	PBP	Pharmacological and Biological Properties	References
Bangio atropurpurea	R-PC	Alleviates allergic airway inflammation	Chang et al. (2011)
Spirulina platensis	C-PC	Antifibrotic	Pattarayan et al. (2017)
Arthospira maxima	C-PC	Reduce intestinal inflammation	Gonzalez et al. (1999)
Spirulina platensis	APC	Reduce intestinal damage	Chueh (2002); Shih et al. (2003)
Spirulina (dry powder)	C-PC	Anti-tumor	Gupta and Gupta (2012)
Spirulina platensis	C-PC	Anti-tumor	Gantar et al. (2012a, b); Saini et al. (2012); Saini and Sanyal (2014); Li et al. (2015a, b); Deniz et al. (2016)
Oscillatoria tenuis	C-PC	Anti-tumor	Thangam et al. (2013)
Anacystis nidulans UTEX625	Recombinant APC	Inhibition of cancer cells	Ge et al. (2005a, b)
Gracilaria lemaneiformis	PE	Inhibition of cancer cells	Ruobing et al. (2007)
Spirulina platensis	C-PC	Inhibition of cancer cells	Pan et al. (2015); Ying et al. (2016)
Spirulina powder	Digestion by pepsin releases biologically active chromopeptides from C-PC	Antioxidant, inhibition of cancer cells	Minic et al. (2016)
Oscillatoria tenuis	C-PC	Antioxidant, inhibition of cancer cells	Thangam et al. (2013)

in the reduction of cardiovascular disease by preventing lipid metabolism, mitochondrial damage and oxidative stress is being played by PC. Inflammatory damage induced by oxidative stress in atherosclerosis is also reduced by PC and also level of COX-2 is increased, which leads to an increase in antioxidative enzymes and thus maintaining the blood lipid level (Riss et al., 2007). Expression of CD59 and reduction of blood fat, muscle cell proliferation and apoptosis of endothelial cell indirectly inhibiting progression of atherosclerosis is done by PC (Li et al., 2013). Many age-related disorders are caused due to increasing levels of oxidative stress in cells. PC can also reduce the formation of one such age-related disease that is cataract in rat model (Kumari et al., 2013). Maintenance of lens transparency, regulation at the level of transcription of lens crystalline redox, is also done by PC (Kumari et al., 2015).

Strasky et al. (2013) stated that PC could enhance the expression of encoding heme oxygenase-1 to increase the enzyme content in the atherosclerotic lesion of mice with apolipoprotein E gene deletion to lessen the severity of the disease. As due to the fact that heme oxygenase-1 can break heme to produce bilirubin which has good antioxidant ability. Regulation of oxidative stress and endothelial cell dysfunction marker proteins such as endothelial nitric oxide synthase and NAD oxidase, and reduction in atherosclerotic lesions is also done by PC. These studies show that PC can potentially treat atherosclerosis by decreasing the oxidative stress. Xia et al. (2016) has reported the protective effects of PC on alcoholic fatty liver and found that PC has shown reduction in the serum levels of ALT, aspartate aminotransferase, triglyceride (TG), total cholesterol and low-density lipoprotein (LDL), and increased the content of SOD and MDA in the liver, thus reducing oxidative stress. Ou et al. (2010) reported that PC could reduce CCl_4 induced liver damage by scavenging ROS and increasing the activity of SOD and glutathione peroxidase (GSH-Px).

This PC-induced inhibition of the TLR2-MyD88-NF-κB signaling pathway in the early stages was very important for protection against pulmonary fibrosis (Li et al., 2017a, b). Cervantes-Llanos et al. (2018) have reported the use of PC as a neuroprotector in rodent models with autoimmune encephalomyelitis, where this blue pigment PC has the ability to reduce oxidative stress and immune response. Gdara et al. (2018) demonstrated that PC decreased liver injuries by reducing the activity of enzymes such as liver transaminases and alkaline phosphatase, which are in an active state by the oxidative stress (Table 12.3).

TABLE 12.3 Antioxidant Activity of Phycobiliproteins and Phycobilins from Blue-Green Algae

Bioactive Compound	Cyanobacterial Origin	Mechanism of Action	References
Phycocyanin	*Arthospira maxima*	Scavenge free radical or free radical less reactive	González et al. (1999)
	Spirulina sp.	Scavenge free radical or free radical less reactive	Zhou et al. (2005)
Phycoerythrin	*Halomicronema* sp.	Scavenge radicals from DPPH reduction of the Fe^{3+}	Patel et al. (2018)
Phycocyanobilin	*Spirulina platensis*	Scavenge radicals from AAPH	Hirata et al. (2000)
	Spirulina platensis	Inhibition of ONOO−- mediated DNA degradation	Bhat and Madyastha (2001)
	Spirulina sp.	Partial inhibition of NADPH oxidase	McCarty (2007)
	Aphanizomenon flos-aquae	Quenching of peroxyl radicals	Benedetti et al. (2010)

Note: AAPH-2,2′: azobis (2-amidinopropane) dihydroxychloride; ONOO: peroxynitrite.

12.5.6 OPTICAL APPLICATIONS

PBPs can be widely used in photodynamic therapy and other fields because PBPs can emit strong fluorescence after being irradiated with a laser (Table 12.4).

12.5.7 USES OF PHYCOBILIPROTEIN (PBP) IN PHOTODYNAMIC THERAPY

Dye-sensitized solar cells (DSSC) are based on the usage of colored dyes as photosensitizers, which are capable of absorbing the energy coming from the solar radiation in the visible range (Govindaraj et al., 2015). PBPs could be of substantial use as natural substitutes to chemical sensitizing dyes in low-cost photovoltaic devices, such as dye-sensitized solar cells (Bora et al., 2012; Schrantz et al., 2017; Sharma et al., 2018; Li et al., 2019a). Some workers have evaluated the use of PC from *Spirulina platensis* in DSSC.

TABLE 12.4   Applications of PBPs in Optics

Protein or Gene Source	PBP	Application	References
Gastroclonium coulteri (Rhodymeniales)	R-PE	Fluorescence probe	Oi et al. (1982)
	PBP	Fluorescence probe	Siiman et al. (1999)
	APC	Photodynamic therapy	Suping et al. (2001)
	R-PE	Photodynamic therapy	Huang et al. (2002)
Microcystis	PC	Photodynamic therapy	Wang et al. (2012)
	R-PE	Fluorescence probe	Gao et al. (2015)
Spirulina platensis	C-PC	Photodynamic therapy	Bharathiraja et al. (2016)
Spirulina	C-PC	Highly efficient fluorescence sensors	Wang et al. (2016)
	APC	Fluorescence probe	Tang et al. (2017); Li et al. (2018)
Streptomyces avidinii	Recombinant fusion PBP (SLA-PEB and SLA-PCB)	Fluorescence probe	Wu et al. (2017)
Synechococcus PCC 6803	Streptavidin-PBPs (SA-PBPs)	Immunoassay technologies	Ge et al. (2017)
Chroococcidiopsis thermalis	ApcF2 (the phycobilisome core subunit)	Fluorescent markers	Ding et al. (2017)
	C-PC	Fluorescence probe	Han et al. (2018)

DSSC assemblies have reported that R-PE signifies a highly suitable photosensitizer promising a high stability with a relatively high solar energy to electricity conversion efficiency ($\eta = 0.11\%$) when being compared to other recombinant proteins ($\eta = 0.30\%$) (Yanuk et al., 2019). This PDT therapy is a new oncology therapy based on enriching a lesion area with photosensitizers, thereby causing oxidative damage to tumor tissue by generating free radicals and active oxygen species upon irradiance. Therefore, choice of photosensitizers with high efficiency, low toxicity and good selectivity is important for effective photodynamic therapy. PC extracted from *Microcystis* (MC-PC) has been studied for its photosensitization effect on HepG2 cells. Cancer cells were incubated with MC-PC following laser irradiation and MTT assay analyzed the cell viability. Results indicated that MC-PC at a dosage 200 μg/ml inhibited the growth of HepG2 cell line and induced apoptosis after 24 h, which led to the conclusion of a new source of PC from *Microcystis* as a nontoxic and effective photosensitizer (Wang et al., 2012).

C-PC incubation with breast cancer cells MDAMB-231 was done and the results indicated that C-PC showed no photo-toxicity without laser irradiation, but when irradiated with a 625 nm laser, the cells were capable to produce oxygen free radicals and ROS, leading to apoptosis and killing of MDA-MB-231 breast cancer cells (Bharathiraja et al., 2016). Upon laser irradiation, APC is also able of generating triplet states and free radical cations, showing that APC can perform photoexcitation and photoionization all together, and so can also be used as type I and type II photosensitizers (Suping et al., 2001). PBPs have a stronger affinity for tumor cells than for normal cells, but the mechanism underlying this mechanism is still uncertain. Additionally, also because PBPs can also be used as a health food supplement to boost immunity level (Levi et al., 2018), some researchers believe that PBPs may have the effect of inhibiting tumor growth through a variety of synergistic effects.

12.5.8   USES OF PHYCOBILIPROTEIN (PBP) IN TEXTILES DYES

The textile industry is primarily dependent on synthetic dyes, which are made from chemicals, petroleum by-products, and minerals. These can cause negative impact on the environment of synthetic dyes, particularly in terms of water pollution (Kant, 2012). However, natural dyes have many advantages in terms of sustainability, including a reduction in carbon footprint for the textile industry (Moldovan et al., 2017a). Recent studies have reported that cotton fabrics colored with PE-based dyes from the mesophilic red algae *Gracilaria cornea* and *Gracilaria gracilis* do fulfill with the European accepted standards (UNE-EN ISO) for textiles in terms of color and fastness to laundering and rubbing tests (e.g., the measurement of the durability of a fabric) (Moldovan et al., 2017a, b). In addition, PC from *A. platensis* and PE from the mesophilic red macroalga *Gracilaria vermiculophylla* has also been used productively for dyeing cotton and wool fabrics (Ferrandiz et al., 2016; Gorman et al., 2017).

12.6   FUTURE PERSPECTIVES

Though PBPs and phycobilins have previously shown a great potential of applicability in the industry due to their wide range of bioactivities, part of this potential is not yet included due to limitations in the purification and characterization of these molecules. It is likely that in the next few years, more studies about the bioactivities of PBPs and phycobilins could be carried out and so that the potential application of these biliproteins can be achieved. Also, that with additional screening of cyanobacteria strains for the presence of phycobilins with appropriate characteristics will result in additional discoveries beyond the fast-growing strains already used in the market. Different reports have suggested that the productions of PBPs are primarily dependent on environmental factors and light intensity (Chadkar, 2012; Khatoon et al., 2018). Also, metabolic, and genetic engineering methods have been done to enhance the production of PBPs. For the study of genetic and metabolic engineering, *Synechocystis* sp. was used as a typical organism (Lin et al., 2017). Some specific cyanobacterial strains, for instance, *A. platensis*, *Synechococcus* sp., *Porphyridium* sp., were subjected for studies of genetic and metabolic engineering approaches to increase the production of PBPs (Saini et al., 2018). PBPs are being used widely for different purposes, such as in pharmaceutical, food, and cosmetic industries. The present market value of C-phycocyanin is between the range of US \$10 and 50 million annually, with the price of food-grade (OD_{620}/OD_{280} = <1.0). On the whole, PC has been the major objective of the market, and since it was accepted by the FDA in the United States, its use has been increased. In Europe, the usage of PC as dye has gained interest due to its natural, non-toxic, and biodegradable features (Venil et al., 2013; de Morais et al., 2018). The actual cost of PBPs production varies according to their purity grade, ranging from 130 USD to 15,000 USD per gram. While the usage of PBPs in food industries, the purity is not a constraint and the price can be easily reduced, but, when used for scientific and pharmaceutical applications, the purity needs to be greater, and the cost could be 100 times higher (de Morais et al., 2018). The market size for PBPs and other pigments is tough to guess, due either to the lack of the statistics of the regional, low technology products (Venil et al., 2013). However, Borowitzka (2013) have assessed the total market of PBPs as bigger than 60 million USD in 2013. A recent report from future market insights has calculated a total market of 112.3 million USD for 2018 with a future of doubling that number until 2028. This report also states that Western Europe is the biggest consumer of this product (around 33%) and that 80% of PC manufactured is used in the food industry. The biggest limitation is the applicability of cyanobacteria for high added-value products and how to reduce the cost in order to gain the interest of the market. A promising way to do that is optimizing the concentration of this product to the biomass, and also developing new applications for it, namely in nutraceutical, cosmetical, and pharmaceutical industries.

KEYWORDS

- allophycocyanin
- anticancer
- anti-inflammatory
- antioxidant
- carotenoids
- cosmetics
- cyanobacteria
- fluorescent
- gram-negative
- phycobiliproteins
- phycobilisomes
- phycocyanin
- phycoerythrin
- prokaryotes

REFERENCES

Arasaki, S., & Arasaki, T., (1983). *Low Calorie, High Nutrition Vegetables from the Sea to Help You Look and Feel Better* (p. 196). Japan Publications, Tokyo.

Arashiro, L. T., Boto-Ordóñez, M., Van, H. S. W., Ferrer, I., Garfí, M., & Rousseau, D. P., (2020). Natural pigments from microalgae grown in industrial wastewater. *Bioresour Technol., 303*, 122894.

Arashiro, L. T., Ferrer, I., Rousseau, D. P. L., Van, H. S. W. H., & Garfí, M., (2019). The effect of primary treatment of wastewater in high-rate algal pond systems: Biomass and bioenergy recovery. *Bioresour Technol., 280*, 27–36.

Ariede, M. B., Candido, T. M., Jacome, A. L. M., Velasco, M. V. R., De Carvalho, J. C. M., & Baby, A. R., (2017). Cosmetic attributes of algae: A review. *Algal Res., 25*, 483–487.

Batista, A. P., Raymundo, Sousa, I., & Empis, J., (2006). Rheological characterization of colored oil-in-water food emulsions with lutein and phycocyanin added to the oil and aqueous phases. *Food Hydrocoll., 20*, 44–52.

Benedetti, S., Benvenut, F., & Pagliarani, S., (2004). Antioxidant properties of a novel phycocyanin extract from the blue-green alga *Aphanizomenon flos-aquae. Life Sci., 75*, 2353–2362.

Benedetti, S., Benvenuti, F., Scoglio, S., & Canestrari, F., (2010). Oxygen radical absorbance capacity of phycocyanin and phycocyanobilin from the food supplement *Aphanizomenon flos-aquae. J. Med. Food., 13*, 223–227.

Benvin, A. L., Creeger, Y., Fisher, G. W., Ballou, B., Waggoner, A. S., & Armitage, B. A., (2007). Fluorescent DNA nanotags: Supramolecular fluorescent labels based on intercalating dye arrays assembled on nanostructured DNA templates. *J. Am. Chem. Soc., 129*, 2025–2034.

Bermejo, P., Piñero, E., & Villar, Á. M., (2008). Iron-chelating ability and antioxidant properties of phycocyanin isolated from a protean extract of *Spirulina platensis. Food Chem., 110*, 436–445.

Bermejo, R., Acién, F. G., Ibáñez, M. J., Fernández, J. M., Molina, E., & Alvarez-Pez, J. M., (2003). Preparative purification of B-phycoerythrin from the microalga *Porphyridium cruentum* by expanded-bed adsorption chromatography. *J. Chromatogr. B., 790*, 317–325.

Bharathiraja, S., Seo, H., Manivasagan, P., Santha, M. M., Park, S., & Oh, J., (2016). *In vitro* photodynamic effect of phycocyanin against breast cancer cells. *Molecules, 21*(11), 1470.

Bhat, V. B., & Madyastha, K. M., (2000). C-Phycocyanin: A potent peroxyl radical scavenger *in vivo* and *in vitro. Biochemical and Biophysical Research Communications, 275*(1), 20–25.

Bora, D. K., Rozhkova, E. A., Scrantz, K., Wyss, P. P., Braun, A., Graule, T., & Constable, E. C., (2012). Functionalization of nanostructured hematite thin film electrodes with the light harvesting membrane protein c-phycocyanin yields an enhanced photocurrent. *Advanced Functional Materials, 22*, 490–502.

Borowitzka, M. A., (1999). Commercial production of microalgae: Ponds, tanks, tubes, and fermenters. *J. Biotechnol., 70*, 313–321.

Borowitzka, M. A., (2013). High-value products from microalgae - their development and commercialization. *J. Appl. Phycol., 25*, 743–756.

Cai, X. H., He, L., Jiang-Jialun, J., Xu, X., & Zheng, S., (1995). The experimental study of application of phycocyanin in cancer laser therapy. *Chin. Mar. Drug., 1*, 15–18.

Cama, V., Piniella-Matamoros, B., & Pentón-Rol, G., (2018). Beneficial effects of oral administration of C-phycocyanin and phycocyanobilin in rodent models of experimental autoimmune encephalomyelitis. *Life Sci., 194*, 130–138.

Castangia, I., Manca, M. L., Caddeo, C., Bacchetta, G., & Pons, R., (2016). Santosomes as natural and efficient carriers for the improvement of phycocyanin reepithelizing ability *in vitro* and *in vivo. Eur. J. Pharm. Biopharm.*, 149–158.

Cervantes-Llanos, M., Lagumersindez-Denis, N., Marín-Prida, J., Pavón-Fuentes, N., Falcon-Choi, W. Y., & Lee, H. Y., (2018). Effect of ultrasonic extraction on production and structural changes of C-Phycocyanin from marine *Spirulina maxima. Int. J. Mol. Sci., 19*.

Chaffey, N., (2014). Raven biology of plants. *Annals of Botany, 113*(7), 7. https://doi.org/10.1093/aob/mcu090.

Chaiklahan, R., Chirasuwana, N., & Bunnag, B., (2012). Stability of phycocyanin extracted from *Spirulina* sp.: Influence of temperature, pH, and preservatives. *Process Biochem., 47*, 659–664.

Chang, C., Yang, Y., Liang, Y., Chiu, C., Chu, K., & Chou, H., (2011). A novel phycobiliprotein alleviates allergic airway inflammation by modulating immune responses. *Am. J. Respir. Crit. Care Med., 183*, 15–25.

Chen, C. Y., (2010). Cultivation, photobioreactor design and harvesting of microalgae for biodiesel: A critical review. *Bioresour. Technol., 102*, 71–81.

Chen, H., Yang, T., Chen, M., Chang, Y., Wang, E. I. C., & Ho, C., (2014). Purification and immunomodulating activity of C-phycocyanin from *Spirulina platensis* cultured using power plant flue gas. *Process Biochem., 49*, 1337–1344.

Chojnacka, K., & Noworyta, A., (2004). Evaluation of *Spirulina* sp. growth in photoautotrophic, heterotrophic and mixotrophic cultures. *Enzym. Microb. Technol., 34,* 461–465.

Chrismadha, T., & Borowitzka, M. A., (1994). Effect of cell density and irradiance on growth, proximate composition and eicosapentaenoic acid production of *Phaeodactylum tricornutum* grown in a tubular photobioreactor. *J. Appl. Phycol., 6,* 67–74.

Chueh, C., (2002). *Method of Allophycocyanin Inhibition of Enterovirus and Influenza Virus Reproduction Resulting in Cytopathic Effect.* US Patent US 6346408B1.

Ciniglia, C., Yang, E. C., Pollio, A., Pinto, G., Iovinella, M., Vitale, L., & Yoon, H. S., (2014). Plastid rbcL gene elucidates origin and dispersal of extremophilic *Galdieria sulphuraria* and *G. maxima. Phycologia., 53*(6), 542–551.

Costa, J. A. V., Moreira, J. B., Lucas, B. F., Braga Da, V. S., Cassuriaga, A. P. A., & Morais De, M. G., (2018). 2018 Recent advances and future perspectives of PHB production by cyanobacteria. *Ind. Biotechnol.,14,* 249–256.

Deniz, I., Ozen, M. O., & Yesil-Celiktas, O., (2016). Supercritical fluid extraction of phycocyanin and investigation of cytotoxicity on human lung cancer cells. *J. Supercrit. Fluids,* 13–18.

Dillon, J. C., Phuc, A. P., & Dubacq, J. P., (1995). Nutritional value of the alga *Spirulina. World Rev. Nutr. Diet., 77,* 32–46.

Ding, W., Miao, D., Hou, Y., Jiang, S., Zhao, B., & Zhou, M., (2017). Small monomeric and highly stable near-infrared fluorescent markers derived from the thermophilic phycobiliprotein, ApcF2. *Mol. Cell Res., 1864,* 1877–1886.

Dong, C., & Wang, W., (2006). Headspace solid-phase micro extraction applied to the simultaneous determination of sorbic and benzoic acids in beverages. *Anal. Chim. Acta., 562,* 23–29.

Doust, A. B., Wilk, K. E., Curmi, P. M. G., & Scholes, G. D., (2006). The photophysics of cryptophyte light-harvesting. *J. Photochem. Photobiol. A: Chem., 184,* 1–17. doi. org/10.1016/j.jphotochem.2006.06.006.

Dufosse, L., Galaup, P., Yaron, A., Arad, S. M., Blanc, P., Murthy, K. N. C., & Ravishankar, G. A., (2005). Microorganisms and microalgae as sources of pigments for food use: A scientific oddity or an industrial reality? *Trends Food Sci. Technol., 16,* 389–406.

Dumay, J., & Morancais, M., (2016). Proteins and pigments. In: Fleurence, J., & Levine, I., (eds.), *Seaweed in Health and Disease Prevention* (pp. 275–318), Academic Press, San Diego.

Elmorjani, K., Thomas, J. C., & Sebban, P., (1986). Phycobilisomes of wild type and pigment mutants of the cyanobacterium *Synechocystis* PCC 6803. *Arch Microbiol., 146,* 186–191.

Eriksen, N. T., (2008). Production of phycocyanin - a pigment with applications in biology, biotechnology, foods, and medicine. *Appl. Microbiol. Biotechnol., 80,* 1–14.

Esbona, K., Inman, D., Saha, S., Jeffery, J., Schedin, P., Wilke, L., & Keely, P. (2016). COX-2 modulates mammary tumor progression in response to collagen density. *Breast Cancer Res., 18*(1), 35.

Fan, C., Jiang, J., Yin, X., Wong, K., Zheng, W., & Chen, T., (2012). Purification of selenium containing allophycocyanin from selenium-enriched *Spirulina platensis* and its hepatoprotective effect against t-BOOH-induced apoptosis. *Food Chem., 134,* 253–261.

Farooq, S. M., Boppana, N. B., Devarajan, A., Sekaran, S. D., & Shankar, E. M., (2014). C-phycocyanin confers protection against oxalate mediated oxidative stress and mitochondrial dysfunctions in MDCK cells. *PLoS One, 9,* e103361.

Farooq, S., Asokan, D., Sakthivel, R., Kalaiselvi, P., & Varalakshmi, P., (2004). Salubrious effect of C-phycocyanin against oxalate-mediated renal cell injury. *Clin. Chim. Acta, 348,* 199–205.

Fernández-Rojas, B., Hernández-Juárez, J., & Pedraza-Chaverri, J., (2014). Nutraceutical properties of phycocyanin. *J. Funct. Foods., 11,* 375–392.

Ferrándiz, M., Moldovan, S., Mira, E., Pinchetti, J. L. G., Rodriguez, T., Abreu, H., Rego, A. M., et al., (2016). Phycobiliproteins - new natural dyes from algae as a sustainable method. *Fibres and Textiles, 3,* 56.

Fischer, W. F., (2008). Biogeochemistry: Life before the rise of oxygen. *Nature, 455,* 1051–1052.

Foley, J., & Ramankutty, N., (2011). Solutions for a cultivated planet. *Nature, 478,* 337–342.

Fukui, K., Saito, T., Noguchi, Y., Kodera, Y., Matsushima, A., Nishimura, H., & Inada, Y., (2011). Relationship between color development and protein conformation in the phycocyanin molecule. *Dyes Pigments, 63,* 89–94.

Galetović, A., Seura, F., Gallardo, V., Graves, R., Cortés, J., Valdivia, C., & Gómez-Silva, B., (2020). Use of phycobiliproteins from Atacama cyanobacteria as food colorants in a dairy beverage prototype. *Foods, 9*(2), 244. https://doi.org/10.3390/foods9020244.

Gantar, M., Dhandayuthapani, S., & Rathinavelu, A., (2012). Phycocyanin induces apoptosis and enhances the effect of topotecan on prostate cell line LNCaP. *J. Med. Food, 15,* 1091–1095.

Gantar, M., Simovic, D., Djilas, S., Gonzalez, W. W., & Miksovska, J., (2012). Isolation, characterization and antioxidative activity of C-phycocyanin from *Limnothrix* sp. strain 37-2-1. *J. Biotechnol., 159,* 21–26.

Gantt, E., & Conti, S. F., (1966). Granules associated with the chloroplast lamellae of *Porphyridium cruentum. J. Cell Biol., 29,* 423–434.

Gantt, E., & Lipschultz, C. A., (1972). Phycobilisomes of *Porphyridium cruentum.* I. Isolation. *J. Cell Biol., 54,* 313–324.

Gantt, E., (1975). Phycobilisomes: Light-harvesting pigment complexes. *Bioscience, 25,* 781–788.

Gantt, E., (1990). Pigmentation and photoacclimation. In: Cole, K. M., & Sheath, R. G., (eds.), *Biology of the Red Algae* (pp. 203–219). Cambridge University Press, Cambridge.

Gantt, E., Lipschultz, C. A., Grabowski, J., & Zimmerman, B. K., (1979). Phycobilisomes from blue green and red algae. *Plant Physiol., 63,* 615–620.

Gao, Y., Pallister, J., Lapierre, F., Crameri, G., Wang, L., & Zhu, Y., (2015). A rapid assay for Hendra virus IgG antibody detection and its titer estimation using magnetic nanoparticles and phycoerythrin. *J. Virol. Methods, 222,* 170–177.

Gardeva, E., Toshkova, E., Yossifova, L., Minkova, K., Ivanova, N., & Gigova, L., (2014). Antitumor activity of C-phycocyanin from *Arthronema africanum* (Cyanophyceae). *Braz. Arch Biol. Technol., 57,* 675–684.

Gdara, N. B., Belgacem, A., Khemiri, I., Mannai, S., & Bitri, L., (2018). Protective effects of phycocyanin on ischemia/reperfusion liver injuries. *Biomed. Pharmacother., 102,* 196–202.

Ge, B., Lin, X., Chen, Y., Wang, X., Chen, H., & Jiang, P., (2017). Combinational biosynthesis of dual-functional streptavidin-phycobiliproteins for high-throughput compatible immunoassay. *Process Biochem., 58,* 306–312.

Ge, B., Qin, S., Han, L., Lin, F., & Ren, Y., (2006). Antioxidant properties of recombinant allophycocyanin expressed in *Escherichia coli. J. Photochem. Photobiol., 84,* 175–180.

Ge, B., Tang, Z., Lin, L., Ren, Y., Yang, Y., & Qin, S., (2005). Pilot-scale fermentation and purification of the recombinant allophycocyanin over-expressed in *Escherichia coli. Biotechnol. Lett., 27,* 783–787.

Ge, B., Tang, Z., Zhao, F., Ren, Y., & Yang, Y., (2005). Scale-up of fermentation and purification of recombinant allophycocyanin over-expressed in *Escherichia coli*. *Process Biochem., 40*, 3190–3196.

Glauser, M., Bryant, D. A., Frank, G., Wehrli, E., Rusconi, S. S., & Sidler, W., (1992). Phycobilisome structure in the cyanobacteria *Mastigocladus laminosus* and *Anabaena* sp. PCC 7120. *Eur. J. Biochem., 205*, 907–915.

Glazer, A. N., (1984). Phycobilisome: A macromolecular complex optimized for light energy transfer. *Biochim. Biophys. Acta, 768*, 29–51.

Glazer, A. N., (1985). Light harvesting by phycobilisome. *Annu. Rev. Biophys. Chem., 14*, 47–77.

Glazer, A. N., (1989). Light guides. *J. Biol. Chem., 264*, 1–4.

Gonzalez, R., Gonzalez, A., & Remirez, D., (2003). Protective effects of phycocyanin on galactosamine-induced hepatitis in rats. *Biotecnol. Appl., 20*, 107–110.

Gonzalez, R., Rodriguez, S., Romay, C., Ancheta, O., Gonzalez, A., & Armesto, J., (1999). Anti-inflammatory activity of phycocyanin extract in acetic acid-induced colitis in rats. *Pharmacol. Res., 39*, 55–59.

Gorman, L., Kraemer, G. P., Yaish, C., Boo, S. M., & Kim, J. K., (2017). The effects of temperature on the growth rate and nitrogen content of invasive *Gracilaria vermiculophylla* and native *Gracilaria tikvahiae* from Long Island Sound, USA. *Algae, 32*, 57–66.

Govindaraj, R., Pandian, M. S., Ramasamy, P., & Mukhopadhyay, S., (2015). Sol-gel synthesized mesoporous anatase titanium dioxide nanoparticles for dye sensitized solar cell (DSSC) applications. *Bulletin of Materials Science, 38*(2), 291–296.

Graverholt, O. S., & Eriksen, N. T., (2007). Heterotrophic high cell density fed batch and continuous-flow cultures of *Galdieria sulphuraria* and production of phycocyanin. *Appl. Microbiol. Biotechnol., 77*, 69–75.

Greenwell, H. C., Laurens, R. J., Shields, R., Lovitt, W., & Flynn, K. J., (2010). Placing microalgae on the biofuel's priority list: A review of the technological challenges. *J. Roy. Soc. Interf., 7*, 703–726.

Grossman, A., Schaefer, M. R., Chiang, G. G., & Collier, J. L., (1993). The phycobilisomes, a light-harvesting complex responsive to environmental conditions. *Microbiol. Rev., 57*, 725–749.

Guan, X. Y., Zhang, W. J., Zhang, X. W., Li, Y. X., Wang, J. F., & Lin, H. Z., (2009). A potent anti-oxidant property: Fluorescent recombinant alpha-phycocyanin of *Spirulina*. *J. Appl. Microbiol., 106*, 1093–1100.

Guan, X., Qin, S., Zhao, F., Zhang, X., & Tang, X., (2007). Phycobilisomes linker family in cyanobacterial genomes: Divergence and evolution. *Int. J. Biol. Sci., 3*, 434–445.

Guglielmi, G., Cohen-Bazire, G., & Bryant, D. A., (1981). The structure of *Gloeobacter violaceus* and its phycobilisomes. *Arch Microbiol., 129*, 181–189.

Gupta, N. K., & Gupta, K. P., (2012). Effects of C-Phycocyanin on the representative genes of tumor development in mouse skin exposed to 12-O-tetradecanoyl-phorbol-13-acetate. *Environ Toxicol. Pharmacol., 34*, 941–948.

Häder, D. P., Williamson, C. E., Wängberg, S., Rautio, M., Rose, K. C., Gao, K., Helbling, E. W., et al., (2015). Effects of UV radiation on aquatic ecosystems and interactions with other environmental factors. *Photochem. Photobiol. Sci., 14*, 108–126.

Han, X., Lv, L., Yu, D., Wu, X., & Li, C., (2018). Coordination induced supramolecular assembly of fluorescent C-Phycocyanin for biologic discrimination of metal ions. *Mater. Lett., 215*, 238–241.

Hemalatha, M., Sravan, J. S., Min, B., & Venkata, M. S., (2019). Microalgae-biorefinery with cascading resource recovery design associated to dairy wastewater treatment. *Bioresour. Technol., 284*, 424–429.

Hirata, T., Tanaka, M., Ooike, M., Tsunomura, T., & Sakaguchi, M., (2000). Antioxidant activities of phycocyanobilin prepared from *Spirulina platensis. Journal of Applied Phycology, 12*(3/5), 435–439.

Ho, M. Y., Niedzwiedzki, D. M., MacGregor-Chatwin, C., Gerstenecker, G., Hunter, C. N., Blankenship, R. E., & Bryant, D. A., (2019). Extensive remodeling of the photosynthetic apparatus alters energy transfer among photosynthetic complexes when cyanobacteria acclimate to far-red light. *Biochimica et Biophysica Acta (BBA)-Bioenergetics, 148064*, 727.

Ho, M. Y., Soulier, N. T., Caniffe, D. P., Shen, G., & Bryant, D. A., (2017). Light regulation of pigment and photosystem biosynthesis in cyanobacteria. *Current Opinion in Plant Biology, 37*, 24–33.

Hsiao, G., Chou, P. H., Shen, M. Y., Chou, D. S., Lin, C. H., & Sheu, J. R., (2005). C-phycocyanin, a very potent and novel platelet aggregation inhibitor from *Spirulina platensis. J. Agric. Food Chem., 53*, 7734–7740.

Huang, B., Wang, G. C., Zeng, C. K., & Li, Z. G., (2002). The experimental research of R-phycoerythrin subunits on cancer treatment: A new photosensitizer in PDT. *Cancer Biother. Radiopharm., 17*, 35–42.

Huang, Z., Guo, B. J., Wong, R. N. S., & Jiang, Y., (2007). Characterization and antioxidant activity of selenium-containing phycocyanin isolated from *Spirulina platensis. Food Chem., 100*, 1137–1143.

Hulspas, R., Dombkowski, D., Preffer, F., Douglas, D., Kildew-Shah, B., & Gilbert, J., (2009). Flow cytometry and the stability of phycoerythrin-tandem dye conjugates. *Cytometry. A, 75*, 966–972.

Hussein, M. M. A., Ali, H. A., & Ahmed, M. M., (2015). Ameliorative effects of phycocyanin against gibberellic acid induced hepatotoxicity. *Pestic Biochem Physiol.,119*, 28–32.

Ihssen, J., Braun, A., Faccio, G., Gajda-Schrantz, K., & Thöny-Meyer, L., (2014). Light harvesting proteins for solar fuel generation in bioengineered photoelectrochemical cells. *Curr. Protein Pept. Sci., 15*(4), 374–384.

Izydorczyk, K., Tarczynska, M., Jurczak, T., Mrowczynski, J., & Zalewski, M., (2005). Measurement of phycocyanin fluorescence as an online early warning system for cyanobacteria in reservoir intake water. *Environ. Toxicol., 20*, 425–430.

Jensen, G. S., Ginsberg, D. I., & Drapeau, C., (2001). Blue-green algae as an immuno-enhancer and biomodulator. *J. Am. Nutraceut Ass., 3*, 24–30.

Jespersen, L., Strømdahl, L. D., Olsen, K., & Skibsted, L. H., (2005). Heat and light stability of three natural blue colorants for use in confectionery and beverages. *Eur. Food Res. Technol., 220*, 261–266.

Jiménez, C., Cossio, B. R., Labella, D., & Niell, F. X., (2003). The feasibility of industrial production of *Spirulina* (*Arthrospira*) in southern Spain. *Aquaculture, 217*, 179–190.

Joshi, S., Kumari, R., & Upasani, V. N., (2018). Applications of algae in cosmetics: An overview. *Int. J. Innov. Res. Sci. Eng. Technol., 7*, 1269–1278.

Juin, C., Chérouvrier, J. R., Thiérym, V., Gagez, A. L., Bérard, J. B., Joguet, N., Kaas, R., et al., (2015). Microwave-assisted extraction of phycobiliproteins from *Porphyridium purpureum. Appl. Biochem. Biotechnol., 175*, 1–15.

Kannaujiya, V. K., & Sinha, R. P., (2015). Impacts of varying light regimes on phycobiliproteins of *Nostoc* sp. HKAR-2 and *Nostoc* sp. HKAR-11 isolated from diverse habitats. *Protoplasma, 252*, 1551–1561.

Kannaujiya, V. K., & Sinha, R. P., (2016). Thermokinetic stability of phycocyanin and phycoerythrin in food grade preservatives. *J. Appl. Phycol., 28*, 1063–1070.

Kannaujiya, V. K., Kumar, D., Pathak, R. J., Sonker, A. S., Rajneesh, Singh, V., Sundaram, S., & Sinha, R. P., (2017). Recent advances in production and the biotechnological significance of phycobiliproteins. In: Sinha, R. P., & Richa, (eds.), *New Approaches in Biological Research* (pp. 1–34), Nova Science Publisher, New York.

Kannaujiya, V. K., Rastogi, R. P., & Sinha, R. P., (2014). GC constituents and relative codon expressed amino acid composition in cyanobacterial phycobiliproteins. *Gene, 546*, 162–171.

Kannaujiya, V. K., Sundaram, S., & Sinha, R. P., (2017b). *Phycobiliproteins: Recent Developments and Future Applications*. Springer, Singapore.

Kant, R., (2012). Textile dyeing industry an environmental hazard. *Natural Science, 4*, 22–26.

Karan, H., Funk, C., Grabert, M., Oey, M., & Hankamer, B., (2019). Green bioplastics as part of a circular bioeconomy. *Trends Plant Sci., 24*, 237–249.

Kefayat, A., Ghahremani, F., Safavi, A., Hajiaghababa, A., & Moshtaghian, J., (2019). C-phycocyanin: A natural product with radio sensitizing property for enhancement of colon cancer radiation therapy efficacy through inhibition of COX-2 expression. *Scientific Reports, 9*(1).

Kehrer, J. P., (1993). Free radicals as mediators of tissue injury and disease. *Crit. Rev. Toxicol., 23*, 21–48.

Khan, M., Varadharaj, S., Ganesan, L. P., Shobha, J. C., Naidu, M. U., Parinandi, N. L., Tridandapani, S., et al., (2006). C-phycocyanin protects against ischemia-reperfusion injury of heart through involvement of p38 MAPK and ERK signaling. *Am. J. Physiol. Heart Circ. Physiol., 290*, 2136–2145.

Khan, S. A., Sharma, G. K., Malla, F. A., Kumar, A., Rashmi, & Gupta, N., (2019). Microalgae based biofertilizers: A biorefinery approach to phycoremediate wastewater and harvest biodiesel and manure. *J. Clean Prod., 211*, 1412–1419.

Khatoon, H., Kok, L. L., Abdu, R. N., Mian, S., Begum, H., Banerjee, S., & Endut, A., (2018). Effects of different light source and media on growth and production of phycobiliprotein from freshwater cyanobacteria. *Bioresource Technology, 249*, 652–658.

Kilpatrick, K. A., (1985). *The Development of a Method for Measure Marine Cyanobacterial Phycoerythrin Extracted in Solvents (Dissertation)*. Texas AM University.

Kissoudi, M., Sarakatsianos, I., & Samanidou, V., (2018). Isolation and purification of food grade C-phycocyanin from *Arthrospira platensis* and its determination in confectionery by HPLC with diode array detection. *Journal of Separation Science, 41*(4), 975–981.

Kothadia, A. D., & Shabaraya, S. A. M., (2011). Evaluation of cataract preventive action of phycocyanin. *J. Pharm. Sci., 3*, 42–44.

Kuddus, M., Singh, P., Thomas, G., & Al-Hazimi, A., (2013). Recent developments in production and biotechnological applications of C-phycocyanin. *Biomed Res. Int., 742*859.

Kumar, C. S., Ganesan, P., Suresh, P. V., & Bhaskar, N., (2008). Seaweeds as a source of nutritionally beneficial compounds: A review. *J. Food Sci. Technol., 45*, 1–13.

Kumari, R. P., & Anbarasu, K., (2014). Protective role of C-phycocyanin against secondary changes during sodium selenite mediated cataractogenesis. *Nat. Prod. Bioprospect., 4*, 81–89.

Kumari, R. P., Ramkumar, S., & Thankappan, B., (2015). Transcriptional regulation of crystallin, redox, and apoptotic genes by C-phycocyanin in the selenite-induced cataractogenic rat model. *Mol. Vis., 21*, 26–39.

Kumari, R. P., Sivakumar, J., Thankappan, B., & Anbarasu, K., (2013). C-phycocyanin modulates selenite-induced cataractogenesis in rats. *Biol. Trace Elem. Res., 151*, 59–67.

Kupka, M., & Scheer, H., (2008). Unfolding of C-phycocyanin followed by loss of non-covalent chromophore-protein interactions-1. Equilibrium experiments. *BBA Bioenergetics., 1777*, 94–103.

Lee, Y. K., (1997). Commercial production of microalgae in the Asia-Pacific rim. *J. Appl. Phycol., 9*, 403–411.

Leema, J. T., Kirubagaran, R., Vinithkumar, N. V., Dheenan, P. S., & Karthikayulu, S., (2010). High value pigment production from *Arthrospira* (*Spirulina*) *platensis* cultured in seawater. *Bioresour. Technol., 101*, 9221–9227.

Leney, A. C., Tschanz, A., & Heck, A., (2018). Connecting color with assembly in the fluorescent 0B-phycoerythrin protein complex. *FEBS J., 285*, 178–187.

Leung, P. O., Lee, H. H., Kung, Y. C., Tsai, M. F., & Chou, T. C., (2013). Therapeutic effect of C-phycocyanin extracted from blue green algae in a rat model of acute lung injury induced by lipopolysaccharide. *Evid Based Complement. Alternate Med., 916*590.

Li, B., Chu, X. M., Xu, Y. J., Yang, F., Lv, C. Y., & Nie, S. M., (2013). CD59 underlines the anti-atherosclerotic effects of C-phycocyanin on mice. *Bio. Med. Res. Int., 729*413.

Li, B., Chu, X., Gao, M., & Li, W., (2010). Apoptotic mechanism of MCF-7 breast cells *in vivo* and *in vitro* Induced by photodynamic therapy with C-phycocyanin. *Acta Biochem. Biophys. Sin., 42*, 80–89.

Li, B., Gao, M., Chu, X., Teng, L., Lv, C., & Yang, P., (2015). The synergistic antitumor effects of all-trans retinoic acid and C-phycocyanin on the lung cancer A549 cells *in vitro* and *in vivo. Eur. J. Pharmacol., 749*, 107–114.

Li, C., Yu, Y., Li, W., Liu, B., Jiao, X., & Song, X., (2017). Phycocyanin attenuates pulmonary fibrosis via the TLR2-MyD88-NF-kappaB signaling pathway. *Sci. Rep., 7*, 5843.

Li, L., Li, L., & Song, K., (2015). Remote sensing of freshwater cyanobacteria: An extended IOP inversion model of inland waters (IIMIW) for partitioning absorption coefficient and estimating phycocyanin. *Remote Sens. Environ., 157*, 9–23.

Li, W., Pu, Y., Gao, N., Tang, Z., & Song, L., (2017). Efficient purification protocol for bioengineering allophycocyanin trimer with N-terminus histag. *Saudi J. Biol. Sci., 24*, 451–458.

Li, W., Su, H. N., Pu, Y., Chen, J., Liu, L. N., Liu, Q., & Qin, S., (2019). Phycobiliproteins: Molecular structure, production, applications, and prospects. *Biotechnology Advances, 37*(2), 340–353.

Liao, G., Gao, B., Gao, Y., Yang, X., Cheng, X., & Ou, Y., (2016). Phycocyanin inhibits tumorigenic potential of pancreatic cancer cells: Role of apoptosis and autophagy. *Sci. Rep., 6*, 34564.

Lin, P. C., Saha, R., Zhang, F., & Pakrasi, H. B., (2017). Metabolic engineering of the pentose phosphate pathway for enhanced limonene production in the cyanobacterium *Synechocystis* PCC 6803. *Scientific Reports, 7*(1).

Lino, C. M., & Pena, A., (2010). Occurrence of caffeine, saccharin, benzoic acid and sorbic acid in soft drinks and nectars in Portugal and subsequent exposure assessment. *Food Chem, 121*, 503–508.

Lissi, E. A., Pizarro, M., Aspee, A., & Romay, C., (2000). Kinetics of phycocyanin bilin groups destruction by peroxyl radicals. *Free Radic. Biol. Med., 28*, 1051–1055.

Liu, Q., Wang, Y., Cao, M., Pan, T., Yang, Y., & Mao, H., (2015). Anti-allergic activity of R-phycocyanin from *Porphyra haitanensis* in antigen-sensitized mice and mast cells. *Int. Immunopharmacol., 25*, 465–473.

Liu, Y., Xu, L., & Cheng, N., (2000). Inhibitory effect of phycocyanin from *Spirulina platensis* on the growth of human leukemia k562 cells. *J. Appl. Phycol., 12*, 125–130.

Lundell, D. J., Williams, R. C., & Glazer, A. N., (1981). Molecular architecture of a light-harvesting antenna. *In vitro* assembly of the rod substructures of *Synechococcus* 6301 phycobilisomes. *J. Biol. Chem., 256*, 3580–3592.

Lynch, M. A., (1998). Age-related impairment in long-term potentiation in hippocampus: A role for the cytokine, interleukin-1? *Prog. Neurobiol., 56*, 571–589.

Macedo, D., Bertolin, T. E., Oro, T., Backes, L. T. H., & Bras, I. C., (2017). Phycocyanin protects against alpha-synuclein toxicity in yeast. *J. Funct. Foods.*, 553–560.

Madhyastha, H. K., Sivashankari, S., & Vatsala, T. M., (2009). C-phycocyanin from *Spirulina fusiformis* exposed to blue light demonstrates higher efficacy of *in vitro* antioxidant activity. *J. Biochem. Eng., 43*, 221–224.

Marín-Prida, J., Pentón-Rol, G., & Rodrigues, F. P., (2012). C-Phycocyanin protects SH-SY5Y cells from oxidative injury, rat retina from transient ischemia and rat brain mitochondria from Ca^{2+}/phosphate-induced impairment. *Brain Res. Bull., 89*, 159–167.

Marquez, F. J., Sasaki, K., Kakizono, T., Nishio, N., & Nagai, S., (1993). Growth characteristics of *Spirulina platensis* in mixotrophic and heterotrophic conditions. *J. Ferment Bioeng.*, (5), 408–410.

Martinez, S. E., Chen, Y., Ho, E. A., Martinez, S. A., & Davies, N. M., (2015). Pharmacological effects of a C-phycocyanin-based multicomponent nutraceutical in an *in-vitro* canine chondrocyte model of osteoarthritis. *Can J. Vet. Res., 79*, 241–249.

Masojídek, J., Koblížek, M., & Torzillo, G., (2013). Photosynthesis in microalgae. In: Richmond, A., & Hu, Q., (eds.), *Handbook of Microalgal Culture: Applied Phycology and Biotechnology* (pp. 20–39). Wiley Blackwell, Oxford.

McCarty, M. F., (2007). Clinical potential of *Spirulina* as a source of phycocyanobilin. *J. Med. Food, 10*, 566–570.

Min, S. K., Park, J. S., Luo, L., Kwon, Y. S., Lee, H. C., & Shim, H. J., (2015). Assessment of C-phycocyanin effect on astrocytes-mediated neuroprotection against oxidative brain injury using 2D and 3D astrocyte tissue model. *Sci. Rep., 5*, 14418.

Minic, S. L., Stanic-Vucinic, D., Mihailovic, J., Krstic, M., Nikolic, M. R., & Velickovic, T. C., (2016). Digestion by pepsin releases biologically active chromopeptides from C-phycocyanin, a blue-colored biliprotein of microalga *Spirulina*. *J. Proteome., 147*, 132–139.

Minkova, K. M., Tchernov, A. A., Tchorbadjieva, M. I., Fournadjieva, S. T., Antova, R. E., & Busheva, M. C., (2003). Purification of C-phycocyanin from *Spirulina* (*Arthrospira*) *fusiformis*. *J. Biotechnol., 102*, 55–59.

Mishra, S. K., Shrivastav, A., Maurya, R. R., Patidar, S. K., Haldar, S., & Mishra, S., (2012). Effect of light quality on the C-phycoerythrin production in marine cyanobacteria *Pseudanabaena* sp. isolated from Gujarat coast India. *Protein Expr. Purif., 81*, 5–10.

Mishra, S. K., Shrivastav, A., Pancha, I., Jain, D., & Mishra, S., (2010). Effect of preservatives for food grade C-phycoerythrin, isolated from marine cyanobacteria *Pseudanabaena* sp. *Int. J. Biol. Macromol., 47*, 597–602.

Mitra, S., Siddiqui, W. A., & Khandelwal, S., (2015). C-phycocyanin protects against acute tributyltin chloride neurotoxicity by modulating glial cell activity along with its anti-oxidant and anti-inflammatory property: A comparative efficacy evaluation with N-acetyl cysteine in adult rat brain. *Chem. Biol. Interact., 238*, 138–150.

Mogany, T., Kumari, S., Swalaha, F. M., & Bux, F., (2019). Extraction and characterization of analytical grade C-phycocyanin from *Euhalothece* sp. *Journal of Applied Phycology, 31*, 1661–1674.

Mogany, T., Swalaha, F. M., Kumari, S., & Bux, F., (2018). Elucidating the role of nutrients in C-phycocyanin production by the halophilic cyanobacterium *Euhalothece* sp. *Journal of Applied Phycology, 30*, 2259–2271.

Moldovan, S., Ferrandiz, M. A., & Bonet, M. A., (2017). Natural cotton printing with red macroalgae biomass of *Gracilaria gracilis* and *Gracilaria cornea*. *Annals of the University of Oradea, Fascicle of Textiles Leatherwork, 20*, 61–66.

Morais, De, M. G., Fontoura Da, P. D., Moreira, J. B., Duarte, J. H., & Costa, J. A. V., (2018). Phycocyanin from microalgae: Properties, extraction, and purification with some recent applications. *Industrial Biotechnology, 14*(1), 30–37.

Morweiser, M., Kruse, O., Hankamer, B., & Posten, C., (2010). Developments and perspectives of photobioreactors for biofuel production. *Appl. Microbiol. Biotechnol., 87*, 1291–1301.

Munier, M., Moranc, M., Dumay, J., Jaouen, P., & Fleurence, J., (2015). One-step purification of R-phycoerythrin from the red edible seaweed *Grateloupia turuturu*. *J. Chromatogr. B, 992*, 23–29.

Nagaoka, S., Shimizu, K., & Kaneko, H., (2005). A novel protein C-phycocyanin plays a crucial role in the hypocholesterolemic action of *Spirulina platensis* concentrate in rats. *J. Nutr., 135*, 2425–2430.

Nagaraj, S., Arulmurugan, P., Rajaram, M. G., Karuppasamy, K., Jayappriyan, K. R., & Sundararaj, R., (2012). Hepatoprotective and antioxidative effects of C-phycocyanin from *Arthrospira maxima* SAG 25780 in CCl_4-induced hepatic damage rats. *Biomed Prev. Nutr., 2*, 81–88.

Nakagawa, K., Ritcharoen, W., Sri-Uam, P., Pavasant, P., & Adachi, S., (2016). Antioxidant properties of convective-air-dried *Spirulina maxima*: Evaluation of phycocyanin retention by a simple mathematical model of air-drying. *Food Bioprod. Process., 100*, 292–302.

Nwoba, E. G., Parlevliet, D. A., Laird, D. W., Alameh, K., & Moheimani, N. R., (2019). Sustainable phycocyanin production from *Arthrospira platensis* using solar-control thin film coated photobioreactor. *Biochemical Engineering Journal, 141*, 232–238.

Oi, V. T., Glazer, A. N., & Stryer, L., (1982). Fluorescent phycobiliprotein conjugates for analyses of cells and molecules. *J. Cell Biol., 93*, 981–986.

Ou, Y., Ren, Z., Wang, J., & Yang, X., (2016). Phycocyanin ameliorates alloxan-induced diabetes mellitus in mice: Involved in insulin signaling pathway and GK expression. *Chem. Biol. Interact., 247*, 49–54.

Ou, Y., Zheng, S., Lin, L., Jiang, Q., & Yang, X., (2010). Protective effect of C-phycocyanin against carbon tetrachloride-induced hepatocyte damage *in vitro* and *in vivo*. *Chem. Biol. Interact., 185*(2), 94–100.

Pagels, F., Guedes, A., Amaro, H. M., Kijjoa, A., & Vasconcelos, V., (2019). Phycobiliproteins from cyanobacteria: Chemistry and biotechnological applications. *Biotechnology Advances, 37*, 422–443.

Pak, W., Takayama, F., Mine, M., Nakamoto, K., Kodo, Y., & Mankura, M., (2012). Antioxidative and anti-inflammatory effects of *Spirulina* on rat model of non-alcoholic steatohepatitis. *J. Clin. Biochem. Nutr., 51*, 227–234.

Pan, R., Lu, R., Zhang, Y., Zhu, M., Zhu, W., & Yang, R., (2005). *Spirulina* phycocyanin induces differential protein expression and apoptosis in SKOV-3 cells. *Int. J. Biol. Macromol., 81*, 951–959.

Patel, S. N., Sonani, R. R., Jakharia, K., Bhastana, B., Patel, H. M., Chaubey, M. G., Singh, N. K., & Madamwar, D., (2018). Antioxidant activity and associated structural attributes of *Halomicronema* phycoerythrin. *Int. J. Biol. Macromol., 111*, 359–369.

Patil, G., Chethana, S., Sridevi, A. S., & Raghavarao, K. S. M. S., (2006). Method to obtain C-phycocyanin of high purity. *J. Chromatogr. A, 1127*, 76–81.

Pattarayan, D., Rajarajan, D., Ayyanar, S., Palanichamy, R., & Subbiah, R., (2017). C-phycocyanin suppresses transforming growth factor-β1-induced epithelial mesenchymal transition in human epithelial cells. *Pharmacol. Rep., 69*, 426–431.

Perez-Garcia, O., Escalante, F. M. E., de-Bashan, L. E., & Bashan, Y., (2011). Heterotrophic cultures of microalgae: Metabolism and potential products. *Water Res., 45*, 11–36.

Pleonsil, P., Soogarun, S., & Suwanwong, Y., (2013). Anti-oxidant activity of holo- and apo-c phycocyanin and their protective effects on human erythrocytes. *Int. J. Biol. Macromol., 60*, 393–398.

Prasanna, R., Sood, A., Suresh, A., & Kaushik, B. D., (2007). Potentials and applications of algal pigments in biology and industry. *Acta Bot. Hungar., 49*, 131–156.

Reddy, C. M., Bhat, V. B., Kiranmai, G., Reddy, M. N., Reddanna, P., & Madyastha, K. M., (2000). Selective inhibition of cyclooxygenase-2 by C-phycocyanin, a biliprotein from *Spirulina platensis. Biochem Biophys Res. Commun., 277*, 599–603.

Remirez, D., Ledón, N., & González, R., (2002). Role of histamine in the inhibitory effects of phycocyanin in experimental models of allergic inflammatory response. *Mediat. Inflamm., 11*, 81–85.

Renuka, N., Guldhe, A., Prasanna, R., Singh, P., & Bux, F., (2018). Microalgae as multifunctional options in modern agriculture: Current trends, prospects, and challenges. *Biotechnol. Adv., 36*, 1255–1273.

Richa, K. V. K., Kesheri, M., Singh, G., & Sinha, R. P., (2011). Biotechnological potentials of phycobiliproteins. *Int. J. Pharm. Bio. Sci., 2*, 446–454.

Rimbau, V., Camins, A., Romay, C., González, R., & Pallàs, M., (1999). Protective effects of C-phycocyanin against kainic acid-induced neuronal damage in rat hippocampus. *Neurosci. Lett., 276*, 75–78.

Riss, J., Décordé, K., & Sutra, T., (2007). Phycobiliprotein C-phycocyanin from *Spirulina platensis* is powerfully responsible for reducing oxidative stress and NADPH oxidase expression induced by an atherogenic diet in hamsters. *J. Agric. Food Chem., 55*, 7962–7967.

Riss, J., Décordé, K., Sutra, T., Delage, M., Baccou, J. C., Jouy, N., Brune, J. P., et al., (2007). Phycobiliprotein C-phycocyanin from *Spirulina platensis* is powerfully responsible for reducing oxidative stress and NADPH oxidase expression induced by an atherogenic diet in hamsters. *J. Agric. Food Chem., 55*(19), 7962–7967.

Romay, C. H., Gonzalez, R., & Ledon, N., (2003). C-phycocyanin a biliprotein with antioxidant, anti-inflammatory, and neuroprotective effects. *Curr. Protein Pept. Sci., 4*, 207–216.

Romay, C., Armesto, J., Remirez, D., Gonzalez, R., Ledon, N., & Garcia, I., (1998). Antioxidant and anti-inflammatory properties of C-phycocyanin from blue-green algae. *Inflamm. Res., 47*, 36–41.

Romay, C., Ledon, N., & Gonzalez, R., (1998). Further studies on anti-inflammatory activity of phycocyanin in some animal models of inflammation. *Inflamm. Res., 47*, 334–338.

Romay, C., Ledon, N., & Gonzalez, R., (1999). Phycocyanin extract reduces leukotriene B4 levels in arachidonic acid-induced mouse ear inflammation test. *J. Pharm. Pharmacol., 51*, 641–642.

Roy, K. R., Arunasree, K. M., Reddy, N. P., Dheeraj, B., Reddy, G. V., & Reddanna, P., (2007). Alteration of mitochondrial membrane potential by *Spirulina platensis* C-phycocyanin induces apoptosis in the doxorubicin-resistant human hepatocellular-carcinoma cell line HepG2. *Biotechnol. Appl. Biochem., 47,* 159–167.

Ruobing, W., Zhenghong, S., Xuecheng, Z., Shuang, Z., & Song, Q., (2007). Expression of the phycoerythrin gene of *Gracilaria lemaneiformis* (Rhodophyta) in *E. coli* and evaluation of the bioactivity of recombinant PE. *J. Ocean Univ., 6,* 373–377.

Schluchter, W. M., Shen, G., Alvey, R. M., Biswas, A., Saunée, N. A., Williams, S. R., & Bryant, D. A., (2010). Phycobiliprotein biosynthesis in cyanobacteria: Structure and function of enzymes involved in post-translational modification. *Advances in Experimental Medicine and Biology,* 211–228.

Schmidt, R. A., Wiebe, M. G., & Eriksen, N. T., (2005). Heterotrophic high cell-density fed-batch cultures of the phycocyanin producing red alga *Galdieria sulphuraria. Biotechnol Bioeng., 90,* 77–84.

Schrantz, K., Wyss, P. P., Ihsen, J., Toth, R., Bora, D. K., Vitol, E. A., Rozhkova, E. A., et al., (2017). Hematite photoanode co-functionalized with self-assembling melanin and C-phycocyanin for solar water splitting at neutral pH. *Catalysis Today, 284,* 44–51.

Sekar, S., & Chandramohan, M., (2008). Phycobiliproteins as a commodity: Trends in applied research, patents, and commercialization. *J. Appl. Phycol., 20,* 113–136.

Senthilkumar, N., Kurinjimalar, C., Thangam, R., Suresh, V., Kavitha, G., & Gunasekaran, P., (2013). Further studies and biological activities of macromolecular protein R Phycoerythrin from *Portieria homemannii. Int. J. Biol. Macromol., 62,* 107–116.

Senthilkumar, N., Suresh, V., Thangam, R., Kurinjimalar, C., Kavitha, G., & Murugan, P., (2013). Isolation and characterization of macromolecular protein R Phycoerythrin from *Portieria hornemannii. Int. J. Biol. Macromol., 55,* 150–160.

Setyaningsih, I., Bintang, M., & Madina, N., (2015). Potentially antihyperglycemic from biomass and phycocyanin of *Spirulina fusiformis* Voronikhin by *in vivo* test. In: Setyobudi, R. H., Scheer, H., Limantara, L., Shioi, Y., Fiedor, L., Brotosudarmo, T., & Prihastyanti, M., (eds.), *Procedia Chemistry* (Vol. 14, pp. 211–215).

Sharma, K., Sharma, V., & Sharma, S. S., (2018). Dye-sensitized solar cells: Fundamentals and current status. *Nanoscale Research Letters, 13,* 381.

Shcherbo, D., Merzlyak, E. M., Chepurnykh, T. V., Fradkov, A. F., Ermakova, G. V., & Solovieva, E. A., (2007). Bright far-red fluorescent protein for whole-body cyanobacterial thylakoid membranes. *Biochem Biophys Acta, 1857,* 256–265.

Sheu, M., Hsieh, Y., Lai, C., Chang, C., & Wu, C., (2013). Anti-hyperlipidemic and antioxidant effects of C-phycocyanin in golden Syrian hamsters fed with a hypercholesterolemic diet. *J. Tradit. Complement Med., 3,* 41–47.

Shih, S. R., Tsai, K. N., Li, Y. S., Chueh, C. C. & Chan, E. C., (2003). Inhibition of enterovirus 71-induced apoptosis by allophycocyanin isolated from a blue green alga *Spirulina platensis. J. Med. Virol., 70,* 119–125.

Sidler, W. A., (1994). Phycobilisome and phycobiliprotein structures. In: Bryant, D. A., (ed), *The Molecular Biology of Cyanobacteria* (pp. 139–216). Kluwer Academic Publishers, Dordrecht.

Siiman, O., Wilkinson, J., Burshteyn, A., Roth, P., & Ledis, S., (1999). Fluorescent neoglycoproteins: Antibody-aminodextran-phycobiliprotein conjugates. *Bioconjug. Chem., 10,* 1090–1106.

Silva, B. A. M., Torzillo, G., Kopecký, J., & Masojídek, J., (2013). Productivity and biochemical composition of *Phaeodactylum tricornutum* (Bacillariophyceae) cultures grown outdoors in tubular photobioreactors and open ponds. *Biomass and Bioenergy, 54,* 115–122.

Simis, S. G. H., Peters, S. W. M., & Gons, H. J., (2005). Remote sensing of the cyanobacterial pigment phycocyanin in turbid inland water. *Limnol. Oceanogr., 50*, 237–245.

Sinha, R. P., Lebert, M., Kumar, A., Kumar, H. D., & Häder, D. P., (1995). Spectroscopic and biochemical analyses of UV effect on phycobiliprotein of *Anabena* sp. and *Nostoc carmium*. *Bot. Acta, 108*, 87–92.

Six, C., Thomas, J. C., Garczarek, L., Ostrowski, M., Dufresne, A., Blot, N., Scanlan, D. J., & Partensky, F., (2007). Diversity and evolution of phycobilisomes in marine *Synechococcus* spp.: A comparative genomics study. *Genome Biol., 8*, R259.

Sode, K. J., Horikoshi, K., Takeyama, J., Nakamura, N., & Matsunga, T., (1991). Online monitoring of marine cyanobacterial cultivation based on phycocyanin fluorescence. *J. Biotechnol., 21*, 209–218.

Sonani, R. R., Patel, S., Bhastana, B., Jakharia, K., Chaubey, M. G., & Singh, N. K., (2017). Purification and antioxidant activity of phycocyanin from *Synechococcus* sp. R42DM isolated from industrially polluted site. *Bioresour. Technol., 245*, 325–331.

Sonani, R. R., Singh, N. K., Kumar, J., Thakar, D., & Madamwar, D., (2014). Concurrent purification and antioxidant activity of phycobiliproteins from *Lyngbya* sp. A09DM: An antioxidant and anti-aging potential of phycoerythrin in *Caenorhabditis elegans*. *Process Biochem., 49*, 1757–1766.

Spolaore, P., Joannis-Cassan, C., Duran, E., & Isambet, A., (2006). Commercial applications of microalgae. *J. Biosci. Bioeng., 101*(2), 87–96.

Stanier, R. Y., & Cohen-Bazire, G., (1977). Phototrophic prokaryotes: The cyanobacteria. *Annu. Rev. Microbiol., 31*, 225–274.

Stewart, D. E., & Farmer, F. H., (1984). Extraction, identification, and quantitation of phycobiliproteins pigments from phototropic biomass. *Limnol. Oceanogr., 29*, 392–397.

Strasky, Z., Zemankova, L., Nemeckova, I., Rathouska, J., Wong, R. J., Muchova, L., Subhanova, I., et al., (2013). *Spirulina platensis* and phycocyanobilin activate atheroprotective heme oxygenase-1: A possible implication for atherogenesis. *Food Funct., 4*, 1586–1594.

Sun, G. Y., Liang, H., & Xu, Q. Y., (2010). Study on antitumor activity of phycocyanin and its antioxidant function. *Prog. Modern. Biomed., 10*, 243–245.

Sun, L., Shumei, W., Chen, L., & Gong, X., (2003). Promising fluorescent probes from phycobiliproteins. *IEEE J. Sel. Top. Quant., 9*, 177–188.

Sun, Y. X., Zhang, J., Yu, G. C., Yan, Y. J., Chen, W. W., & Chi, M. F., (2012). Experimental study on the therapeutic effect of C-phycocyanin against pulmonary fibrosis induced by paraquat in rats. *Zhonghua Lao Dong Wei Sheng Zhi Ye Bing Za Zhi., 30*, 650–655.

Sun, Y., Zhang, J., Yan, Y., Chi, M., Chen, W., & Sun, P., (2011). The protective effect of C-phycocyanin on paraquat-induced acute lung injury in rats. *Environ. Toxicol. Pharmacol., 32*, 168–174.

Suping, Z., Jingxi, P., Zhenhui, H., Jingquan, Z., Side, Y., & Lijin, J., (2001). Generation and identification of the transient intermediates of allophycocyanin by laser photolytic and pulse radiolytic techniques. *Int. J. Radiat. Biol., 77*, 637–642.

Tang, Y., Xie, M., Jiang, N., Huang, F., Zhang, X., & Li, R., (2017). Icarisid II inhibits the proliferation of human osteosarcoma cells by inducing apoptosis and cell cycle arrest. *Tumor Biol., 39*, 1393383919.

Tavanandi, H. A., Mittal, R., Chandrasekhar, J., & Raghavarao, K. S. M. S., (2018). Simple and efficient method for extraction of C-phycocyanin from dry biomass of *Arthospira platensis*. *Algal Res., 31*, 239–251.

Telford, W. G., Moss, M. W., & Moreseman, J. P., (2001). Cyanobacterial stabilized phycobilisomes as fluorochromes for extracellular antigen detection by flow cytometry. *J. Immunol. Methods., 254*, 13–30.

Thajuddin, N., & Subramanian, G., (2005). Cyanobacterial biodiversity and potential application in biotechnology. *Current Science, 89*(1), 47–57.

Thangam, R., Suresh, V., Asenath, P. W., Rajkumar, M., Senthilkumar, N., & Gunasekaran, P., (2013). C-Phycocyanin from *Oscillatoria tenuis* exhibited an antioxidant and *in vitro* antiproliferative activity through induction of apoptosis and G0/G1 cell cycle arrest. *Food Chem., 140*, 262–272.

Thomas, J. C., & Passaquet, C., (1999). Characterization of a phycoerythrin without α-subunits from a unicellular red alga. *Journal of Biological Chemistry, 274*(4), 2472–2482.

Thoren, K. L., Connell, K. B., Robinson, T. E., Shellhamer, D. D., Tammaro, M. S., & Gindt, Y. M., (2006). The free energy of dissociation of oligomeric structure in phycocyanin is not linear with denaturant. *Biochemistry, 45*, 12050–12059.

Vandamme, D., Foubert, I., & Muylaert, K., (2013). Flocculation as a low-cost method for harvesting microalgae for bulk biomass production. *Trends Biotechnol., 31*, 233–239.

Venil, C. K., Zakaria, Z. A., & Ahmad, W. A., (2013). Bacterial pigments and their applications. *Process Biochem., 48*, 1065–1079.

Viskari, P. J., & Colyer, C. C., (2003). Rapid extraction of phycobiliproteins from cultured cyanobacteria samples. *Anal. Biochem., 319*, 263–271.

Wang, C. Y., Wang, X., Wang, Y., Zhou, T., Bai, Y., & Li, Y. C., (2012). Photosensitization of phycocyanin extracted from *Microcystis* in human hepatocellular carcinoma cells: Implication of mitochondria-dependent apoptosis. *J. Photochem. Photobiol. B, 117*, 70–79.

Wang, X., Yu, J., Kang, Q., Shen, D., & Li, J., (2016). Molecular imprinting ratiometric fluorescence sensor for highly selective and sensitive detection of phycocyanin. *Biosens. Bioelectron., 77*, 624–630.

Wu, H. L., Wang, G. H., Xiang, W. Z., Li, T., & He, H., (2016). Stability and antioxidant activity of food-grade phycocyanin isolated from *Spirulina platensis*. *International Journal of Food Properties, 19*, 2349–2362.

Wu, J., Chen, H., Zhao, J., & Jiang, P., (2017). Fusion proteins of streptavidin and allophycocyanin alpha subunit for immunofluorescence assay. *J. Biochem. Eng., 125*, 97–103.

Wyman, M., (1992). An *in vivo* method for the estimation of phycoerythrin concentration in marine cyanobacteria (*Synechococcus* spp.). *Limnol. Oceanogr., 37*, 1300–1306.

Xia, D., Liu, B., Xin, W., Liu, T., Sun, J., Liu, N., Qin, S., & Du, Z., (2016). Protective effects of C-phycocyanin on alcohol-induced sub-acute liver injury in mice. *J. Appl. Phycol., 28*, 765–772.

Xie, Y., Jin, Y., Zeng, X., Chen, J., Lu, Y., & Jing, K., (2015). Fed-batch strategy for enhancing cell growth and C-phycocyanin production of *Arthrospira* (*Spirulina*) *platensis* under phototrophic cultivation. *Bioresour. Technol., 180*, 281–287.

Yadav, S., Sinha, R. P., Tyagi, M. B., & Kumar, A., (2011). Cyanobacterial secondary metabolites. *International Journal of Pharma and Bio Sciences., 2*(2), 144-167.

Yañuk, J. G., Cabrerizo, F. M., Dellatorre, F. G., & Cerdá, M. F., (2019). Photosensitizing role of R-phycoerythrin red protein and β-carboline alkaloids in dye sensitized solar cell. Electrochemical and spectroscopic characterization. *Energy Reports*. doi: 10.1016/j.egyr.2019.10.045.

Ying, J., Wang, J., Ji, H., Lin, C., Pan, R., & Zhou, L., (2016). Transcriptome analysis of phycocyanin inhibitory effects on SKOV-3 cell proliferation. *Gene, 585*, 58–64.

Young, I., Chuang, S., Hsu, C., Sun, Y., & Lin, F., (2016). C-phycocyanin alleviates osteoarthritic injury in chondrocytes stimulated with H_2O_2 and compressive stress. *Int. J. Biol. Macromol., 93*, 852–859.

Zhang, X., Li, J. Y., & Gong, X. G., (2010). Isolation of C-PC subunits from *Spirulina platensis* and inhibitory effect on SPC-A-1 cell line. *J. Zhej. Uni., 37*, 319–323.

Zhao, L., Peng, Y., Gao, J., & Cai, W., (2014). Bioprocess intensification: An aqueous two-phase process for the purification of C-phycocyanin from dry *Spirulina platensis. Eur. Food Res. Technol., 238*, 451–457.

Zheng, J., Inoguchi, T., Sasaki, S., Maeda, Y., McCarty, M. F., Fujii, M., Ikeda, N., et al., (2013). Phycocyanin and phycocyanobilin from *Spirulina platensis* protect against diabetic nephropathy by inhibiting oxidative stress. *AJP Regul. Integr. Comp. Physiol., 304*, R110–R120.

Zhou, Z. P., Liu, L. N., Chen, X. L., Wang, J. X., Chen, M., Zhang, Y. Z., & Zhou, B. C., (2005). Factors that affect antioxidant activity of c-phycocyanin from *Spirulina platensis. J. Food Biochem., 29*, 313–322.

INDEX

For Product Safety Concerns and Information please contact our EU
representative GPSR@taylorandfrancis.com
Taylor & Francis Verlag GmbH, Kaufingerstraße 24, 80331 München, Germany

www.ingramcontent.com/pod-product-compliance
Lightning Source LLC
Chambersburg PA
CBHW060753220326
41598CB00022B/2424